D0699264

HIGH-TECH HANDICAPPING IN THE INFORMATION AGE

HIGH-TECH HANDICAPPING IN THE INFORMATION AGE

An Information Management Approach to the Thoroughbreds

James Quinn

WILLIAM MORROW AND COMPANY, INC.
New York

Grateful acknowledgment is made for permission to reprint the following:

"Comments on Lee Lawrence's 'Handicappers' Guide to Fiscal Fitness,'" © 1984 by Phillips Racing Newsletter, August 1984 edition. Reprinted by permission of Phillips Racing Newsletter.

Past performance and results charts copyright © 1985 by *Daily Racing Form, Inc.* Reprinted by permission of copyright owner.

Library of Congress Cataloging-in-Publication Data

Quinn, James, 1943–
High-tech handicapping in the information age.

Bibliography: p.
1. Horse race betting—Data processing.
2. Information storage and retrieval systems—Horse race betting. I. Title.
SF331.Q548 1986 798.4'01'0285 85-15404
ISBN 0-688-05388-2

Printed in the United States of America

First Edition

1 2 3 4 5 6 7 8 9 10

BOOK DESIGN BY BERNARD SCHLEIFER

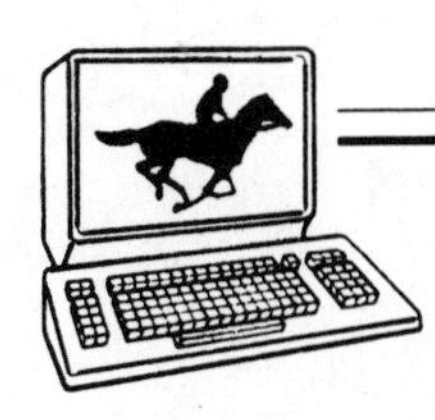

ACKNOWLEDGMENTS

MY FRIEND and writing colleague Dick Mitchell is first among three associates whose advice and help with the manuscript I am pleased to credit.

During the formative planning and development phases I was grateful to sit repeatedly as a guest at Mitchell's dinner table, discussing—in the main I listened—the role and function of computer technology in modern handicapping with an authentic expert in mathematics and computer science, and always under the delightful conditions provided by Dick and his wife, Cynthia. From these conversations I gleaned the decisive, centralizing role of *databases* in the developing schemes of the high-tech, information society. Eventually, Mitchell passed muster on the book's two key chapters on databases for handicappers. I was glad to get his stamp of approval.

Conversations and correspondence with handicapping colleague George Kaywood, of Omaha, Nebraska, formed the background for the book's review of the contemporary high-tech marketplace in handicapping. Kaywood read that chapter, and another on the fundamentals of computer technology, or just about everything that potential high-tech handicappers will need to know. A founder of the handicapping special-interest subgroup within American Mensa, Kaywood is fast becoming a leading developer of what this book refers to as "information programs" for handicappers and racegoers.

Dina Fallon, office administrator of Peopleware Systems, Inc., of Laguna Beach, California, guided the transmission of a severely edited first draft onto the company's word processor. Following that, she carefully managed the word processing of the hundreds of corrections, changes, and additions I insisted upon. It was a massive task, obviously outside of the mainstream of office business, well done. My deep gratitude, and respects.

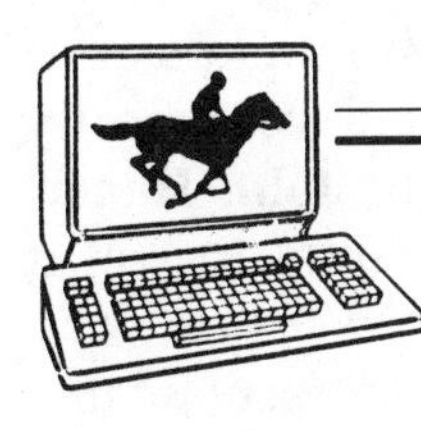

PREFACE

Knowledge is the capital of tomorrow... information is the principal resource of the high-tech, information society....

"A Nation at Risk," a report by the President's Commission on Education in America, 1983

TO UNDERSTAND AT ONCE the value of information, let's begin with the greatest overlay of all time, the Arlington Budweiser 'Million of 1983, and parlay that to the most misunderstood race on the American calendar, the Kentucky Derby, in particular the 1984 running.

The two races not only illustrate several of this book's propositions on thoroughbred handicapping, they dramatize the splendid opportunities awaiting those confidently progressive handicappers fully prepared to enter the age of information. As will be seen immediately and convincingly, I do not exaggerate by suggesting that handicappers having (a) access to the pertinent information, (b) the know-how to interpret it correctly, and (c) the experience to use it smartly could readily have walked away with a full year's income after placing prudent, normal-sized wagers on these two highly visible races.

The favorite at 7 to 5 for the 1983 Arlington Budweiser 'Million was the great gelding **John Henry,** at age eight a conspicuous underlay in the straight pools. At 2 to 1 the second choice was Charles Whittingham's five-year-old turf handicap star **Erin's Isle,** winless since taking Santa Anita's San Juan Capistrano (Gr. 1) four months back, and whose form had clearly deteriorated in that interim.

No other horse in the 1983 'Million was priced below 10 to 1. Was the 'Million that lopsided, truly a two-horse race? It was not. One horse was actually a fantastic overlay, sent to the post at 38 to 1 when it should have been approximately 4 to 1. The difficulty was that hardly anyone in the United States realized it. Certainly not the experts, those public selectors and media correspondents whose job it is precisely to inform and to help us interpret the information. Interviewed on national television just prior to the race, most explained why **John Henry** figured to win. My colleague Andrew Beyer, an astute speed and trip handicapper, who also understands the difference between underlays and overlays, said vehemently that a California horse would win, "but it won't be **John Henry. Erin's Isle** has this field at his mercy, and I expect him to win it easily. . . ."

Beyer did not mention **Tolomeo**. The American correspondents discounted the foreign runners as a group; these were not even the best of the overseas candidates. None suspected the three-year-old **Tolomeo** had a decent chance. Perhaps the public selectors and media correspondents showed us mainly a lack of information about international racing, not to mention the relative probabilities of nonclaiming three-year-olds beating older horses in late August. They clearly hold their own.

In merry old England, to be sure, the three-year-old homebred **Tolomeo** went off at 4 to 1.

What did the British players know the Americans did not? Plenty.

Tolomeo had recently placed in the Gr. 1 Sussex Stakes there, losing to an older horse in a prestigious race having the sixth highest purse in the country. The American equivalent of the Sussex Stakes purse is $115,500. Moreover, the colt had won and placed in other important events in England and its full record was practically prototypical of the well-bred, lightly raced, nicely developing three-year-old that might be any kind.

Now does **Tolomeo** deserve to be 38 to 1 in the 'Million? Of course not. But the American horseplayer did not have ready access to vital information. The *Daily Racing Form* does not supply the purse values and eligible ages of foreign stakes that are graded; it provides no useful information about races that are ungraded. How to know which are the relatively important stakes? How to compare one foreign stakes with another? With United States stakes?

But suppose handicappers had access to a catalogue of that information. Or, suppose they could query a computerized database that stored the information, either on their personal micro or on a mini that was the hub of a regional information system for handicappers? The catalogue now exists; the computerized database will be forthcoming.

In the database situation, handicappers might dial a centrally located minicomputer from residential telephone hook-ups, or if they had constructed a personal database, they might use their microcomputer keyboards to submit the "query":

> List the Gr. 1 races in England open to 3up according to their purse values.

Within seconds, handicappers would find this kind of information splashed across a terminal screen:

England Gr. 1 stakes, 3up

King George VI and Queen Elizabeth Diamond Stakes	$244,200
Benson and Hedges Gold Cup	165,000
Dubai Champion Stakes	165,000
Sussex Stakes	115,500
Coral Eclipse Stakes	107,250
Gold Cup	89,100
King's Stand Stakes	66,000

Now handicapperrs realize only **seven** of these important races are carded throughout England and Tolomeo has just finished second in the fourth richest. Does that information help handicappers better estimate the English three-year-old's chances in the 'Million?

Armed with that information, handicappers at Arlington Park might have decided to wager on Tolomeo at 38 to 1. Those who did would have been generously rewarded when the colt beat John Henry by a neck and returned $78.40 to win. Is that an overlay?

Handicappers might also have chosen to box Tolomeo and John Henry in the Exacta, which costs $2 at Arlington Park. Those who decided to do that would have won $439.20 for each deuce. A $20 Exacta box returned $4,392! What do handicappers without information about foreign horses and international racing think of that?

Now, the kicker. The Arlington Budweiser 'Million is simulcast to numerous racetracks throughout North America; local pari-mutuel wagering is conducted.

Comes a letter from one Winford Mulkey, of Siloam Springs, Arkansas: "Did you know that on the simulcast of the Arlington 'Million to Louisiana Downs the winner paid over $200 to win and the Exacta ($3) to John Henry paid **six thousand dollars!"**

I did not know. Further, had I attended at Louisiana Downs on

Chart of Arlington Budweiser Million

NINTH RACE
Arlington
AUGUST 28, 1983

1 ¼ MILES.(turf). (1.59⅘) 3rd Running THE ARLINGTON BUDWEISER MILLION (Grade I). Purse $1,000,000 Guaranteed. 3-year-olds and upward. Weight for age. By subscription of $1,000 each horse which should accompany the nominations and a second eligibility payment of $2,500 due on Monday, May 2, 1983. A further payment of $2,500 (refundable if the horse is not give the opportunity to run) will be due on Wednesday, July 20, 1983. An International Board of Handicappers will select a field of 14 starters and 10 also eligibles in order of their preference by Saturday, July 30. For those that failed to nominate by Tuesday, March 15, 1983 a supplementary nomination may be made for a fee of $40,000 not later than Wednesday, July 20. If a supplementary nominee is not selected to run a refund of $35,000 will be made. All selected nominees to pay $5,000 additional to pass the entry box and $5,000 additional to start with $1,000,000 Guaranteed to be divided $600,000 to the winner, $200,000 to second, $110,000 to third, $60,000 to fourth and $30,000 to fifth. Weight For Age. Northern Hemisphere: 3-year-olds, 118 lbs. Fillies, 114 lbs. 4-year-olds and up, 126 lbs. Fillies and Mares, 122 lbs. Southern Hemisphere: 3-year-olds, 113 lbs. Fillies, 109 lbs. 4-year-olds and upward, 121 lbs. Fillies and Mares, 117 lbs. Closed with 252 nominations. Supplementary nominee: Bold Run.

Value of race $1,000,000, value to winner $600,000, second $200,000, third $110,000, fourth $60,000, fifth $30,000. Mutuel pool $645,542. Perfecta Pool $229,587.

Last Raced	Horse	Eqt.A.Wt	PP	¼	½	¾	1	Str	Fin	Jockey	Odds $1
27Jly83 ^{4}Eng^{2}	Tolomeo	3 118	5	2^{hd}	$3^{1/2}$	$3^{1/2}$	3^{hd}	$3^{1\frac{1}{2}}$	1^{nk}	Eddery P	38.20
4Jly83 ^{8}Hol^{1}	John Henry	8 126	13	$3^{1/2}$	$2^{1/2}$	$2^{1\frac{1}{2}}$	2^{1}	2^{1}	$2^{1/2}$	McCarron C J	1.40
7Aug83 ^{8}Sar^{5}	Nijinsky's Secret	5 126	7	1^{1}	1^{1}	1^{hd}	$1^{1/2}$	$1^{1/2}$	3^{2}	Velez J A Jr	10.30
24Jly83 ^{8}Bel^{2}	Thunder Puddles	4 126	9	$6^{1/2}$	$6^{1/2}$	6^{1}	$6^{1\frac{1}{2}}$	6^{1}	4^{nk}	Cordero A Jr	13.10
24Jly83 ^{8}Bel^{3}	Erins Isle	5 126	2	$8^{1\frac{1}{2}}$	10^{hd}	8^{1}	8^{2}	8^{4}	5^{nk}	Pincay L Jr	a-2.40
8Aug83 ^{8}Sar^{1}	Hush Dear	5 122	6	9^{1}	7^{1}	5^{hd}	4^{hd}	4^{1}	6^{2}	Vasquez J	19.90
14Aug83 ^{4}Fra^{6}	(S)Bold Run	b 4 126	8	$4^{1/2}$	4^{1}	4^{hd}	$5^{1/2}$	5^{hd}	7no	Starkey G	f-36.60
27Jly83 ^{4}Eng^{5}	Muscatite	3 118	3	5^{1}	5^{hd}	7^{hd}	7^{hd}	7^{hd}	$8^{3/4}$	Fires E	24.00
8Aug83 ^{7}Sar^{2}	Trevita	6 122	4	7^{hd}	9^{hd}	$11^{1\frac{1}{2}}$	11^{1}	10^{5}	$9^{2\frac{1}{2}}$	Velasquez J	38.30
23Jly83 ^{8}AP^{3}	Rossi Gold	7 126	1	$11^{1/2}$	13^{3}	13^{4}	13^{2}	13^{1}	10^{1}	Day P	21.90
5Jly83 ^{2}Eng^{9}	Be My Native	4 126	11	12^{hd}	12^{1}	10^{1}	9^{hd}	$9^{1/2}$	$11^{2\frac{1}{2}}$	Piggott L	17.40
24Jly83 ^{8}Bel^{1}	Majesty's Prince	4 126	12	14	14	14	14	11^{2}	12^{10}	Maple E	10.40
7Aug83 ^{8}Dmr^{2}	The Wonder	5 126	14	13^{4}	8^{hd}	$9^{1/2}$	$10^{1/2}$	14	13^{nk}	Shoemaker W	a-2.40
27Jly83 ^{4}Eng^{6}	The Noble Player	3 118	10	$10^{1/2}$	$11^{1\frac{1}{2}}$	$12^{1\frac{1}{2}}$	12^{2}	12^{2}	14	Cauthen S	f-36.60

a—Coupled: Erins Isle and The Wonder.
f—Mutuel field.
(S) Supplementary nomination.

OFF AT 4:43 Start good for all but HUSH DEAR, Won driving. Time, :24⅖, :50⅗, 1:15⅘, 1:41⅕, 2:04⅖ Course good.

$2 Mutuel Prices:				
	5–TOLOMEO	78.40	33.20	17.00
	11–JOHN HENRY		4.80	3.40
	7–NIJINSKY'S SECRET			6.00

$2 PERFECTA 5–11 PAID $439.20.

B. c, by Lypheor—Almagest, by Dike. Trainer Cumani Luca M. Bred by Corduff Stud (Ire).

TOLOMEO rated just off the early pace, rallied along rail leaving second turn, took command nearing the sixteenth pole and was all out to hold safe JOHN HENRY. The latter rushed up wide to force the pace of NIJINSKY'S SECRET, held on gamely in drive and just missed. NIJINSKY'S SECRET made the early pace while under wraps, held on willingly but was not good enough in final sixteenth. THUNDER PUDDLES was wide early, rallied mildly between horses. ERINS ISLE circled rivals on second turn but lacked needed closing response. HUSH DEAR stumbled at the break, made a strong run outside horse late on second turn, tired between rivals. BOLD RUN had early speed while wide, remained forwardly placed between horses on second turn, gave way in drive. MUSCATITE saved ground to no avail. TREVITA raced evenly. ROSSI GOLD struck the rail first time in upper stretch, passed tired horses. BE MY NATIVE showed little. MAJESTY'S PRINCE was always outrun. THE WONDER was through early. THE NOBLE PLAYER failed to reach contention.

Owners— 1, D'Alessio C; 2, Dotsam Stable; 3, McDougal Mrs J A; 4, Rockwood Stable; 5, Sweeney B; 6, Whitney C V; 7, Fradkoff & Seltzer; 8, Kais Al-Said; 9, Augustin Stable; 10, Combs Leslie II; 11, Hsu K; 12, Marsh J D; 13, DuBreil Alaine-Marquise DeMoratall; 14, Sangster R E.

Scratched—Craelius (25Jly83 ^{8}Hol^{1}).

August 28, 1983, I could not have seized the moment. I would have lacked the relevant information. I would not have been sufficiently informed to use a $30 Exacta box between **Tolomeo** and **John Henry**, collect $60,000 in one gigantic swoop, and thereby share bragging rights to the greatest overlay ever in thoroughbred racing.

To imagine that an Exacta with the champion **John Henry** in the two spot from a winner at 4-to-1 odds in England could return $6,000 for $3 (2000 to 1) at a major American racetrack dramatizes as little else could the importance of new and unfamiliar information at the track. Those who have got it indeed will have won the day.

A central argument of this book is that the most successful, most prosperous handicappers of today and for the near future, that is, the information age, will not be the method players—not the class-form experts, not the speed-pace analysts, not the adjusted final time "figure" addicts, not the trainer-patterns specialists, not the trip fanciers, and not devotees of classical comprehensive handicapping principles—but rather the information managers, those handicappers who combine access to the greatest amount of information with the capacity to manage it efficiently and the ability to interpret it smartly. The information managers will be the relatively big winners of the information age. They will become consistently successful handicappers who have mastered what I have conveniently called the information management approach to this difficult, challenging pastime.

To illuminate our directions further, nowhere is the information management approach to handicapping more applicable than to the nation's most popular race, the Kentucky Derby. The problem of analyzing the Kentucky Derby is unique in American racing. To wit, how to evaluate the prospects of still-developing three-year-olds competing at a classic distance (1¼ miles) none have ever run over a racing surface few have yet experienced?

Until 1981 only conventional handicapping applied. Experts identified the most impressive horses graduating from the multitude of preliminary stakes, and distinguished these utilizing various methods emphasizing class appraisal, pace analysis, speed figures, form cycles, trainer-jockey patterns, and on and on. Wrong. None of the conventional methods could fairly project the Derby winners. They failed, abysmally.

That point was proved beyond dispute in 1981 when a bloodlines scholar named Steven A. Roman published his continuous research on "dosage," the relative influences of speed and stamina in a horse's immediate four-generation pedigree, and showed that since 1940 no horse having a ratio of speed to stamina in its bloodlines greater than a given index figure (4.0) had ever won the Derby.

Dosage information was new, unfamiliar information. It applied with such astonishing reliability to the Kentucky Derby and the Belmont Stakes that it has altered for all time the handicapping of those two classics. To unravel each year's Derby, handicappers need first of all a set of dosage figures for all the starters. As a first step, they can confidently eliminate any horse having a dosage index greater than 4.0. But accessing the dosage information can be a sticky, laborious problem.

Moreover, Roman's research promotes a method for playing the Derby that relates pedigree and performance, but depends absolutely on information management skills. Roman's method bets on all Derby starters acceptable on dosage and weighted within ten pounds of the Experimental Handicap leader (that industry-based handicap ranks the previous season's two-year-old crop by weights). A few horses each season are selected by the method, and since 1972 a bet on all of them would have netted profits comfortably upwards of 90 percent on investment. A variation of the Roman method, suggested by the author, invests on all starters acceptable on dosage that have won one of eleven important Gr. 1 stakes that are key preliminaries to the Derby. Five are for two-year-olds: Champagne Stakes, Laurel Futurity, Young America Stakes, Hollywood Futurity, and The Remsen. Six are for three-year-olds: Flamingo Stakes, Florida Derby, Santa Anita Derby, Wood Memorial, Arkansas Derby, and the Blue Grass Stakes. All other stakes winners are eliminated, presumably outclassed.

Both methods combine pedigree and performance, but the graded stakes inventory allows handicappers to consider important three-year-olds which either did not start at two or might have been improving impressively lately while only placing at two.

These approaches relate pedigree to performance, yet the vital information cannot be located in the *Daily Racing Form* past performance tables on Derby day. The information needs to be collected, stored, and processed, and retrieved efficiently whenever handicappers need it.

At the appointed Derby hour out go additional queries to the handicapper's computerized database.

Chart of the Kentucky Derby

EIGHTH RACE
Churchill
MAY 5, 1984

1 ¼ MILES. (1.59⅖) 110th Running KENTUCKY DERBY (Grade I) Scale weight Stakes. $250,000 Added. 3-year-olds. With a subscription fee of $200 each, an entry fee of $10,000 each and a starting fee of $10,000 each. All fees to be paid to the winner. $250,000 shall be paid by Churchill Downs Incorporated (the "Association") as the Added Purse. Second place shall receive $100,000, third place shall receive $50,000 and fourth place shall receive $25,000 from the Added Purse. (The Added Purse and fees to be divided equally in the event of a dead heat.) Starters shall be named through the entry box on Thursday, May 3, 1984, at the usual time of closing. The maximum number of starters shall be limited to twenty and each shall carry a weight of 126 lbs., fillies, 121 lbs. In the event that more than twenty entries pass through the entry box at the time of closing, the starters shall be determined at that time with preference given to those horses that have accumulated the highest earnings, excluding earnings won in a restricted race. For purposes of this preference, a "restricted race" shall mean a state-bred restricted race (a race where entries are restricted to horses qualifying under state breeding programs), a sales restricted race (a race where entries are restricted by the origin of purchase), and a money restricted sweepstake (a race where entries are restricted by the amount of money previously earned). The owner of the winner shall receive a gold trophy. No supplementary nominations shall be accepted. (Closed with 312 nominations.)

Value of race $712,400, value to winner $537,400, second $100,000, third $50,000, fourth $25,000. Mutuel pool $5,428,787.

Last Raced	Horse	Eqt.	A.	Wt	PP	¼	½	¾	1	Str	Fin	Jockey	Odds $1
17Apr84 ^{7}Kee^{2}	Swale		3	126	15	$3^{1\frac{1}{2}}$	3^{1}	2^{1}	1^{2}	1^{5}	$1^{3\frac{1}{4}}$	Pincay L Jr	3.40
1Apr84 ^{9}Lat^{6}	Coax Me Chad		3	126	19	17^{3}	$17^{1\frac{1}{2}}$	$14^{1\frac{1}{2}}$	7^{hd}	2^{hd}	2^{2}	McCauley W H	f-9.90
21Apr84 ^{9}OP^{4}	At The Threshold		3	126	14	$11^{\frac{1}{2}}$	$9^{1\frac{1}{2}}$	$6^{\frac{1}{2}}$	$4^{1\frac{1}{2}}$	$3^{\frac{1}{2}}$	3^{nk}	Maple E	37.70
21Apr84 ^{9}OP^{3}	[D]Gate Dancer	b	3	126	20	$19^{\frac{1}{2}}$	18^{3}	15^{1}	$9^{1\frac{1}{2}}$	6^{1}	$4^{\frac{3}{4}}$	Delahoussaye E	18.90
8Apr84 ^{8}SA^{4}	Fali Time	b	3	126	7	$7^{\frac{1}{2}}$	$7^{\frac{1}{2}}$	$7^{1\frac{1}{2}}$	$6^{\frac{1}{2}}$	5^{hd}	5^{1}	Hawley S	18.70
21Apr84 ^{9}OP^{2}	Pine Circle		3	126	18	$14^{\frac{1}{2}}$	14^{4}	$11^{\frac{1}{2}}$	$12^{\frac{1}{2}}$	$7^{\frac{1}{2}}$	$6^{\frac{3}{4}}$	Smith M E	b-6.00
26Apr84 ^{7}Kee^{4}	Fight Over		3	126	6	6^{hd}	5^{1}	3^{1}	$2^{1\frac{1}{2}}$	4^{1}	7^{hd}	Vergara O	78.90
8Apr84 ^{8}SA^{5}	Life's Magic		3	121	5	$13^{1\frac{1}{2}}$	13^{1}	$9^{\frac{1}{2}}$	8^{1}	8^{3}	8^{1}	Brumfield D	a-2.80
26Apr84 ^{7}Kee^{2}	Silent King		3	126	11	20	20	20	19	11^{2}	9^{3}	Shoemaker W	4.80
26Apr84 ^{7}Kee^{5}	Rexson's Hope	b	3	126	9	$18^{1\frac{1}{2}}$	19^{4}	19^{3}	$18^{\frac{1}{2}}$	$10^{\frac{1}{2}}$	10^{nk}	Gaffglione R	f-9.90
26Apr84 ^{7}Kee^{7}	So Vague		3	126	4	$15^{\frac{1}{2}}$	$15^{1\frac{1}{2}}$	18^{3}	$16^{\frac{1}{2}}$	$12^{\frac{1}{2}}$	11^{2}	Cooksey P J	f-9.90
28Apr84 ^{8}CD^{2}	Biloxi Indian	b	3	126	17	$5^{1\frac{1}{2}}$	4^{hd}	4^{hd}	3^{hd}	$9^{1\frac{1}{2}}$	$12^{1\frac{1}{4}}$	Patterson G	f-9.90
26Apr84 ^{7}Kee^{1}	Taylor's Special		3	126	10	9^{1}	8^{hd}	8^{hd}	10^{2}	13^{2}	13^{3}	Maple S	6.80
21Apr84 ^{8}Aqu^{2}	Raja's Shark	b	3	126	2	$8^{2\frac{1}{2}}$	$11^{1\frac{1}{2}}$	12^{1}	14^{hd}	14^{hd}	14^{hd}	Wilson R	59.10
26Apr84 ^{8}Hol^{1}	Bedouin		3	126	8	$16^{1\frac{1}{2}}$	16^{1}	$17^{\frac{1}{2}}$	$17^{\frac{1}{2}}$	16^{2}	15^{nk}	Sibille R	f-9.90
31Mar84 ^{9}OP^{1}	Vanlandingham		3	126	12	$4^{\frac{1}{2}}$	6^{hd}	$10^{1\frac{1}{2}}$	$11^{\frac{1}{2}}$	$15^{\frac{1}{2}}$	16^{2}	Day P	b-6.00
28Apr84 ^{8}CD^{3}	Secret Prince		3	126	13	$10^{\frac{1}{2}}$	$12^{\frac{1}{2}}$	$13^{\frac{1}{2}}$	$13^{\frac{1}{2}}$	$17^{\frac{1}{2}}$	$17^{2\frac{1}{4}}$	Perret C	f-9.90
21Apr84 ^{8}Aqu^{3}	Bear Hunt	b	3	126	3	$2^{\frac{1}{2}}$	$2^{\frac{1}{2}}$	5^{1}	$15^{1\frac{1}{2}}$	$18^{1\frac{1}{2}}$	$18^{7\frac{1}{2}}$	MacBeth D	57.40
21Apr84 ^{9}OP^{1}	Althea		3	121	1	$1^{\frac{1}{2}}$	1^{1}	1^{hd}	$5^{\frac{1}{2}}$	19	19	McCarron C J	a-2.80
21Apr84 ^{8}GG^{2}	Majestic Shore		3	126	16	12^{5}	10^{hd}	$16^{\frac{1}{2}}$	—	—	—	Lively J	f-9.90

Majestic Shore, Eased.
[D]-Gate Dancer Disqualified and placed fifth.
a-Coupled: Life's Magic and Althea; b-Pine Circle and Vanlandingham.
f—Mutuel field.
OFF AT 5:40. Start good for all but GATE DANCER. Won driving. Time, :23⅗, :47⅖, 1:11⅘, 1:36⅘, 2:02⅖ Track fast.

$2 Mutuel Prices:				
	10-SWALE	8.80	4.80	3.40
	12-COAX ME CHAD (f-field)		8.00	4.00
	9-AT THE THRESHOLD			13.80

Dk. b. or br. c, by Seattle Slew—Tuerta, by Forli. Trainer Stephens Woodford C. Bred by Claiborne Farm (Ky).

SWALE, away in good order, went after ALTHEA from the outside approaching the end of the backstretch, took over soon after starting the turn, drew off quickly entering the stretch and was under pressure to hold sway. COAX ME CHAD, outrun to the far turn, moved through along the inside leaving the far turn, came out slightly during the drive and finished well to best the others. AT THE THRESHOLD rallied racing into the far turn, remained a factor into the stretch while racing between horses but wasn't good enough, and leaned in slightly brushing with FIGHT OVER. GATE DANCER broke poorly from his outside post positon, raced wide into the stretch while advancing but failed to sustain his bid while lugging in and bumping FALI TIME several times. GATE DANCER WAS DISQUALIFIED AND PLACED FIFTH FOLLOWING A STEWARDS INQUIRY AND A FOUL CLAIMED BY THE RIDER OF FALI TIME. FALI TIME, never far back, remained a factor into the stretch while racing between horses and was bothered by GATE DANCER during the drive. PINE CIRCLE made up some ground approaching the final furlong but lacked a late response. FIGHT OVER, well placed into the backstretch, moved through along the inside nearing the far turn, raced forwardly into the stretch, then brushed with AT THE THRESHOLD while tiring. A CLAIM OF FOUL AGAINST COAX ME CHAD BY THE RIDER OF FIGHT OVER, FOR ALLEGED INTERFERENCE THROUGH THE STRETCH,

WAS NOT ALLOWED. LIFE'S MAGIC moved up along the inside nearing the end of the backstretch but was finished soon after going seven furlongs. SILENT KING, badly outrun for a mile, passed tired horses while racing very wide. REXSON'S HOPE was without speed. SO VAGUE failed to be a serious factor. BILOXI INDIAN gave way after racing forwardly for a mile. TAYLOR'S SPECIAL, wide throughout, raced within striking distance to the far turn but lacked a further response. RAJA'S SHARK saved ground to no avail. BEDOUIN raced extremely wide. VANLANDINGHAM was finished racing into the far turn and came back sore. SECRET PRINCE tired. BEAR HUNT prompted the pace into the backstretch, held on well until near the end of the backstretch and had nothing left. ALTHEA had speed from the outset, made the pace to the far turn while saving ground but was finished after going seven furlongs. MAJESTIC SHORE dropped back steadily while racing wide approaching the end of the backstretch and was eased when unable to keep up.

Owners— 1, Claiborne Farm; 2, Miller E E; 3, Partee W C; 4, Opstein K; 5, Mamakos & Stubrin; 6, Loblolly Stable; 7, Bwamazon Farm-Sabarese; 8, Mel Hatley Racing Stable; 9, Hawksworth Farm; 10, Rose Elsie A Stable Inc; 11, Hyperion Thoroughbreds; 12, Sundance Stable; 13, Lucas W F; 14, Feiner I; 15, Elmendorf; 16, Loblolly Stable; 17, Brodsky Elaine M; 18, Taylors' Purchase Farm; 19, Alexander-Aykroyd-Groves; 20, Oldknow & Phipps.

Trainers— 1, Stephens Woodford C; 2, Warren Ronnie G; 3, Whiting Lynn; 4, Van Berg Jack C; 5, Jones Gary; 6, McGaughey Claude III; 7, Parisella John; 8, Lukas D Wayne; 9, Delp Grover G; 10, Rose Harold J; 11, Russell Gerry M; 12, Carpenter Diane; 13, Mott William I; 14, Campo Salvatore; 15, Mandella Richard; 16, McGaughey Claude III; 17, Terrill William V; 18, Laurin Roger; 19, Lukas D Wayne; 20, Rettele Loren.

Sort the three-year-old graded stakes winners having a dosage index (DI) higher and lower than 4.0; list the horses and DIs.

Get the 1983 Experimental Handicap.

List the previous year's two-year-olds that were Gr. 1 stakes winners; get the Gr. 1 races won.

List this year's three-year-old Gr. 1 stakes winners; get the Gr. 1 races won.

The information can be retrieved manually if not by computer, but hardly as quickly or efficiently. Technology notwithstanding, information management skills will be the key to success when analyzing the Derby, the Belmont Stakes, and all Gr. 1 races at classic distances or farther.

As always, when critical information is not well distributed, the few in possession of it are at decisive advantage. The 1984 run for the roses is a glorious case in point.

The pre-Derby favorite was the brilliant Alydar filly Althea, recent winner of the mile and one-eighth Arkansas Derby by five lengths in track record time. No one expected, therefore, that Althea would not get the Derby distance comfortably, yet the filly's dosage index was 5.55, far too unbalanced on the speed wing, and too high to represent a Derby probable, let alone the race's favorite. At 5 to 2 Althea could be summarily dismissed.

Another popular pre-Derby favorite, Blue Grass Stakes and Louisiana Derby winner **Taylor's Special,** had a dosage index of "infinity," or zero influences on the stamina wing of its pedigree. Throw out Taylor's Special.

If handicappers had approached the 1984 Derby by asking which of the twenty starters were acceptable on dosage and had been weighted within ten pounds of the Experimental Handicap leader, as Roman recommends, only two horses could be considered: Swale, D1 of 1.93, and Fali Time at 2.50. Coincidentally, the same two survived if handicappers preferred horses acceptable on dosage which had won an important Gr. 1 preliminary. The eighteen others were eliminated. How convenient! Ah, but only for handicappers having the right information.

Swale and Fali Time finished first and fourth, respectively. The frontrunning favorite, Althea, under pressure steadily from Swale, tired badly before reaching the quarter pole and staggered in nineteenth of twenty. Taylor's Special did not get a call.

Even more. I was not caught unaware on this memorable day, as at Arlington Park in 1983. At Longacres, in Seattle, to participate in handicapping seminars during Derby week of 1984, your author, while doping the Derby on dosage, noted that of the fifteen starters acceptable on dosage, five constituted part of a wagering "field," and sent forth the advice that Longacres handicappers should not only back Swale strongly to win but also box that selection in Exactas with Fali Time, the longshot Silent King, and the "field."

Swale paid $8.80 at Longacres in the simulcast there, and the $5 Exacta from Swale to the "field" horse Coax Me Chad, a dosage probable, returned $247.25.

The lesson for approximately sixty Longacres handicappers who had not fixed on it previously was the value of having dosage information when three-year-olds stretch out to classsic distances against first-rank competition. Whether they store it in computer memory or not, in future years those sixty players will collect dosage indexes on the two- and three-year-old Gr. 1 stakes winners. At Derby time, they will not proceed without that information, and they will wish fervently to find prerace favorites like Althea, horses with high dosage indexes and therefore relatively little chance to win.

Interestingly, most of the 1984 pre-Derby journalism centered on the brilliant champion Devil's Bag's capacity to get the Derby distance. The colt had stopped mysteriously at 1-to-5 in the Flamingo Stakes and had not won around two turns until seven days before the Derby, in the Derby Trial mile at Churchill Downs, in which he finished well enough, but apparently weary. Amost unanimously, the experts remained skeptical that Devil's Bag would get a mile and one-quarter comfortably, but none questioned Althea's stamina in the slightest. In 1984 the experts thereby not only once again missed the crucial considerations in their pre-Derby analyses, that year they turned reality upside down.

Devil's Bag has a dosage index of 1.0, a perfect blend of speed and endurance. He would have entered the starting gate at Churchill Downs the quintessential classic prospect. That is, the first American racehorse to challenge the classics while combining the champion's performance record and the perfect pedigree.

With a dosage index of 5.55, Althea would have been fortunate to finish in-the-money at Churchill, and the experts should have told us so. So much for the conventional wisdom of this widely misunderstood sport.

To cement the point, consider briefly the 1983 pre-Derby favorite **Marfa,** trained not so coincidentally by leading southern California horseman D. Wayne Lukas, trainer too of Althea, and the personification of a contemporary horseman whose announced objective annually is to win the Derby, but whose eight starters in the race from 1981 to 1984, ironically, with two exceptions, had high dosage indexes. In its two starts prior to the Derby **Marfa** had won the Jim Beam Stakes at Latonia and the Santa Anita Derby, the California race, impressively.

Each year at Derby time the *Los Angeles Times* publishes for weeks its Triple Crown ratings, a ranking of the Kentucky Derby candidates by the panel of racing secretaries F. E. (Jimmie) Kilroe, of Santa Anita, in southern California, Tommy Trotter, of Arlington Park, in Chicago, and Lenny Hale, of Aqueduct and Belmont Park, in New York, the three most distinguished racetrack handicappers in the United States. The racing secretaries selected **Marfa** to win the 1983 Derby. Almost all leading newspapers in the nation followed suit. In consequence, bettors risked most of their money on **Marfa,** sending the colt into the gate at Churchill Downs their 2-to-1 favorite.

But Marfa had a dosage index of 6.60 and should not have been selected to win by anyone in the know. Positioned perfectly, in striking position at the top of the long Louisville homestretch, Marfa perished in that stretch, as have so many others, true to its pedigree prospects. It finished an obscure fifth, unplaced, and won no money. A year later, the colt had not won since, and if sent to the post in 1984 at a classic distance against graded stakes competition, **Marfa** would have been a longshot, even to place or show.

The types of handicapping information essential to analyzing these two races, about international racing and dosage, are relatively new and unfamiliar. The methods they entail are thus unconventional—not commonly used or popularly understood. So much the better for handicappers in the know. Numerous races are susceptible to similar analyses. Handicappers tied too tightly to conventional methods cannot do much about those races. They will wander afield repeatedly.

Moreover, and more to the thrust of this book, the familiar and trusted information about handicapping, of class appraisal, form cycles, speed figures, pace analyses, trainer patterns, body language, breeding, trips, and money management, has multiplied so abundantly across the past twenty years that recognizing the appropriate interpretations and applications has become seriously problematic for many regular handicappers. Newcomers are often overwhelmed. In 1983–84, no less than four major books were published. Research on thoroughbred handicapping and pari-mutuel wagering has intensified, even at the university level. Those trends will surely continue. Even as information systems have been changing so many contemporary industries and organizations in drastic ways, and they have and will, this book attempts to show contemporary handicappers how to adapt most effectively to that great upheaval.

Handicappers who learn how to handle all types of information appropriately, manage information efficiently, and use it to make handicapping decisions that are new and different, will find themselves and their methods changing in fundamental ways. They probably will find themselves relying on information and methods they have not even considered before now.

The information age is a brand-new world for all of us. Its promises are many and large. Before the promises can be fulfilled, however, an essential attitude and point of view must be accepted by everyone, even the most hardened and veteran of handicappers. It's a time for all of us to change.

CONTENTS

Chapter 1

NEW PERSPECTIVES

The more things change, the more they remain the same.

A FRENCH CLICHÉ

THE AMBITION OF this book is to guide thoroughbred handicappers, however diverse their ideas and methods, into the age of information. In that purpose the book is evolutionary. It will take theories, knowledge, and practices that already exist and are popularly understood, and adapt them to new directions and changing conditions, notably the general societal movement toward information management. As a result, the conventional practice of handicapping should improve. Procedure should be enhanced, knowledge broadened and transformed, and handicappers' decision-making should become better informed.

Those intentions are noble enough. If that were all, however, handicappers of the information age would not be expected to change how they think or what they do in any essential way. Nuances of interpretation and analytical insights might emerge, perhaps the use of prior knowledge would be altered slightly, and probably some new information would be incorporated into the standard conduct of numerous practiced handicappers.

But as the familiar social science saying puts it, the more things change, the more they remain the same. That cliché becomes particularly powerful whenever fundamental change is more technological than intellectual. The computer, the instrument of contemporary information management, is fundamentally a technological change.

In a deeper sense this book is intended to be revolutionary. It seeks as its fundamental purpose to change all handicappers in basic, irreversible ways. Its boldest intent is to use the ideas associated with information management and the computer to alter the fundamental character of how handicappers make decisions and how handicapping is practiced throughout this country.

Let me hasten to become clear as to what I believe six of the most profound changes should be.

First, a moving away from strict adherence to the *selection* purpose of handicapping and toward the *decision* purpose of handicapping. That is, make decisions, not selections.

Handicappers of the information age are urged to embrace a more dynamic analytical approach to the past performances. Concepts such as contenders, selections, and methods give way to outcome scenarios, decision alternatives, and computerized databases. The notion of making selections is replaced by the notion of making decisions. A selection is not a decision. The one represents an opinion, the other a choice. Yet most handicappers treat the two concepts as if they were interchangeable. The reasoning is practically syllogistic. If a selection, then a decision. In practice, they thus deny themselves the tremendous possibilities and promises held out by the information age.

Methods aside, the conventional handicapping process consists of (a) identifying contenders and eliminating noncontenders, (b) distinguishing the contenders variously, and (c) making selections.

Traditionally, handicappers work off-track and take their selections to the races with them. The homework has been dutifully finished. In essence, handicappers have made up their minds before they arrive. If track conditions warrant, body language permits, and the price looks acceptable, the bets are placed.

A worst-case scenario for handicappers arriving at the track with purposeful and carefully determined selections is that the selections should be ignored or changed, and the original horses win. This is psychologically debilitating; to be avoided at all costs. Much less agonizing, but upsetting, is the loss of money on horses that were not really **selections,** in races that were previously thought **unplayable.** Thus, many a hardened veteran can be spotted in the customary boxes, anticipating the key races of an afternoon and clinging to those few key-race selections as if they were irreplaceably prized possessions. Notwithstanding new information that might be meaningful or distracting, in arched tones and with great indignity they will assure you they are not about to change their minds.

This behavior has got to go.

The problem with the **selection** philosophy is informational. First, so many important and useful types of information become available to handicappers only at the racetrack, after the routine paperwork has been completed. Hardly a day will pass that (a) new **selections** not previously crystallized become intensely real or (b) off-track **selections** suddenly lose some of their glow or (c) the alternatives to several off-track **selections** become decidedly superior investment opportunities at the price. On most days, all three eventualities will come to pass.

Handicappers who find themselves arriving at the track with selections cast in gold, to decide from that point only whether to pass or play, and forsaking other options will be asked here to pay more attention to identifying the on-track information needs for making final handicapping decisions. The same races can be analyzed effectively in multiple ways. A proliferation of legitimate handicapping information has put many handicappers into precisely this kind of box. Each approach might identify a dissimilar "selection," or in this book's terminology, decision alternatives. What's what? Which horses deserve how much support?

Second, the suspension of belief in the utility of a particular methodology, model, or handicapping paradigm for all purposes; a correspoding attention to all types of information that might bear in meaningful ways on race outcomes.

Method players are not being asked here to detach themselves from their favorite handicapping activities. In fact, the organization of ideas, knowledge, and procedures into systematic methods is indispensable to effective race analysis and rational decision-making. In this sense class handicapping and speed handicapping can be expected to work equally well. Comprehensive methods and hybrid methods that intermingle pace analysis, form analysis, class, speed, distance, trainer, jockey, weight, post position, and other factors, the emphases and priorities varying from method to method, might also perform comparably well.

The case can be forwarded that various methods apply more appropriately to different kinds of races, and the same race might be analyzed effectively in different ways. Opposite logics do apply. Therefore, and to the degree that it is true, handicappers who become generalists as opposed to specialists will ultimately seize the advantage. That is precisely the point of view dispensed here. Speed handicappers are urged to become experts on class. Class handicappers are urged to calculate adjusted final times, at least, perhaps to go all the way to making speed figures. Both groups are urged to master well-known techniques of pace analysis and form analysis. A comprehensive working knowledge of the handicapping literature as a whole and in all its diversity is strongly recommended to all.

The reasons are easily underscored. Speed handicapping works especially well with claiming races, notably claimers rising in class. But it also works well at times with nonwinners allowance horses. Class handicapping works best with nonclaiming races, notably the longer races for better horses. But it cannot be ignored in claiming contests. Maiden races are not greatly susceptible to either method. Turf races cannot be analyzed as dirt races can. Sprints differ markedly from routes. Races at classic distances among superior horses belong in a category by themselves. Racetracks differ markedly from one another, and the same racing surface differs variously from one day to the next. Human errors by trainers and jockeys affect race outcomes every day. So do accidental mishaps and bad luck encountered during the races. Rain distorts normal performance. So does intense heat or cold.

Thus it can be seen that the main problem with too much reliance on specialized methods for all purposes is also informational. By definition, systematic methods screen out the information not encompassed by the methods. Too much meaningful information is thereby lost too much of the time. Or, when considered at all, any information extraneous to the method is most frequently used to evaluate further the method's key selection and not the race in its entirety. The methodical analysis is therefore skewed. The results are biased to favor the method. Selections are practically preordained. Systematic methods produce the same kinds of selections again and again.

An *alternative decisions* approach to handicapping employs various systematic methods for different races and **always** supplements that kind of analysis with a broader **information analysis** of all the circumstances of the race. Handicappers will take several slants on the same race. As a result, two or more **scenarios** having different probable outcomes will emerge. The scenarios clarify the decision alternatives to be evaluated. Even more information, about price, in particular, will be needed to conclude the decision-making process.

By this point handicappers should begin to sense that an alternating *decisions model* of handicapping intensifies the role of information and introduces the importance of information management. Fortunately, two conditions now coexist which promote the handicapping approach recommended by this book. First, the information has been generated and is available. Second, information technology makes it amazingly accessible. The final burden thus rests with information management. What decisions need to be made about this race? What information is needed for making informed decisions?

Third, the development of handicapping databases.

A database is an electronic record-keeping system. It is a high-tech term and without question the most important concept of information management. Computerized databases facilitate information processing in ways never before possible. Anyone who relies on information to make decisions can take it as a rule of thumb that the better the database, the better the decisions. In fact, databases are so critical to managers of the information society that within the near future just about everyone who is anyone will have one. If constructed properly, as this book tells, and **not** as commonly constructed today, personal databases will make high-tech handicapping an indispensable feature of state-of-the-art handicapping for this generation. Thus handicappers, however averse to computers, should understand in no uncertain terms that those who become the first on the block to develop a database will automatically leap ahead of the racetrack crowd by years.

Handicappers should understand as well that they do not need computers to construct and access databases. Databases can be manual record-keeping systems. When they go to the files of manual systems, handicappers will be doing what administrative, secretarial, and clerical personnel have been doing forever. By so doing they will also suffer the limitations and constraints of manual labor. They will sacrifice speed, efficiency, and valuable time. Much more important, computerized databases afford handicappers a **processing** capability manual databases cannot approach. The possibilities are virtually endless. Thus the most treasured attribute of electronic databases is productivity—information productivity. Handicappers without computerized databases lose that productivity, and that they cannot afford to lose.

* * *

Fourth, a reliance on **intuitive reasoning** skills to make the majority of handicapping and wagering decisions.

Even more than high-tech handicapping, the most revolutionary proposition of this book urges handicappers to rely on **intuitive thinking skills** to make the most of their future racetrack decisions. The proposition will not be startling to handicappers familiar with information technology and information management. It will not prove surprising either to those who have ever had to cope with large amounts of new or unfamiliar information for problem-solving or decision-making, either on the job, in the home, or at playing games. How to handle all that information at once?

As will be shown, extensive evidence indicates that intuitive reasoning outperforms logical deductive thinking when trying to solve problems whose solutions demand relatively large amounts of information processing. Techniques are available to help handicappers develop the skills. Does anyone not believe the art of handicapping qualifies?

To explain.

Deductive reasoning is reductive; it proceeds from the general to the particular. Its logic connects general principles, as with the general principles of handicapping, to specific instances of those principles, which may or may not apply. In other words, it proceeds from handicapping guidelines, tenets, or rules to races or horses which fit those guidelines, tenets, or rules. When handicapping is methodical, its logic is deductive and the purpose is to identify the horse that best satisfies the method's application. Method handicappers do this all the time. They should continue to do so, but not nearly as decisively; they should include it as part of a broader routine.

Intuitive reasoning is additive; it proceeds from the particular to the general. Its logic connects specifics, as with facts and data items about horses and races, to broad or general conclusions. In other words, it proceeds from numerous specific pieces of information about horses in a field and about races to conclusions about which horses figure to win, and which should be bet.

When handicapping is intuitive, that is, when it utilizes relatively large amounts of information and various types of information, its logic is inductive or additive, and the purpose is to identify horses that fit **all** the circumstances of today's race well. Method handicappers rarely do this. But they can begin the practice soon and should expect to make more and more intuitive decisions as the information age progresses.

Intuitive reasoning is a learned skill. It is hardly instinctive or impulsive in the stereotypical conceptions of woman's intuition. Further, as with any skill, it is developed and refined through technique and practice. As is true of skill training generally, a tremendous practice effect occurs.

If women who are handicappers become better at intuitive handicapping than do men, it will not be because they have keener instincts or because they keep in touch with feelings more intensely, but rather because they have practiced thinking inductively more frequently throughout their lives. Until recent years adult women have been socialized and acculturated in the home and family, where most of the thinking is intuitive inductive, and not in the professions, corporations, and work places, where the thinking, however ill-informed and incorrect, is usually logical deductive.

Carnegie-Mellon University in Pittsburgh carries out much research on inductive reasoning and intuitive problem-solving. Handicappers will be more than impressed with the rates of learning that can be demonstrated, even by subjects not especially susceptible to the training and not very confident of their ability to learn how to do it. So far as I know, the Carnegie researchers have yet to explore racetracks as social laboratories for their experiments. Unknown to them, they are missing out on a population and problem-solving, decision-making environment arguably matchless for testing their hypotheses.

Granted, method players and many racetrack veterans are not very susceptible to change or to new training of any kind, but other experimental conditions are nearly perfect. Method selections aside, the problems of handicapping demand inductive (intuitive) thinking repeatedly throughout the day. All manner of facts, information items, and opinions must be assimilated and sorted. Even more accommodating, the size of the bet in preprogrammed wagering methods aside, betting problems and decisions are entirely intuitive. Which horses to bet? What kinds of bets look most rewarding? How much to bet? What odds to accept? What odds not to accept? Whether to bet at all?

The classic problem is the setting of prerace betting odds that can distinguish underlays and overlays; that is, establishing a handicapper's morning line. As crucial as this objective is, no method, scientific, mathematical, or statistical, is available which handicappers can use to establish odds lines that reflect the true relative chances of horses.

Attempts to set accurate odds lines are made in a few exceptional cases relatively rigorously by statistical estimation, and in all other cases decidedly less rigorously by individual speculation.

If intuitive reasoning can assist in handicapping where players grapple with large amounts of information, it should improve the handicapper's typical betting behavior dramatically.

Fifth, an unprecedented emphasis on exotic wagering.

An information management approach that presumes more than a single way to skin a race utilizes opposite logics, multiple types of information, and intuitive reasoning, and results in outcome scenarios, decision alternatives, and final decisions do not pay its dividends as often from the straight pools. Wagering to win is appropriate to the selection model of handicapping. That practice should endure, unless the selections become underlays, that is, and must be passed, which happens more today than ever.

The essential corollary to this book's promotion of information management in handicapping and intuitive reasoning in decision-making is exotic wagering. Optimal results should be expected and can be obtained only from the combination pools. Many times the results will be fantastic, and seem fantastically easy to achieve, notably when the information from which they spring is not widely distributed. The promise of the information management approach is more solidly backed plays and more winners at better prices.

On a more practical level, exotic wagering is increasingly recognized as the only approach by which recreational handicappers can beat the races impressively. As opposed to professionals, who play strictly for profits that represent income, recreational handicappers play for fun and profit concurrently, and with discretionary money, not capital. The recreation is equally as important as the profit. Thus it plays an equally decisive role in the betting.

In seminars at tracks across the country, the author begins by asking the participants who rely on handicapping profits for 100 percent of their income to identify themselves. No one yet has raised a hand. Just a few have identified themselves as expecting approximately 50 percent of their income from handicapping profits.

This book's extensive treatment of **intuitive wagering** will be welcomed by most handicappers. Its guidelines encourage them to do roughly what they have preferred to do since the Exacta was introduced to the American horseplayer fifteen years ago; bet in exotic ways that promise the richest rewards from the smallest investments. If recreational handicapping and wagering consists of whatever one enjoys, as it does, information management, intuitive decision-making, and exotic wagering support those preferences.

To become successful, however, handicappers must entertain two changes in traditional recreational betting styles: one, structure a preestablished capital investment in planful, multifaceted ways; two, restrict all wagers to horses having a rational basis for support at fair to generous odds.

On the wagering front too, and not unimportant, the personal lifestyles of regular handicappers and incontrovertible facts of modern racing also argue persuasively for an increased dependency on exotic wagering.

Recreational handicappers often transform themselves into professionals for short to extended periods each season. For thirty days, sixty days, ninety days, or perhaps six months, the recreational handicapper attends the track regularly, sets achievement goals, and intends to maximize profits. This is healthy procedure. It is strongly recommended to handicappers who never seem to get off the daily treadmill. Moreover, it's the only real way to improve one's performance; to grow, if you will.

Regarding profits, straight wagering simply cannot optimize the returns during betting periods of six months or less. Weekend players have scarcely a hope to win significant money in the straight pools. In fact, computer simulations have revealed only one method of racetrack investment capable of returning important win profits when betting conventional amounts, say five percent of $1,000 to begin, at attainable results levels, perhaps 32 percent winners at average odds of 2.6 to 1. It is optimal betting, or the Kelly criterion method, where a fixed percentage amount equal to the player's advantage over the game is bet each time. All other well-defined, grind-it-out, fixed percentage, progressive methods of win wagering involve an unacceptable risk of going bust.

When Bill Quirin simulated a 700-race season for which 30.4 percent winners at average odds of 2.75 to 1 represented actual results, and repeated that computer experiment one thousand times for each of five money-management methods, randomizing the win-loss distribution each time, Kelly returned average profits of $4,054 and $3,075 when wagering 4 percent or 5 percent of the bankroll. The method never tapped out, but the profit margin across a 700-race season should not cause too much jubilation among handicappers. If three win bets a day are made, that's 700 bets in 233 racing days.

And that's the fatal flaw with grind-it-out methods. They progress slowly, very, very slowly. Kelly takes several seasons to return significant money. It's a long-term investment strategy, absolutely. Few handicappers wish to invest that kind of time and effort before profit-taking. For that reason, method play is almost completely abandoned.

In practice, the conventional short-term alternative strategy for making money in the straight pools is to vary the size of one's bets in accord with the strength of one's opinions. That is, bet more when more confident, less when less confident, employing some kind of subjective formula that regulates the size of the bet. This is precisely what horseplayers have done for all time, but without much resemblance to invoking any kind of exacting formulae. Unfortunately, except in the very short run, this does not work very well either. The same statistical realities that support methodical percentage betting for the long term penalize this kind of uneven unit betting "long term," a period as brief as three to six months. As time goes by, the central tendencies of the sport and of each player's level of skill, that is, win percentage, average odds on winners, win streaks, and loss streaks, gather their increasingly inexorable force. Over time, handicappers thus will lose an increasing number of their bigger, more confident bets. Huey Mahl has cautioned us about this phenomenon as much as anybody. Mahl likes to say that handicappers who lose the big ones and win the little ones, "are deep in mashed potatoes again."

An antidote has been proposed by Andrew Beyer, a leading proponent of sizing the bets to match the relative strength of handicapping opinions. Beyer urges handicappers to alert themselves to the possibility of "killing" a given race, presumably every so often during the season, and definitely in the exotics, when the key horse and all the attending circumstances of the race appear to be ideal. By this sudden kill, argues Beyer, regular recreational handicappers seize their best opportunities to beat the races in a significant way.

Within a framework of intuitive wagering that invests varying amounts from race to race, depending on rational considerations and not merely subjective opinions, this book promotes Beyer's concept of the sudden kill but extends its application beyond a single occasional race to the short term, perhaps two to three weeks or a month, when solid handicapping form and good fortune have joined to sustain a winning pattern. As Beyer has urged, the "kill" occurs mainly in the exotics. But by this book's approach, the size of the wagers, though dramatically increased for the short duration, will not, when lost, deliver seasonal knockout blows of their own. Needless to say, this kind of exaggerated intuitive wagering to maximize profits short term will be among the most difficult skills for handicappers to master. It is arguably the most difficult skill of all.

Sixth, participation in information networks.

In the organization, computerized information systems facilitate the logical and physical flow of data and information to executives, managers, and other decision-makers. Such systems are high-tech innovations presently happening everywhere in the corporate world, and normally extend far beyond the boundaries of the physical organization into the much larger corporate environment. The computer technology that supports these communication systems is called a **datacom network,** or data communications networks. Modern handicappers who are information managers will want to use datacom networks. Teamwork beats individual play in the information age.

The obviously persuasive reason for information networks among handicappers is that few individuals can collect, encode, store, process, retrieve, and use all the information appropriate to daily race analysis, and do so quickly, efficiently, and effectively. Trip information is difficult to collect, even more difficult to interpret for handicappers who have not attended the races. Pace figures are not kept (stored) by numerous speed handicappers who rely on final times. Class handicappers do not often calculate speed figures. Speed handicappers do not often possess class ratings. Body language information can be devilish: difficult to collect, interpret, and update. Personal databases might not be updated sufficiently for practical reasons. Members of an information network might divide the responsibility for accessing the various types of information. Decision-making remains individualized.

Information networks also can control greatly for the diversity and continuity of handicappers' information needs by eliminating the constraints of time and place confronting individual handicappers. The information systems can be as complicated as high-tech datacom networks, involving more communications software and technology than an individual's computerized work station would warrant, or they can be loosely arranged human networks of information collectors and distributors. Most communications networks will probably be hybrid systems, where one or two members own microcomputers. Their individual databases might contain some different data items, thereby contributing different types of decision support information.

The purpose remains the same. Enhance the information flow to decision-makers (handicappers), in order to improve the quality of their handicapping decisions. This book argues that handicappers who participate in information networks will discover more winners at higher prices than handicappers who stay the course alone. And because handicapping information is so diverse and so diffuse, participants in the same information networks will often combine the same types of information to make different decisions. Each decision will be better informed nonetheless, and results should improve all around.

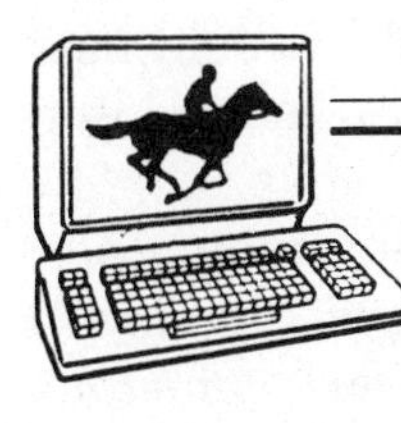

Chapter 2

HANDICAPPING'S INFORMATION EXPLOSION IS NOW

Can you please explain what we are supposed to do with all this different information? I mean, how are we supposed to handle it all? What's important? What's not important? I mean, there's just so much of it now. No horse satisfies everything; every horse figures on some of it. I mean, we can throw them all out, or leave them all in. How are we supposed to decide what to do?

MIKE MASON
Seminar on Thoroughbred Handicapping
Pasadena, CA, 1982

DURING HOLLYWOOD PARK'S spring-summer session of 1984 the grass racing defied its history. Suddenly the frontrunners were winning consistently over a course on which they had never before lasted. This phenomenon repeated itself day after day, race after race. Ordinary horses approached track record times that had stood for years. Lee Rousso, age 25, two-time winner of the national handicapping contests held in Reno, Nevada, which netted him $147,000, and one of the most astute, skillful handicappers on the American scene, commented he had never seen a track bias so powerful.

"What's so incredible," said Rousso, "is that this course has always had probably the strongest come-from-behind bias to be found anywhere. It's just a frontrunners' graveyard. No matter how slow or uncontested the fractions, even top horses die on this grass."

It's not difficult to imagine that many southern California method handicappers took a pounding from the Hollywood turf that season. Class handicappers surely did. Form cycle analysts did. Trainer intentions specialists did. So did classical comprehensive handicappers and so, paradoxically, did numerous pace analysts, whichever refused to begin and end the deliberations with speculations about the early speed. To be sure, systematic handicapping methods did not count for much in the face of the Hollywood turf's speed bias throughout that season. A less severe bias would have mattered less, but method players would have chased many losers nonetheless.

Prior to the mid-seventies the literature of handicapping contained scarcely a reference to track bias. Then Steve Davidowitz promoted the concept strongly. So did Andrew Beyer. Numerous, more recent discussions of trip handicapping deal extensively with track biases, such that they have by now become part and parcel of the general handicapping methodology. Contemporary handicappers pay attention to speed biases and off-pace biases, to inside biases and outside biases, and to combination biases, such as speed biases that favor the rail or speed biases that favor the outside paths or off-pace biases that favor the rail or off-pace biases that favor the outside paths. These change variously from one part of the season to the next, and it can get annoyingly confusing to the uninformed.

In concert with the literary trends, the author as recently as 1980 would have lost large amounts to a speed bias on the turf at Hollywood. I would have neither recognized nor understood what was happening. Track biases apparently are not as frequent or meaningful on the normally glib surfaces of southern California tracks as they are elsewhere.

When Andrew Beyer played at Santa Anita during 1983, he mentioned two months into the season that he had hardly noticed any kind of bias handicappers might exploit. By now, in 1985, it's clear that when turf courses experience prolonged periods of intense heat or intense cold, they get terrifically firm. During Santa Anita's 1984 winter season preceding Hollywood's, it hardly rained a drop. At times the heat became intense indeed, particularly just prior to Hollywood's opening. The familiar southern California storm season of January/February/March had passed us by. The weather conditions affected the Hollywood Park turf course tremendously.

Southern California handicappers possessing a working knowledge of track biases were equipped to adapt quickly to the Hollywood turf in 1984; others were lost. Most in the know presumably were spared serious financial losses, and several probably converted the speed bias into considerable winnings. Modern players like Lee Rousso make a killing when normal circumstances change in this unpredictable way. They are well informed, up to date. They collect numerous types of information, can interpret all of it properly, and know precisely how to cope with changing conditions.

The Hollywood turf situation of 1984 supports a central theme of this book. Information managers have much more handicapping ammunition than do systematic method players. Times change. Playing conditions change. So do the information needs of thoroughbred handicappers.

Only yesterday, from September 1983 until July 1984, ten months, four major works on handicapping were published. Each extended the knowledge base of this specialized field; each complicated the methodology. Houghton Mifflin gave us Andrew Beyer's discourse on trip handicapping, emphasizing the use of trip notes in relation to speed figures and his pointed recommendations about modern money management; William L. Scott published under his own imprint, Amicus Press, a major investigation of form analysis, liberalizing still further the prevailing standards of recent action and identifying at least three "form defects" that might reliably eliminate many low-priced horses and favorites, without eliminating many winners; William L. Quirin contributed through William Morrow a comprehensive state-of-the-art reprise on analyzing the past performances, and in the process extended his own speed methodology to include pace figures and speed figures in combination, as well as providing par variants for every North American racetrack; and Harcourt Brace Jovanovich introduced two university scholars, William T. Ziemba, professor of management science at British Columbia University, in Vancouver, Canada, and Donald Hausch, a doctoral candidate at Northwestern University, who published jointly the Z-System, a fail-proof method of place and show wagering that entails the use of high technology—handheld computers—at the racetrack. The traditional guidelines on place and show betting were thus summarily superseded.

Also in 1984, in April, in anticipation of the Kentucky Derby, *The Thoroughbred Record* published "An Analysis of Dosage," by Dr. Steven A. Roman, the most completely accessible explanation of dosage and the dosage index to reach a popular audience since *Daily Racing Form* "Bloodlines" columnist Leon Rasmussen first introduced Roman's stimulating, provocative, persuasive, and amazingly reliable research back in May 1981. Applications of the dosage index to handicapping, many of them ill-informed and tangential, have since been developed by individuals at racetracks across the country.

The glut of handicapping information thus continues to accumulate unabated. It will continue to accumulate for the forseeable future. The struggle to legitimize handicapping instruction, intellectually, has finally been won. All can move on now to still deeper investigations of the subtleties, complexities, and paradoxes of the whole process of handicapping. The racetrack is yet a vast open laboratory. How fortunate and appropriate for handicappers that they should be finding out so much of what they need to know just as the high-tech age of information has been gathering its steam. The relationship is symbiotic; the two go hand in hand. But many handicappers are troubled. What to do with all that information? It's the operative question of the day.

THE INFORMATION PROBLEM

So far the information problems of handicapping have not been properly formulated. The question is not what handicappers should do with all that diverse information beating down upon them. That issue is not the real problem. For the proper perspective on that, handicappers have to backtrack.

Historically, exactly the converse of the situation today has been the real problem for handicappers. Tom Ainslie bemoaned the lack of scholarly literature on the pastime when he felt his initial stirrings toward handicapping as an intellectual pursuit. Ainslie wrote that he could hardly find a single worthwhile book.

Other, later searches of the literature, including my own, found the single regularly published author for the thirties and forties was Robert Saunders Dowst, whose class-consistency discoveries dominated his thinking, as well as his several books. Of form analysis, there was nothing. Of speed methods, there was little. Of pace analysis, the same. Trainer intentions, trips, biases, body language, and other modern specialties went unattended. In 1983 Andrew Beyer told the audience at the first national conference on handicapping in Los Angeles that Dowst himself recognized the humble state of the art of his times and suggested that Dowst would be astonished to witness the intellectual evolution well underway now.

The few other handicappers who found a serious audience, writers such as Robert Rowe, Colonel E. R. Bradley, and Ray Taulbot concerned themselves with systems and methods that usually surveyed a specific slice of handicapping, such as pace, but ignored the wider context. Since the full context of handicapping was hardly understood at all, practitioners would generally learn that systems and methods, even substantial ones, did not work well enough. Taking up the slack, of course, were Runyonesque tales of racetrack characters and fast-buck, simple-minded systems of no worth. In an unfortunate way, these carried the day and guaranteed that the stereotypical perceptions of handicapping instruction, which characterized the field, would loom as major impediments to progress once the real work began in the late sixties. The rip-off systems and fast-buck artists slowly have been losing their charms to hard-nosed handicapping information ever since.

Under these circumstances the seeming avalanche of information raining down on handicappers in recent years has proceeded as if from a historical imperative. It has simply been filling the void. If handicappers of today think they perceive an oversupply of theories, a multiplicity of methods, and too much new or unfamiliar information, that perception is largely an illusion. Handicappers instead might appreciate the need for solid information of all kinds. Almost nothing came before. Handicapping, as both an art and science, has been evolving into a discipline. That evolution requires a tremendous knowledge base—information. So the response to the most common question about the current amount of information is to reformulate the issues. The amount of information is not the real information problem. Handicappers have instead still far to travel on the information front.

For many regular handicappers, far too many, a legitimate information problem is access. Recreational handicappers simply do not have the appropriate information at hand when they need it. This thwarts effective decision-making. Most players have chosen to live peacefully with these constraints. That is a major mistake.

Information management specialists have developed a deliberate style of responding to the access issues. They deliver the high-tech argument. Computers. Databases. Communications systems. They first point out that in the purest sense the availability of information is not really the problem and never has been. Most types of information, even the most obscure—remote scientific findings on the most esoteric topics, information specialists insist—have always been available. People can get any information they really want or need. In this tortured sense, availability is tantamount to access. Accessing the information, however, has simply been such an overwhelmingly impractical problem, especially for the individual, that few individuals bother. Access has not been cost effective.

But in the information age, the argument goes, we have computers and data communications systems, including satellites, fiber optics, and all the wonders of high technology, such that large amounts of specialized data and information can be sent across the oceans and circulated among the continents within minutes or seconds. The access problem has thereby been solved. Obtaining information is now both practical and cost effective. In consequence, information is now a primary resource. That conclusion is true. It holds for handicappers.

As we have just noted, the mere existence of the requisite information has been the greater practical obstacle for handicappers. Now that information availability has largely been achieved, what about practical, hands-on, cost-effective access? For most handicappers, access to comprehensive handicapping information remains a problem, but it need not be. For our purposes, access is one of two serious problems handicappers confront in the information age, but by far it is the less serious of the two.

The access issue can be boiled down to the modern handicapper's need for several types of information not contained in the past performance tables. Having access to the *Daily Racing Form* and its results charts, while indispensable, is no longer enough. Handicappers with access to no more information than that will be at a competitive disadvantage. Far too many handicappers can interpret and use that information skillfully; the competition is particularly strong on the major circuits of New York and southern California, but it is relatively strong just about everywhere. Underlays resulting from widely distributed access to conventional information are plentiful.

The access problem solution involves, first, identifying the types of information to be collected, and second, developing an information system that organizes, encodes, stores, processes, retrieves, and displays that information. A quick fix at step two is the microcomputer. The personal computer encodes, stores, processes, retrieves, and displays complex information quickly and efficiently. It dissolves the access problem into manageable homemade solutions.

Before proceding to identify the types of unfamiliar information handicappers should want to access regularly, and therefore should regard as indispensable to personal information systems, it's appropriate to stress again that microcomputers are not essential. Information systems can be manual systems. A file of standard-sized manila folders that contain the various data records of interest can be perfectly suitable for information management purposes, but that is obviously not as quick, efficient, flexible, or convenient. Moreover, the processing capability, or information productivity, of manual systems depends on the individual's willingness and skill in manipulating large amounts of data. When processing capability is desired, manual systems become cumbersome and burdensome. They are no match for computers. Productivity goes down. Decision-making falters. Thus, a running theme here is that handicappers should bite the bullet and buy a micro. Notwithstanding that, handicappers who do not own microcomputers and do not intend to buy one, can also cope. They should commit themselves at once to developing a manual information system.

THE HANDICAPPER'S INFORMATION SYSTEM

An information system is a logical and physical array of informational sources and types that are relevant to routine operations, problem-solving, and decision-making.

The handicapper's information system can become fantastically elaborate. Describing its logical and physical structure is within the scope of this book. Some of the sources and types of handicapping information described below are familiar to practitioners; others are unfamiliar. Most are indispensable to effective handicapping in the eighties; all are useful. Particularly helpful are types of information not trapped by the past performance tables.

It is convenient to organize the types of handicapping information under six headings: theories and methods, 1965–1984; probability studies; statisical compilations; reference materials; soft information; financial records.

MAJOR THEORIES AND METHODS, 1965–1984

If different handicapping methods apply more effectively to different kinds of races, and they do, handicappers are not adequately equipped by adhering strictly to a single approach. This book urges modern players to become generalists rather than specialists. That means handicappers must be prepared to analyze the past performances by utilizing each of the following approaches.

In 1965, and more elaborately in 1968, Tom Ainslie described a methodology he referred to as comprehensive handicapping. Its rationale was that *all* the factors of handicapping played *a part* in predicting race outcomes, the interrelations of the factors varying in priority and emphasis depending on the circumstances of the race.

Ainslie set down specific guidelines for evaluating horses on class, form, distance, and pace, his key fundamentals, and outlined an analytical method for proceeding. In essence, horses first were eliminated on relatively strict considerations of form and distance, the survivors designated the contenders. Contenders next were evaluated in terms of class appraisal and pace analysis. Final contenders were distinguished on a list of thirty-plus factors, which encompassed all the aspects of handicapping. A selection was made, tentatively. That horse was inspected again at the track, in the paddock and post parade. If body language were acceptable, the bets were placed. If horses were not comprehensively acceptable, races were judged unplayable, and passed.

In response to the phenomenal acceptance of this book's guidelines, Ainslie subsequently published a private rating method. From a base ten, and utilizing a checklist of handicapping standards, handicappers added or subtracted points for each horse in stepwise fashion. The horse with the highest point total equalled the final selection. The rating method's application requires access only to the past performance tables. Ainslie revises the comprehensive method periodically, as the knowledge base of handicapping changes or grows.

In the late sixties comprehensive handicapping represented a major historical breakthrough. It provided a general theory and method of handicapping where previously none had existed. Several of the subsequent in-depth investigations of handicapping as theory and method concentrated on the class-form or speed-pace dynamics inherent in comprehensive handicapping, or more broadly, Ainslie's basic propositions about racing and handicapping, which, after all, had been empirically based, but not statistically validated. Several of the first probability studies of the numerous handicapping characteristics, for example, focused on Ainslie's specific guidelines as hypotheses or points of departure.

Comprehensive handicapping persists today as a general theory that all practitioners should study closely, but many of its specific guidelines have been outdated. Its classic guidelines on form analysis have been roundly liberalized. Its handling of the distance factor is known now to be far too conservative and incomplete. In-depth investigations of the speed and class factors have resulted in new knowledge that supersedes the comprehensive handicapping guidelines.

Moreover, additional handicapping factors of real significance, which even comprehensive handicapping omitted, have since been identified and explored, notably speed figures, early speed indexes, dosage, trainer intentions, eligibility conditions, body language, trips, track biases, and turf pedigrees. Most contemporary handicappers rely on methods that encompass several of these new-wave concepts.

In practice, comprehensive handicapping has suffered in ways not peculiar to that method but to any that gains wide distribution and application among practitioners. As thousands of enlightened, born-again handicappers flocked to racetracks attempting to implement comprehensive handicapping, prices on the methods' selections plunged. The method thus began to throw seasonal losses; it supported too many underlays. As comprehensive handicapping relies exclusively on *Daily Racing Form*'s past performances, that problem persists, even as alternative approaches increasingly uncover attractive alternatives to "comprehensive" selections.

Nonetheless, as with any general theory that has been rigorously investigated, confirmed, and validated in its essential thrust, comprehensive handicapping provides an outstanding grounding in the practice of fundamental handicapping. The method can be considered classical. Its rationale, general propositions, and specific guideposts should be understood in part for their own sake and in part as referents by which other approaches might be fairly evaluated.

It should be abundantly clear to all by now that speed handicapping affords handicappers some of the sweetest opportunities of any season. What could be more advantageous than picking horses whose final times appear slower but are actually faster. Adjusted final times beat unadjusted final times repeatedly each season, and notably among horses rising and dropping in claiming class. Skillful speed handicapping permits handicappers to benefit from these unusual or obscured situations. Further, when adjusted final times have been converted to speed figures, performances at different distances can be compared fairly, shippers can be evaluated, and all sorts of convenient comparisons become facile.

Speed handicapping can be laborious and time-consuming, but it is also particularly susceptible to high-tech handicapping. The computer reduces the time and labor traditionally involved, and its processing capability can spit out figures and all the interesting comparisons instantly. If a speed handicapping database has been constructed properly, all kinds of applications programs might be developed to search that database. Handicappers of all persuasions, for example, could obtain speed figures merely by entering a horse's name.

This book believes all handicappers of the information age should embrace speed handicapping for general purposes, at least calculating adjusted final times. Handicappers without microcomputers can stop there, with adjusted final times, but those having microcomputers can begin to construct a database that contains par tables, final times, and daily variants for all races on the circuit; the raw times and beaten lengths of horses; the speed chart and beaten-lengths chart of handicappers. The potential applications of that kind of database are fantastic.

Four speed handicapping methods promoted variously by Andrew Beyer, Bill Quirin, Henry Kuck, and others have now enjoyed general acceptance and use in the field. In ascending order of their labor demands and complexity, they can be characterized as follows:

1. *Speed handicapping with adjusted final times.* The method involves constructing class-distance par times tables, calculating race variants and daily track variants, and adjusting the raw final times of races and horses to reflect the combined influences of class (pars) and track surface (variants).
2. *Speed handicapping with figures.* The method converts adjusted final times into figures that can be used to compare horses that have competed at different distances and class levels. Beyer's speed charts embody the principle of proportional time, as opposed to parallel time, such that when horses' figures are projected to other distances, the figures at the other distances reflect the proportionate value of the original time at the new distances. This approach provides a finer

estimate of true speed than do parallel time charts or speed figure charts, which embody the assumptions of parallel time.

3. *Speed handicapping with projected times.* This method is highly advanced. It ignores pars and concentrates instead on the well-known figures of horses that consistently run the same adjusted final times. Beyer and Kuck both promote the approach. It demands extensive experience with the local horse population and a highly practiced skill at making reliable projections. Few recreational handicappers can use the approach successfully, but several professionals do. Times and speed figures are also adjusted variously according to highly practiced estimates of the effects on running time due to trips, biases, and even high winds. This brand of speed handicapping becomes greatly esoteric.
4. *Speed handicapping with pace pars.* The method relies on class-distance final time pars and the associated pace pars, and it has been developed and promoted vigorously in recent years by Bill Quirin. The pace-speed figures are *not* combined to produce an ultimate figure, but instead are interpreted analytically, in terms of pace analysis. The method involves calculating the pace pars and fractional times in tenths. To evaluate shippers, it utilizes par variants for the respective racetracks. It also provides nine typical "race shapes," defined by expected pairs of figures against which the actual paired figures of horses under study can be interpreted.

Speed handicapping has several useful applications. Beyer asserts it's the surest way to decide whether claiming horses can move ahead in class successfully, often at nice prices. Quirin argues that figures can occasionally provide the telltale clues in maiden races and preliminary nonwinners allowance races, where class differences can be wide. In claiming races for older horses generally, where levels remain relatively stable and the differentiations permitted by class-form analyses among several horses at comparable price levels are often not so severe, speed handicapping often can unmask real differences in ability which do exist.

Unfortunately, speed handicapping has suffered the same sad experience as did comprehensive handicapping earlier: lowered odds on figure selections. Thus, the most progressive speed handicappers of today employ speed methods in combination with related methods, notably trip handicapping, pace analysis, and class appraisal.

Handicappers who will be generalists in the information age need to master the knowledge underlying five additional approaches that have gained widespread allegiance and application. They must also develop a working knowledge of the methods the five entail. The first two deal extensively with class appraisal.

In 1981 the author promoted an understanding of eligibility conditions as the critical determinant of effective class evaluation. As the role of eligibility conditions is precisely to define and limit the class of the field, the idea is to recognize the past performance profiles of horses that are well suited to conditons, as distinguished from those that are poorly suited. The performance profiles of horses that win most frequently under various conditions of eligibility were arranged in descending order of preference. Elimination profiles were also identified. By matching horses to conditions properly, handicappers not only could avoid backing horses unsuited to the class of race but also could regularly discover hidden opportunities not popularly understood, particularly in the garden variety of allowance and stakes races.

In a companion booklet, the author prepared detailed selection and elimination profiles for the various conditions of eligibility carded in major racing. Implementation studies have revealed that the top preference horses win approximately 20 percent of the races reviewed, with profit margins at 10 to 15 percent. High preference horses win much more frequently than do the lower-order profiles. Elimination profiles do not eliminate many winners. Classifying horses as suited or not suited to eligibility conditions has been widely recognized as the most effective technique for getting a fast line on the relative class of the field, particularly under any nonclaiming conditions.

Another approach to class appraisal has been termed ability times by its inventor, author William L. Scott. Proceeding from Scott's determination that speed and class represent the two key determinants of race outcomes, the method calculates the final quarter times of sprints, the second call times of middle-distance races, and modifies these crucial quarter-mile times by considerations of form and early speed. A chart is provided that converts the modified "ability" times into figures. Handicappers calculate several ability times for each horse and rely on the top two. The high-figure horse gets the play. The method also produces *double-advantage* horses, whose two top-ability times figures are each better than either figure of the other horses. Practitioners swear that double advantage horses return powerful results consistently. Bill Quirin's latest probability data confirms that experience as fact. It cites the impact value (I.V.) of double-advantage horses as 2.78 and estimates their dollar return at 37 percent profit. Several devotees of ability times have learned that the method can reveal interesting figure horses when recent races have been sufficiently unrepresentative that they interfere with the reliability of other methods. Ability times is not as sensitive as most systematic methods to recent races.

An approach to handicapping that relies exclusively on early speed consists of a clever rating method proposed years ago by Bill Quirin, who labeled the procedure *speed points*. Quirin developed speed points when his massive probability studies indicated clearly that early speed was the most important factor in handicapping. Approximately 60 percent of all races at all tracks were being won, Quirin found, by horses that had run first, second, or third at the first call in their latest starts.

The speed points method assigns 0, 1, or 2 points to each horse for each of three races, depending on their position and beaten lengths at the first call. Handicappers never rate races prior to the five most recent races. Horses can earn 0 to 8 speed points, and a percentage of a race's total speed points is calculated for each horse. Quirin provides basic interpretation guidelines, and users of the method, this one included, agree readily with Quirin's assertion that speed points serve handicappers as the surest predictor of a race's early speed. Studies indicate profits are more than possible merely by backing the top speed-points horses every time.

In 1983 Andrew Beyer published the first comprehensive explanation and discussion of trip handicapping, which involves the scheduled observation of horses in competition. Beyer's method emphasizes running position and track bias more than trouble, and involves short, simple, but deadly accurate notations at several points of call, including the starting gate, the first turn, the backside, the far turn, entering the stretch, and in the stretch.

Beyer recommends interpreting trip notation in relation to speed figures, a position strongly endorsed by Quirin in his 1984 distillation of the state of the art of handicapping. Trip handicapping has the highly agreeable potential of unlocking overlays, horses whose recent races might actually have looked quite different than the past performance lines suggest.

A fifth major approach, normally used in combination with other analytical methods, and one too often ignored by untutored handicappers, is Bonnie Ledbetter's body language profiles of horses ready to race, or unready to race. Six appearance and behavior profiles are drawn in detail by Ledbetter, of the *sharp* or *frightened* horses, of the *ready* or *dull* horses, of *angry* horses, and of *hurting* horses.

As 90 percent of races are won by horses that look *sharp, ready,* or *dull* on the track, handicappers are strongly advised to recognize these and to avoid the *frightened, angry,* and *hurting* horses. Selections and contenders that look frightened, angry, or hurting should be readily eliminated, certainly at low prices.

A more interesting application is the confident support of especially *sharp* horses that perhaps looked dull last time and lost. Thinking positively, Ledbetter has persuaded thousands of handicappers that contenders can sometimes be distinguished at the paddock.

Handicappers who have developed what the author often calls a working knowledge of these methodologies will find themselves in control of diverse methods of analysis and evaluation, each of which can be used variously, frequently, and profitably, as situations demand. Such handicappers are well along the path to becoming generalists rather than specialists. Thus, they enjoy the generalist's broad analytical scope and edge. But they are still far from qualifying as information managers.

PROBABILITY STUDIES

That last point is worth winning. All handicappers who get their results solely by the expert application of systematic methods are asked to try their hand with the restricted turf stakes included here. It's for three-year-olds, carded at Santa Anita, April 18, 1984.

Do speed handicappers like Barcelona? Do class handicappers lean toward Bedouin? Coming off a troubled trip, does the consistent, improving Bean Bag fit the conditions snugly? Which horse has the best ability times? How do comprehensive handicappers pick among Juliet's Pride, Barcelona, Bedouin, or Bean Bag?

8th Santa Anita

1⅛ MILES. (TURF). (1.45⅗) 4th Running of THE LA PUENTE STAKES. $60,000 added. 3-year-olds which have never won *$20,000. (Allowance.) By subscription of $50 each to accompany the nomination, $600 additional to start, with $60,000 added, of which $12,000 to second, $9,000 to third, $4,500 to fourth and $1,500 to fifth. Weight, 120 lbs. Non-winners of $16,000 at one mile or over, 2 lbs.; of such a race of $13,000, 4 lbs.; of such a race of $11,000, 6 lbs. A trophy will be presented to the owner of the winner. *A race worth $20,000 to the winner. Closed Wednesday, April 11, 1984 with 17 nominations. (Maiden and claiming races not considered.)

Juliet's Pride

Own.—Cole-Freeman-Gold **120**

Dk. b. or br. c. 3, by Table Run—Up to Juliet, by First Balcony
Br.—Rowan & Whitney (Cal) 1984 4 1 0 0 $18,225
Tr.—Stute Melvin F 1983 7 1 3 1 $92,481
Lifetime 11 2 3 1 $110,706

8Apr84-3SA 1 :46^3 1:11^1 1:36^{4}ft 4 116 1^{hd} 1^3 1^{2½} 1^{¾} VlenzulPA^2 [S]Aw30000 84-15 Juliet's Pride, Bean Bag, QuatiPink 7

28Mar84-8SA 1⅛ :46^3 1:10^4 1:48^{3}ft 33 118 5^{4½} 5^{5½} 9^{12} 9^{13} McGrnC^{10} [S]Bradbury 73-10 TsmmShw,MghtyAdvrsry,LotsHony 10

14Mar84-8SA 1 1/16 :47^1 1:11^1 1:43^{3}gd 24 114 10^{13} 10^{18} 10^{19} 10^{18¼} ShmkrW^4 [S]Sta Ctlna 65-26 Tights, Prince True, Gate Dancer 10

14Mar84—Steadied at 7/8

3Mar84-7SA 6f :22 :45^3 1:09^{4}ft 4¾ 114 6^{5½} 6^8 6^9 4^{7¾} McCarronCJ^6 Aw23000 81-14 Barcelona,LordOfTheWind,TkkTkk 7

19Nov83-8BM 1 1/16 :51^1 1:17^4 1:55 sy 3¾ 115 4^{7½} 2^4 2^5 2^{½} VlnzulPA^6 El Cmno Rl — — BoldT.Jy,Juliet'sPride,John'sRuffi 10

5Nov83-8SA 1 1/16 :46^4 1:11^3 1:44^{1}ft 9¾ 118 9^{7½} 8^9 7^6 5^{5¼} VlenzuelPA^{10} Norfolk 75-20 Fali Time, Life's Magic, Artichoke 10

5Nov83—Wide into stretch

16Oct83-8SA 1 1/16 :46^3 1:11^2 1:44 ft *3-2 117 4^{4½} 4^6 3^8 3^{7½} VlenzulPA^1 El Rio Rey 73-17 Artichoke, Slugfest, Juliet's Pride 5

16Oct83—Veered out, bumped stretch

14Sep83-8Dmr 1 :45^4 1:10^2 1:34^{4}ft 15 115 5^4 3^3 2^6 2^{6½} ValenzuelPA^1 Dmr Fut 87-14 Althea, Juliet's Pride, Gumboy 5

31Aug83-8Dmr 1 :45^3 1:10^4 1:37^{1}ft 15 115 5^4 3^4 4^3 2^{1½} ValenzuelaPA^2 Balboa 88-19 Party Leader,Juliet'sPride,Gumboy 7

23Jly83-8Hol 6f :21^4 :44^1 1:09^{2}ft 32 117 8^{9¾} 8^{14} 7^{13} 4^{12½} VlnzlPA^8 Hol Juv Chp 77-21 Althea,RjectedSuitor,AtoCommndr 9

Apr 16 SA 4f ft :48^4 h Apr 4 SA 5f ft 1:01^2 h Mar 26 SA 4f ft :48^2 h Mar 21 SA 5f ft 1:00^4 h

Loft

Own.—Harris C E II (Lessee) **114**

Ch. c. 3, by Roberto—Journey, by What a Pleasure
Br.—North Ridge Farm (Ky) 1984 3 0 0 0 $1,500
Tr.—Jolley Leroy 1983 3 2 1 0 $23,560
Lifetime 6 2 1 0 $25,060

14Mar84-8SA 1 1/16 :47^1 1:11^1 1:43^{3}gd 22 115 5^4 8^9 9^{14} 9^{16¾} Pierce D^3 [S]Sta Ctlna 65-26 Tights, Prince True, Gate Dancer 10

14Mar84—Very rank to place 7/8 turn, took up sharply at 3/4

2Mar84-7SA 1 1/16 :47^1 1:11^3 1:42^{4}ft 7¾ 115 6^3 3^1 5^{5½} 5^{8½} CastanedaM^8 Aw30000 81-18 Distant Ryder, Lotsa Honey,Bozina 8

19Feb84-9SA 7f :22^2 :45^1 1:22^{2}ft 9-5 115 8^5 7^{7½} 5^{12} 5^{12¾} Pierce D^4 Aw30000 75-14 Fifty SixInaRow,SharperOne,Armin 8

14Oct83-6Med 1 :46^3 1:11^1 1:37^{3}ft *1 112 3^3 3^{2½} 2^4 2^5 CordrAJr^5 [S]Montclair 88-18 Shuttle Jet, Loft, Arabian Gift 6

27Sep83-3Med 1 :47^4 1:12^3 1:38^{1}ft *6-5 120 3^{2½} 3^1 1^{1½} 1^3 Cordero AJr^5 Aw12000 90-14 Loft, Tango Victor, RainbowCastle 7

24Aug83-4Bel 5½f :23 :46^4 1:05^{3}ft *6-5 118 3^{1½} 3^{1½} 1^{hd} 1^{¾} Cruguet J^5 Mdn 87-12 Loft, Gold Studs, Fingers Inthe Till 8

●Apr 16 SA ⓣ 3f fm :38^3 h (d) Apr 9 SA ⓣ 7f fm 1:28^4 h (d) Apr 5 SA 4f ft :51 h ●Mar 29 SA 1 ft 1:38^2 h

Swivel

Own.—Holt L **120**

B. c. 3, by Don B—Cuppling, by Dress Up
Br.—Newmann D (Cal) 1984 4 1 0 0 $20,925
Tr.—Holt Lester 1983 6 2 0 1 $24,900
Lifetime 10 3 0 1 $45,825

7Apr84-4SA 1 1/16 :45^4 1:10^4 1:43^{2}ft 16 120 6^{11} 4^{2½} 4^{1½} 1^1 Olivares F^4 Aw33000 84-18 Swivel, Bedouin, Barcelona 6

2Mar84-7SA 1 1/16 :47^1 1:11^3 1:42^{4}ft 30 117 8^{8½} 8^{11} 8^{10} 7^{13¾} Olivares F^6 Aw30000 73-18 Distant Ryder, Lotsa Honey,Bozina 8

20Feb84-4SA 1 1/16 :46^1 1:11^1 1:43^{2}ft 17 120 7^{15} 7^{9½} 6^{6½} 5^{7½} Olivares F^1 Aw30000 76-16 GateDncer,PrinceTrue,HevenlyPlin 7

22Jan84-7SA 1 1/16 :46^3 1:10^4 1:43^{4}ft 4½ 120 5^9 4^{6½} 3^{5½} 4^5 Olivares F^6 Aw27000 77-21 Tights, Bold Batter Up, Bedouin 6

26Dec83-7SA 1 1/16 :46^4 1:11^3 1:44 ft 10 118 8^{9½} 6^6 3^{1½} 1^{1¼} Olivares F^9 Aw21000 81-16 Swivel,KeepOnTalking,CopyMaster 9

26Dec83—Checked, altered course at 1/8

5Nov83-8SA 1 1/16 :46^4 1:11^3 1:44^{1}ft 68 118 10^{7¾} 9^{9½} 8^{6½} 6^{7½} Olivares F^5 Norfolk 73-20 Fali Time, Life's Magic, Artichoke 10

22Oct83-4SA 1 1/16 :47 1:12^2 1:45^{4}ft 3¾ 116 7^{7½} 4^{3½} 3^2 1no Delahoussaye E^1 Mdn 72-19 Swivel, Lotsa Honey, Intimidation 9

8Oct83-6SA 1 :46 1:11^3 1:38^{3}ft 12 112^5 9^{17} 7^{12} 7^8 3^{1½} Estrada J C^9 Mdn 73-16 OneO'ClockJump,LotsHoney,Swivel 9

8Oct83—Extremely wide into stretch

10Sep83-6Dmr 6f :22^3 :46^1 1:11^{1}ft 6½ 112^5 8^{5¾} 7^{8¾} 7^{12} 5^{8½} Estrada J C^1 [S]Mdn 73-15 Slgfst,Fortn'sKngdom,Rosl'sChoc 12

27Aug83-4Dmr 6f :22^2 :46^1 1:11 ft *8-5 117 6^{3¾} 5^{7½} 5^9 5^{10½} Lamance C^5 [S]Mdn 73-15 JohnJerry,Slugfest,Roslie'sChoice 10

27Aug83—Broke slowly

Apr 14 SA 4f ft :50 h Apr 4 SA 3f ft :37 h Mar 30 SA 5f ft 1:00^3 h Mar 25 SA 5f ft 1:00^2 h

***Heavenly Plain**
Own.—Gumprt or Hffmn or Gttsgen 114
Gr. c. 3, by Mount Hagen—Nuageuse, by Prince Regent
Br.—Collinstown Stud Fm Ltd (Ire) 1984 4 0 1 3 $38,400
Tr.—Vienna Darrell 1983 7 2 0 2 $5,123
Lifetime 11 2 1 5 $43,523 Turf 6 2 0 2 $5,123

4Mar84-6SA 1 1/16 :464 1:11 1:431ft *2 1/2 117 14 1/2 12 1/2 2hd 21 3/4 Pincay L Jr1 Aw23000 83-17 MightyAdversry,HevenlyPlin,Czdor 8
20Feb84-4SA 1 1/16 :461 1:111 1:432ft 3 1/2 117 38 1/2 34 1/2 52 3/4 34 1/2 Pincay L Jr4 Aw30000 80-16 GateDncer,PrinceTrue,HevenlyPlin 7
20Feb84—Steadied, altered course leaving 3/16
5Feb84-8BM 1 1/16 :461 1:103 1:422ft 15 120 13 13 21 1/2 32 1/2 BlackK3 Cmno Rl Dby 84-17 FrenchLegionire,GtDncr,HvnlyPlin 11
16Jan84-3SA 1 1/16 :471 1:114 1:441ft 8 1/2 115 13 1/2 12 1/2 31 1/2 34 DelhoussyeE5 Aw22000 76-25 PrinceTrue,KepOnTlking,HvnlyPlin 6
16Jan84—Broke slowly, bumped after start
19Oct83-8SA 7f :221 :45 1:242ft 26 117 46 1/2 59 78 79 1/2 DlhssyE3 Sunny Slope 69-21 Vencimiento, Bozina, Cardell 7
19Oct83—Broke slowly
1Aug83♦5Leopardst'n(Ire) 7f 1:323fm 4 123 ① 31 1/2 McGrtG Ardenode Stud ExecutivePride,ReoRcine,HvnlyPlin 7
29Jly83♦2Galway(Ire) 7f 1:282fm*2-3 124 ① 12 1/2 McGrathG Tuam Plt Heavenly Plain,ActionGirl,Arcanum 5
25Jly83♦2Galway(Ire) 7f 1:293fm 5 126 ① 13 McGrtG Athenry(Mdn) HevenlyPlin,Shubumi,MonrdPrince 7
25Jun83♦2Curragh(Ire) 6f 1:133gd 7 123 ① 62 1/2 McGrath G Tyros Pymster,Productivity,DesysDlight 17
11Jun83♦1Naas(Ire) 6f 1:164gd *2 122 ① 3 3/4 McGrtG Eadestown Plt Paymaster,LateSally,HeavenlyPlin 16
●Apr 16 SA ① 6f fm 1:16 h (d) Apr 10 SA 5f ft 1:001 h Apr 4 SA 5f ft 1:04 h Mar 30 SA 3f ft :364 h

Barcelona
Own.—Beal-French Jr-Lukas 114
Ch. c. 3, by Key to the Mint—Bold Honor, by Bold Ruler
Br.—Jones & Farish III (Ky) 1984 7 2 1 1 $56,820
Tr.—Lukas D Wayne 1983 0 M 0 0
Lifetime 7 2 1 1 $56,820

7Apr84-4SA 1 1/16 :454 1:104 1:432ft 2 3/4 117 2 1/2 2hd 1hd 31 3/4 Pincay L Jr5 Aw33000 82-18 Swivel, Bedouin, Barcelona 6
25Mar84-11FG 1 1/8 :463 1:104 1:493ft 27 118 66 1/2 68 1/2 68 1/2 44 1/2 Frazier R L7 La Dby 91-21 Tylor'sSpecil,SilentKing,FightOver 7
14Mar84-8SA 1 1/16 :471 1:111 1:433gd 14 115 31 35 711 814 1/4 VlnzulPA1 Ⓢ Sta Ctlna 69-26 Tights, Prince True, Gate Dancer 10
3Mar84-7SA 6f :22 :453 1:094ft *4-5 120 2 1/2 2 1/2 2 1/2 1hd ValenzuelPA5 Aw23000 89-14 Barcelona,LordOfTheWind,TkkTkk 7
22Feb84-8SA 6f :214 :451 1:10 ft 8 115 22 1/2 21 1/2 1 1/2 2hd VlnlPA4 Ⓢ Bolsa Chica 88-18 Artichoke, Barcelona, Yukon's Star 6
22Feb84—Lugged in
12Feb84-6SA 7f :223 :454 1:244ft 4 1/2 118 22 1/2 22 1 1/2 1 3/4 Valenzuela P A5 Mdn 76-18 Barcelona, Guards, Grey Missile 8
28Jan84-2SA 6f :213 :444 1:091ft 19 118 54 47 411 412 Sibille R3 Mdn 80-12 NobleFury,LionOfTheDesrt,T.H.Ang 8
28Jan84—Lugged in stretch
Apr 4 SA 5f ft 1:014 h Mar 10 SA 4f ft :50 h Feb 19 SA 6f ft 1:163 h

Cazador
Own.—HancockIII-Bradley-Chndler 114
B. c. 3, by Cougar II—Country Romance, by Halo
Br.—HancockIII-Bradley-Chandler (Ky) 1984 4 1 1 1 $20,225
Tr.—Whittingham Charles 1983 2 M 0 1 $2,850
Lifetime 6 1 1 2 $23,075

18Mar84-3SA 1 1/16 :463 1:103 1:434ft *6-5 118 815 815 813 44 1/2 ShoemkerW1 Aw23000 78-18 Coopers Hill, Tabare, Crystal Court 8
18Mar84—Broke slowly, extremely wide into stretch
4Mar84-6SA 1 1/16 :464 1:11 1:431ft 3 118 815 814 711 38 1/2 ShoemkerW3 Aw23000 77-17 MightyAdversry,HevenlyPlin,Czdor 8
4Mar84—Broke slowly, wide into stretch
12Feb84-4SA 1 1/16 :464 1:11 1:431ft 6 1/2 118 613 58 36 23 ShoemkerW3 Aw23000 82-18 TsunmiSlew,Czdor,MightyAdversry 6
12Feb84—Lugged out down backstretch
15Jan84-6SA 1 1/16 :471 1:122 1:444ft 4 1/2 118 1014 65 32 12 Shoemaker W9 Mdn 77-16 Cazador, DoubleCash,HonorMedal 10
15Jan84—Wide
31Dec83-6SA 1 1/16 :472 1:124 1:44 ft 8 1/2 118 66 64 1/2 45 1/2 38 1/2 Shoemaker W2 Mdn 73-18 PrinceTrue,ExplosivePasser,Cazdor 7
31Dec83—Broke slowly
18Dec83-4Hol 6f :221 :451 1:092ft 16 119 716 815 817 615 Valenzuela P A6 Mdn 75-14 TsunmiSlew,BloomingTyrnt,GretDl 8
18Dec83—Broke slowly
Apr 16 SA ① 3f fm :371 h (d) ●Apr 12 SA 1f ft 1:404 h Apr 7 SA 7f ft 1:28 h Apr 2 SA ① 5f fm 1:03 h (d)

Bedouin *
Own.—Elmendorf 118
Ro. c. 3, by Al Hattab—Lady in Red, by Prince John
Br.—Elmendorf Farm (Ky) 1984 4 0 1 1 $10,650
Tr.—Mandella Richard 1983 6 2 0 0 $101,029
Lifetime 10 2 1 1 $111,679 Turf 1 0 0 0 $1,875

7Apr84-4SA 1 1/16 :454 1:104 1:432ft 4 1/2 114 47 1/2 32 3 1/2 21 Hawley S1 Aw33000 83-18 Swivel, Bedouin, Barcelona 6
28Mar84-7SA 1 1/16 :471 1:113 1:424ft 6 1/2 117 74 65 1/2 612 610 1/4 DelhoussyeE5 Aw30000 77-18 Distant Ryder, Lotsa Honey,Bozina 8
28Mar84—Bumped after start, lugged in stretch
5Feb84-8BM 1 1/16 :461 1:103 1:422ft 5 120 1013 99 812 811 LmcC10 Cmno Rl Dby 76-17 FrenchLegionire,GtDncr,HvnlyPlin 11
5Feb84—Broke in a tangle
22Jan84-7SA 1 1/16 :463 1:104 1:434ft *6-5 117 611 59 46 1/2 34 1/2 DelhoussyeE3 Aw27000 78-21 Tights, Bold Batter Up, Bedouin 6
18Dec83-8Hol 1 1/16 :454 1:101 1:413ft 31 121 1112 98 1/2 88 1/4 47 DelhoussyeE8 Hol Fut 88-14 Fali Time, Bold T. Jay,Life'sMagic 12
25Nov83-8Hol 1 ①:473 1:122 1:373gd 2 1/2 118 811 713 611 56 3/4 PincyLJr2 Hst The Flg 73-22 Precisionist, Fali Time, Tights 8
25Nov83—Run in divisions
13Nov83-6SA 1 :473 1:13 1:392m *9-5 116 63 1/2 43 1hd 15 1/2 Hawley S3 Aw24000 71-26 Bedouin,PenutsTurn,ForeignLegion 7
13Nov83—Broke slowly
28Aug83-4Dmr 1 :461 1:12 1:38 ft 3 116 76 1/2 56 32 1/2 13 1/2 Hawley S2 Mdn 78-16 Bedouin,Intimidtion,OverlndJourny 8
7Aug83-4Dmr 1 :47 1:121 1:383ft 5 1/2 116 812 813 513 410 1/2 McCarron C J7 Mdn 65-18 Accontbly,PlntyCnscs,SpclKndGy 10
27Jly83-6Dmr 5 1/2f :224 :461 1:044ft *8-5 117 68 1/2 711 513 415 McCarron C J2 Mdn 72-20 Bozina, UsuallyReliable,EndDisplay 8
●Apr 16 SA ① 4f fm :483 h (d) Apr 4 SA 4f ft :483 h Mar 29 SA 1f ft 1:391 h Mar 22 SA 5f ft 1:01 h

Bean Bag

Ch. c. 3, by Olympiad King—Chili Beans, by Inverness Drive

Br.—Ring Mrs Connie M (Cal) 1984 4 1 3 0 $24,600

Own.—Ring Mrs Connie M 114 Tr.—Sterling Larry J 1983 0 M 0 0

Lifetime 4 1 3 0 $24,600

8Apr84-3SA 1 :463 1:111 1:364ft *8-5 120 41½ 43½ 22½ 2¾ Pincay L Jr3 [S]Aw30000 83-16 Juliet's Pride, Bean Bag, QuatiPink 7

18Mar84-6SA 1 1/16 :46 1:11 1:444ft *9-5 118 78¾ 64 2hd 11½ Pincay L Jr8 Mdn 77-18 Bean Bag, ParkRow,You'reMyLove 9

18Mar84—Fanned extremely wide 7/8 turn

3Mar84-6SA 6f :22 :452 1:102ft *6-5 118 72¾ 63½ 43 21½ Pincay L Jr1 Mdn 85-14 M.DoubleM.,BnBg,MorningThundr 12

3Mar84—Bobbled start, crowded 3/8 turn, steadied at 1/4

17Feb84-6SA 6f :22 :452 1:092ft 5¼ 118 61¾ 3½ 33½ 22¼ DelhoussyeE11 [S]Mdn 89-17 DebonireJunior,BenBg,Sri'sDlight 12

17Feb84—Wide 3/8 turn

●Apr 16 SA ①4f fm :483 h (d) Apr 13 SA 3f ft :362 h Apr 4 SA 6f ft 1:15 h Mar 29 SA 4f ft :473 h

Bold Batter Up

Ch. c. 3, by Mr Bold Batter—Dress Me Up's Girl, by Aczay

Br.—Hopkins A F (Cal) 1984 6 1 1 2 $29,250

Own.—Hopkins A F 114 Tr.—Arena Joseph 1983 7 2 1 1 $24,375

Lifetime 13 3 2 3 $53,625

29Mar84-7SA 7f :222 :45 1:222ft 7½ 117 46 46 34 34 Hawley S4 Aw31000 84-21 DebonireJunior,Crrizzo,BoldBttrUp 7

16Mar84-3SA 1 1/16 :461 1:101 1:43 ft 5¾ 114 42½ 34 37 34½ Hawley S2 Aw32000 81-20 FftySxInRow,M.DoblM.,BoldBttrUp 6

1Mar84-9SA 1 :471 1:114 1:371ft *2½ 116 31½ 41¾ 2½ 1no Hawley S8 62500 82-21 BoldBtterUp,RefueledII,IrishS'gtti 8

20Feb84-4SA 1 1/16 :461 1:111 1:432ft 13 114 27 22½ 2½ 45 Meza R Q3 Aw30000 79-16 GateDncer,PrinceTrue,HevenlyPlin 7

20Feb84—Bobbled start, lugged in stretch

2Feb84-7SA 6½f :212 :44 1:161ft 13 116 66½ 710 69 56¾ ValenzuelPA2 Aw24000 82-19 Commemorate,Bozin,DeputyGenerl 8

2Feb84—Lugged in throughout, wide into stretch, bumped at 1/8

22Jan84-7SA 1 1/16 :463 1:104 1:434ft 8 114 34 23½ 22½ 22½ Hawley S5 Aw27000 79-21 Tights, Bold Batter Up, Bedouin 6

31Dec83-7SA 6f :22 :452 1:112ft 9½ 115 31 31½ 2½ 11½ ValenzuelPA5 Aw20000 81-18 Bold Batter Up, Teitei, IrishS'getti 6

8Dec83-2Hol 6f :22 :452 1:11 ft 3½ 116 64¾ 87 63 53½ Valenzuela P A5 50000 78-17 FleetJoey,OvrlndJourny,BitOMuff 10

23Nov83-3Hol 6f :221 :453 1:111ft 4½ 115 41½ 32½ 31½ 2hd Valenzuela P A1 40000 81-23 RetGoldDust,BoldBttrUp,BitOMuff 7

6Nov83-2SA 6f :223 :462 1:121ft *4-5 118 1hd 11½ 13 14½ ValenzuelaPA2 M32000 77-18 Bold Batter Up,Morse,EssentialJoe 7

Mar 10 SA 6f ft 1:113 h

Crystal Court

B. c. 3, by Crystal Water—Woman Driver, by Traffic Judge

Br.—Ross A (Cal) 1984 3 1 0 1 $18,850

Own.—Ross A or Mildred 118 Tr.—Anderson Laurie N 1983 5 1 0 1 $14,175

Lifetime 8 2 0 2 $33,025

1Apr84-4SA 1 1/16 :462 1:103 1:434ft 2½ 114 12½ 13½ 16 14½ Meza R Q1 Aw20000 82-16 CrystlCourt,Vigor'sPrince,BnRdmd 6

1Apr84—Hit side of gate, bumped after start

18Mar84-3SA 1 1/16 :463 1:103 1:434ft 23 114 11½ 12½ 15 3¾ Meza R Q4 Aw23000 81-18 Coopers Hill, Tabare, Crystal Court 8

18Feb84-7SA 6½f :213 :441 1:164ft 35 118 46½ 48 713 710½ Meza R Q2 [S]Aw23000 75-21 DistntRydr,StrMtril,LordOfThWind 8

31Dec83-7SA 6f :22 :452 1:112ft 8 120 43 41¾ 31½ 42¾ ShoemkerW1 Aw20000 78-18 Bold Batter Up, Teitei, IrishS'getti 6

16Dec83-6Hol 6f :22 :45 1:10 ft 5½ 119 31½ 43 46½ 45¾ Meza R Q2 Aw25000 81-22 GteDncer,LuckyBuccner,PrincBowr 6

25Nov83-4Hol 6f :233 :464 1:121gd 7½ 118 11½ 13 17 15½ Meza R Q4 [S]Mdn 76-28 Crystal Court, Lothar, Saros Saros 9

25Nov83—Bumped start

21Oct83-6SA 6f :221 :453 1:103ft 50 117 65 55 45 37 Meza R Q2 [S]Mdn 78-28 FaliTime,QuieroDinero,CrystlCourt 7

21Oct83—Bumped and jostled after start, lugged in through stretch

6Oct83-6SA 6f :213 :451 1:111ft 15 117 67¾ 58½ 515 616½ Meza R Q9 [S]Mdn 66-21 FftySxInRw,Rsl'sChc,Frtn'sKngdm 9

6Oct83—Lugged in badly throughout

Apr 16 SA HC 4f fm :474 h (d) Apr 9 SA ①1 fm 1:46 h (d) Mar 26 SA 5f ft 1:01 h Mar 16 SA 4f ft :47 h

Coopers Hill

B. c. 3, by Cox's Ridge—Tooth Fairy, by Impressive

Br.—Allen & Tenny (Ky) 1984 7 2 1 1 $31,675

Own.—Kikis P T 116 Tr.—Fanning Jerry 1983 1 M 0 0 $425

Lifetime 8 2 1 1 $32,100

28Mar84-8SA 1⅛ :463 1:104 1:483ft 14 118 64¾ 65¼ 68¼ 610¼ Fell J9 [R]Bradbury 76-18 TsnmShw,MghtyAdvrsry,LotsHony 10

18Mar84-3SA 1 1/16 :463 1:103 1:434ft 12 118 45 46 25 1nk Fell J3 Aw23000 82-18 Coopers Hill, Tabare, Crystal Court 8

4Mar84-2SA 1 1/16 :471 1:122 1:45 ft 9 118 42½ 52½ 32½ 1nk Fell J9 Mdn 76-17 CoopersHill,Reptrit,LionOfThDsrt 12

26Feb84-2SA 1 :463 1:12 1:374ft 3½ 118 41¾ 32 33 32½ Toro F1 Mdn 76-15 Riva Riva,GreyMissile,CoopersHill 10

26Feb84—Steadied at 3 1/2, crowded 3/8 turn

19Feb84-2SA 1 1/16 :463 1:113 1:441ft 5 118 54 55½ 712 713 Toro F9 Mdn 67-14 WoodlndWy,OcnViw,LionOfThDsrt 9

19Feb84—Veered out start

4Feb84-2SA 6½f :213 :443 1:164ft 4½ 118 67¾ 68½ 54 43½ Toro F1 Mdn 83-14 Appetite, Ocean View,SavorySauce 7

4Feb84—Very wide stretch

22Jan84-6SA 6½f :222 :454 1:18 ft 24 118 42¾ 32¾ 42 2no Toro F7 Mdn 80-21 Carrizzo, Coopers Hill, River Yang 12

13Jly83-6Hol 6f :221 :45 1:101ft 49 116 811 614 614 515½ Toro F2 Mdn 70-19 Precisionist,TriumphntBnnr,FliTim 8

13Jly83—Bobbled start

Apr 14 SA 5f ft 1:00 h Apr 5 SA 5f ft 1:01 h Mar 15 SA 5f ft 1:003 h

In fact, method players have no basis for unlocking the mysteries of this everyday feature. As often happens with better nonclaiming three-year-olds, estimates of true class and speed remain elusive. The colts in the La Puente Stakes are still defining themselves. Traditional systematic methods often do not apply as well. Predicting the probable outcome is thereby reduced to speculation, a guessing game.

Or is it? Consider the information presented below.

An excerpt from *Master Grass Sires List,* 1982, published by the leading handicapping author, mathematician Bill Quirin follows:

Here's our list of the top ten grass sires:

1. Little Current
2. Stage Door Johnny
3. Roberto
4. Tell
5. Rock Talk
6. Advocator
7. One For All
8. Verbatim
9. Nodouble
10. Ambernash

Now handicappers might return to the 1984 La Puente Stakes and find the generous winner rather esily. It is Loft, son of Roberto, the world's third most productive grass sire. When Quirin examined the grass races in which the sons and daughters of Roberto were competing for the first or second times, he discovered Roberto's get won approximately twice their rightful share of those races and returned $3.26 for each $2 bet, a dollar return of 63 percent.

The lesson for handicappers? Prepare to support the sons and daughters of Roberto and other leading grass sires when first they switch to the grass. The higher the odds, the better. If recent dirt form is poor, so much the better as well; the price will be appealing.

Handicappers need to know these statisical facts, and any like them. That is, handicappers need to understand the probabilities of handicapping. Handicappers who appreciated Roberto's status among successful turf sires would hardly have hesitated to bet Loft in the restricted La Puente Stakes. Those who had that information and also remembered trainer Leroy Jolley's strong opinion of the colt in January of 1984 might have chosen to wager seriously in a race unplayable by systemaic methods.

EIGHTH RACE
Santa Anita
APRIL 18, 1984

1 ⅛ MILES.(turf). (1.45⅘) 4th Running of THE LA PUENTE STAKES. $60,000 added. 3-year-olds which have never won *$20,000. (Allowance.) By subscription of $50 each to accompany the nomination, $600 additional to start, with $60,000 added, of which $12,000 to second, $9,000 to third, $4,500 to fourth and $1,500 to fifth. Weight, 120 lbs. Non-winners of $16,000 at one mile or over, 2 lbs.; of such a race of $13,000, 4 lbs.; of such a race of $11,000, 6 lbs. A trophy will be presented to the owner of the winner. *A race worth $20,000 to the winner. Closed Wednesday, April 11, 1984 with 17 nominations. (Maiden and claiming races not considered.)
Value of race $66,250, value to winner $39,250, second $12,000, third $9,000, fourth $4,500, fifth $1,500. Mutuel pool $557,687.

Last Raced	Horse	Eqt.	A.	Wt	PP	St	¼	½	¾	Str	Fin	Jockey	Odds $1
14Mar84 ^{8}SA^{9}	Loft		3	114	2	3	6$^{1\frac{1}{2}}$	6^{1}	8$^{1\frac{1}{2}}$	1^{1}	1$^{4\frac{1}{2}}$	Shoemaker W	6.20
7Apr84 ^{4}SA^{2}	Bedouin		3	118	6	8	9	7^{1}	5^{hd}	6^{hd}	2no	Hawley S	6.00
8Apr84 ^{3}SA^{2}	Bean Bag		3	117	7	6	5$^{\frac{1}{2}}$	5^{hd}	7$^{\frac{1}{2}}$	7^{2}	3^{hd}	Pincay L Jr	2.90
7Apr84 ^{4}SA^{1}	Swivel		3	120	3	7	7^{1}	9	9	9	4$^{2\frac{1}{2}}$	Olivares F	11.30
29Mar84 ^{7}SA^{3}	Bold Batter Up		3	116	8	5	3^{hd}	3^{hd}	3^{hd}	3$^{\frac{1}{2}}$	5^{1}	Black K	22.40
8Apr84 ^{3}SA^{1}	Juliet's Pride	b	3	120	1	1	2$^{1\frac{1}{2}}$	2^{1}	4^{2}	8$^{\frac{1}{2}}$	6$^{\frac{1}{2}}$	Fell J	*9.40
4Mar84 ^{6}SA^{2}	Heavenly Plain	b	3	114	4	2	4^{1}	4$^{2\frac{1}{2}}$	2$^{1\frac{1}{2}}$	2$^{\frac{1}{2}}$	7$^{\frac{1}{2}}$	Garcia J A	3.80
18Mar84 ^{3}SA^{4}	Cazador		3	114	5	9	8^{hd}	8^{hd}	6$^{1\frac{1}{2}}$	5^{hd}	8$^{4\frac{1}{2}}$	McCarron C J	3.70
1Apr84 ^{4}SA^{1}	Crystal Court	b	3	118	9	4	1^{2}	1$^{1\frac{1}{2}}$	1$^{\frac{1}{2}}$	4^{hd}	9	Meza R Q	17.30

OFF AT 5:07. Start good. Won handily. Time, :22⅕, :46⅗, 1:11⅗, 1:36⅗, 1:49⅘ Course firm.

$2 Mutuel Prices:			
2–LOFT	14.40	9.20	5.40
7–BEDOUIN		7.40	4.60
8–BEAN BAG			3.20

Ch. c, by Roberto—Journey, by What a Pleasure. Trainer Jolley Leroy. Bred by North Ridge Farm (Ky).

LOFT saved ground while unhurried for seven furlongs, found room on the rail entering the stretch, quickly overtook the leaders and drew off in the final furlongs. BEDOUIN, outrun early, rallied wide into the stretch, drifted in while responding in the drive and could not overtake the winner. BEAN BAG rallied widest around the final turn, also drifted in during the drive but finished strongly. SWIVEL lacked early speed, swung far wide rallying in the stretch but finished strongly in the middle of the course. JULIET'S PRIDE tired in the stretch, as did HEAVENLY PLAIN. CAZADOR, outrun early, rallied to contention between horses near the rail entering the stretch but tired in the final furlong. CRYSTAL COURT also tired badly in the final furlong. BARCELONA (5) AND COOPERS HILL (11) WERE WITHDRAWN. ALL WAGERS ON THEM IN THE REGULAR POOLS WERE ORDERED REFUNDED AND ALL OF THEIR PICK SIX SELECTIONS WERE SWITCHED TO THE STARTING FAVORITE, BEAN BAG (8).

Owners— 1, Harris C E II (Lessee); 2, Elmendorf; 3, Ring Mrs Connie M; 4, Holt L; 5, Hopkins A F; 6, Cole-Freeman-Gold; 7, Gumprt or Hffmn or Gttsgen; 8, Hancock III-Bradley-Chandler; 9, Ross A or Mildred.

Trainers— 1, Jolley Leroy; 2, Mandella Richard; 3, Sterling Larry J; 4, Holt Lester; 5, Arena Joseph; 6, Stute Melvin F; 7, Vienna Darrell; 8, Whittingham Charles; 9, Anderson Laurie N.

Overweight: Bean Bag 3 pounds; Bold Batter Up 2.

Scratched—Barcelona (7Apr84 ^{4}SA^{3}); Coopers Hill (28Mar84 ^{8}SA^{6}).

Probability studies inform handicappers of the relative power of numerous past performance characteristics. The method of analysis is crucial. Descriptive statistics do not apply; that is, of all grass winners in New York during 1983, only 10 percent were first starters on the turf. The correct procedure divides the percentage of winners having a characterisic by the percentage of starters having the characteristic. The resulting statistic has been called an impact value; the notation is I.V.

An impact value of 1.0 represents the expected percentage of winners having the characteristic, based on the number of starters. Below 1.0 means the characterisic performs below expectation; above 1.0 means the characteristic performs above expectation.

Roberto's impact value as a turf sire through 1982 was 1.82, meaning his get won 82 percent more than their fair share of the grass races they entered. Through the same year, two long-time suspected grass sires, The Axe II and Hawaii, had earned I.V.s of 0.85 and 0.84, respectively; each sire's runners had won less than a fair share of their turf starts by approximately 15 percent. Handicappers need to know, store, and recall these significant statistics. Handicappers who become information managers will be more likely to have the information at hand than handicappers who do not. High-tech handicappers will be able to retrieve the data instantaneously.

Below are listed four probability studies of considerable significance to handicappers, regardless of personal methods. In each case the samples are national, and thus represent the major racing programs throughout the country. Sample sizes are large enough so that the inferences they support are subject only to small, unavoidable error degrees.

1. Percentages Probabilities	Fred Davis	1974
2. Computer Discoveries in Thoroughbred Handicapping	William Quirin	1979, 1984
3. Master Grass Sires List	William Quirin	1982, 1984
4. An Analysis of Dosage	Steve Roman	1981, Annual

Roman's research on dosage does not yield probability values in the strict sense, but it belongs here, as it depends upon inferential statistics and not just descriptive stats. Roman's research with a statistic he terms the dosage index reveals significant differences among the dosage values earned by sprinters and distance horses, and among stakes winners as a group and Group 1 stakes winners at classic distances and beyond.

A perfect blend of speed and stamina in a horse's pedigree yields a dosage index of 1.0. Higher values indicate greater speed than stamina, lower values greater stamina than speed, such that 2.0 means twice as much speed as stamina, and 0.50 means twice as much stamina as speed. When predicting whether horses will get classic distances against top-grade competition, a dosage index of 4.0 is a guideline value distinguishing the statistical probables and improbables with amazing reliability. No horse having a dosage index above 4.0 has ever won the Kentucky Derby; only two have won the Belmont Stakes. Handicappers can fairly conclude that

Table 1
Selected High-Low Handicapping Impact Values (I.V.)

Past Performance Characteristic	*I.V.*
Maiden races, 3-up, 2nd place finish last out.	2.74
Maiden races, 3-up, no starts.	0.44
Claiming races, more than 30 days since last start.	0.40
Claiming races, drop of 30 percent or more in class.	3.78
Claiming races, rise in class by more than 30 percent.	0.19
Sprints, horse with highest average when highest two *Form* speed ratings are combined.	2.38
Handicap stakes, horses having weight of 123 pounds or more.	2.70
Routes, older horses, have never routed before.	0.40
All horses, 1st, 1st call position.	2.50
Six furlongs, 1 to 3, 1st call position.	2.03
Maiden graduates, won maiden race in two lengths faster than par for next start.	2.42

horses having dosage indexes greater than 4.0 should not be expected to win Group 1 races at classic distances.

The handicapper's information system should include all the significant impact values obtained by well-designed national or local probability studies. Table 1 presents selected I.V.s from the Davis and Quirin studies, specifically, several values at 2.0 or greater (wins twice or more its fair share) and 0.50 or lower (wins half or less its fair share). There are many more of less power but of genuine interest.

A final caution on the handicapping probabilities reported by local studies: Samples should include two hundred races at least, and the statistical procedures must compare percentages of winners and percentages of starters. It's important. Handicappers and racing customers in general have long been misled by descriptive statistics presented to imply statistical relationships incorrectly. Descriptive statistics only describe a specific population of events. They cannot describe relationships and do not generalize to other populations.

STATISTICAL COMPILATIONS

Descriptive statistics do provide vital information about a particular population, such as trainers. The data can describe the group of interest by frequencies, by subgroupings, by averages, or by rankings. No attempt should be made to generalize the statistics to larger or comparable populations or events.

In recent years trainer statistics have become a staple of handicapping information services available on several local circuits. Each trainer's performance is recorded into numerous performance categories, and the mass of data is scanned to identify trends and patterns. These are interpreted either strictly, as indicators of good and poor performance, or in terms of trainer intentions. Those handicappers of southern California who have the information know that trainer Richard Mandella is the tops with first starting maidens and that claiming specialist Mike Mitchell performs much better with class rises, notably just after a claim, than he does with class drops.

At Longacres, near Seattle, professional handicapper Paul Braseth annually publishes a Trainer Pattern Analysis of the previous season's entire 125-day meeting, as part of his *Northwest Track Review*'s multiple information services. The trainer analyses are supported by data for twenty-two categories of performance, including each trainer's performance with class drops, class rises, claims, first starters, jockey switches, repeaters, distance changes, track conditions, workout patterns, layoffs, and horses of different age, sex, and class.

Longacres is a minor oval with numerous stables, most having a large quota of cheaper stock, the most difficult to manage effectively over long periods. Trainer performance is therefore a crucial variable.

Having constructed and analyzed his massive trainer database, which he develops manually, by the way, Braseth rarely confronts a training or racing pattern for any horse three-and-up which he cannot comprehend in terms of trainer performance patterns. During the three racing days I accompanied Braseth to Longacres in 1984, he picked twelve winners, no underlays, and half depended largely on recognizing the training and racing patterns of relatively cheap horses. With this supply of trainer performance information, Braseth plays Longacres at tremendous individual advantage, shared only by subscribers to his trainer analysis report.

Several other professionals have done the same elsewhere. Interestingly, developers of these massive trainer statistical compilations have met with consumer resistance when trying to market the information to regular handicappers. As early as 1981, San Francisco professional Ron Cox, publisher of that area's outstanding *Northern California Track Record,* developed a database that examined every trainer on the Bay Meadows-Golden Gate Fields circuit from all conceivable angles. Cox prepared a three-year baseline for many data cells, such as percentage of wins with four-and-up on the first start following layoffs of ninety days or longer. Amazingly, nobody purchased the product, which cost approximately $150.

Cox concluded that recreational handicappers did not want to struggle with a tome of information about a single factor, the trainer. He was proven correct in 1982. That year Cox condensed the trainer data into four lists of names. Based on performance, trainers were classified as A trainers, B trainers, C trainers, or D trainers, with short pointed explanations of each performance category. D trainers, for example, had not won a single race (there were dozens on the list), and Cox merely advised his consumers not only to avoid those trainers' horses absolutely, but to bet more on properly priced selections in races where a D trainer's horse was favored. Cox sold approximately $15,000 worth of the shorter, easier-to-handle lists.

Although statistical compilations are vital components of the handicapper's information system, they tend to be large and messy, putting potential users off. Yet the information management technology of today affords handicappers easy access to and use of the massive, complicated trainer databases developed by professionals such as Braseth, Cox, and others. If trainer stats were stored electronically, the computer could search that database rapidly and respond to any spontaneous questions handicappers might pose.

Imagine a handicapper analyzing a field of ten nonclaiming three-year-olds, entered at a middle distance, with four or five trying the distance for the first time. The handicapper needs to know which trainers make this distance change most effectively. The handicapper queries the computerized trainer database: Get the performance record for trainers (names), for three-year-olds, sprints to routes. Within seconds, the computer displays for each trainer the number of starts, wins, and win percentages; it might show the average odds on winners too. One trainer in the race might have a demonstrated advantage; others might never complete the move successfully.

Because trainers exercise the most influence over a thoroughbred's career and trainer data will be regularly useful to handicappers, trainer performance information leads the list of statistical compilations high-tech handicappers are urged to collect.

In this context, a warning about trainer statistics is in order. Narrow categories of performance information, such as percentage of wins with claiming horses, aged three, dropping in class, stretching out, and switching to a more favorable jockey, must contain a reliable number of instances. Preferably a number approaching thirty. A small number of instances, below ten, for example, can be more misleading than helpful. Thirty instances approaches what statisticians refer to as a "normal distribution" of events. The statistics are far more reliable. Fortunately, broader categories of performance, such as percentage of route wins, are not often jeopardized by inadequate sample sizes.

On the other hand, trainer pattern data is not normally distributed but abnormally skewed. Very small numbers of instances, below five even, can represent the pattern well enough. So for performance data, insist on representative sample sizes; for pattern data, small samples are acceptable.

Handicapper Greg Lawlor, of San Diego, understands the reliability factors with trainer statistics. Lawlor markets trainer information in southern California, and his pattern data are distributed across numerous small cells, but he also has a six-year performance database. The database is computerized and readily updated. The increasing power and reliability of that kind of evolving database is dynamite stuff in the right hands.

Here's a list of useful statistical compilations for inclusion in the handicapper's information system.

- Trainer statistics, all categories of interest
- Trainer patterns analyses
- Jockey statistics, all categories of interest
- Stakes races, by grade designations, purse values, and eligible ages
- Foreign stakes races, same categories as above
- Leading sire lists, for sprints, routes, slop, and grass
- Leader owner lists, nationally or by local circuit

Handicappers who salute the information age by becoming information managers will add to the above recommendations with additional lists of their own. They will be able to not only retrieve the information within seconds but also relate it in numerous ways to data stored in other databases.

REFERENCE MATERIALS

The following sources of information are listed in the order of frequency with which handicappers might choose to consult them.

- Results charts of *Daily Racing Form*
- Past performance histories of all older horses, four-up, on the local circuit
- Periodicals: *The Blood Horse* and *The Thoroughbred Record.* Selected information and data, such as the stud fees of sires or auction prices of yearlings.
- Products of local information services
- *Trackside* diagrams and descriptive data on all North American racetracks

When, if ever, *Daily Racing Form* gets around to designing, constructing, and marketing its massive databanks of racing information as electronic databases, handicappers can subscribe to the computerized services, and with simple telephone calls locate information about the results of past races or the past performance histories of horses anywhere within seconds. That day is still years away. Handicappers will be keeping their old *Forms* and results charts for quite a while longer. I once relished a handicapping acquaintance who had cut out and pasted together the past performance histories of every active horse on the southern California circuit. After a few years of painstaking effort, my friend gave up the ghost. The routine was not sufficiently rewarding. But what an electronic database that will be!

Trackside, a reference guide to the layouts of all North American racetracks, published by Chicagoan Tim Zurick, is a depository of unusually organized information. It provides track diagrams, the starting points of regularly run distances, the distances to turns from starting gates, lengths of stretch runs, leading trainers and jockeys at each track, average purses, selected final time pars, and other useful information not otherwise available in this format.

SOFT INFORMATION

Soft information is subjective. It consists of the points of view, judgments, and opinions of others. It includes, too, the kinds of inside information for which racetrack characters have forever been immortalized. Intelligent authors of handicapping texts rarely fail to denounce inside information as worthless, but all worlds change. In the information age, soft information plays its role. Some of it can be decisively helpful at times.

An illustration is in order.

On January 21, 1984, Santa Anita offered the Gr. 1 San Fernando Stakes. At a mile and one-eighth, it was the second of three stakes in a series uniquely limited to young four-year-olds. Proceeding from the assumption that Gr. 1 races for older horses almost invariably are won by previous Gr. 1 winners, I had decided to pass the San Fernando, as none of these four-year-olds had yet to capture a Gr. 1 event. During the homework phase of handicapping I had checked the three horses whose past performances appear below, but intended to bet none. Interco had not raced in thirty-five days, recorded indifferent workouts, and was switching from soft turf to dirt. Desert Wine had been runner-up in two classics but had not won even a Gr. 2 event and was stretching out following a layoff and a single sprint, the worst of stretch-out patterns statistically. Glacial Stream had looked sharp winning the Gr. 2 Malibu, but its essential class fell a cut below Gr. 1, and with leading rider Chris McCarron aboard was certain to be an underlay.

On that crisp Sunday I happened to join my handicapping friend Ruth Eilken, wife of Santa Anita racing secretary Louis Eilken, in walking ring ceremonies for the San Fernando outsider Hula Blaze and had even decided to string along with the party and bet $20 on the colt, which did look the most intriguing of the longshots. On leaving the paddock, wouldn't you know but Interco's trainer Ted West sidled over to Mrs. Eilken, exchanged greetings, and said, confidently, while pointing to the totalizator board, "My horse should be 8 to 5 against this field; bet a little."

I shortly learned the Eilkens and Wests have been close friends for years. I also knew something about Ted West. He keeps both feet on the ground. Trainers notoriously oversell their top horses, to themselves and to others, but West likes to handicap a little, and he

8th Santa Anita

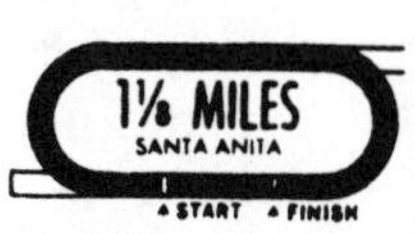

1⅛ MILES. (1.45⅘) 32nd Running of THE SAN FERNANDO STAKES (Grade I). $125,000 added (plus $50,000 Breeders' Cup Premium Awards). 4-year-olds (foals of 1980). (Allowance.) By subscription of $100 each to accompany the nomination, $500 to pass the entry box and $1,250 additional to start, with $125,000 added, of which $25,000 to second $18,750 to third, $9,375 to fourth and $3,125 to fifth. Weight, 126 lbs. Non-winners of two races of $100,000 or three of $50,000 in 1983-84 allowed 3 lbs.; of a race of $100,000 since April 25 or two of $50,000 in 1983, 6 lbs.; of a race of $40,000 since December 25, or two of $30,000 or one of $50,000 in 1983 or a race of $100,000 in 1982, 9 lbs.; of a race of $30,000 in 1983 or one of $30,000 at one mile or over in 1982, 12 lbs. Starters to be named through the entry box by the closing time of entries. A trophy will be presented to the owner of the winner. Closed Wednesday, January 11, 1984 with 26 nominations.

Coupled—Load The Cannons and Estupendo.

Glacial Stream — B. g. 4, by Crystal Water—Cascapedia, by Chieftain
Br.—Ridder B J (Cal) — 1984 1 1 0 0 $43,150
Own.—B J Ridder Estate **120** Tr.—Campbell Gordon C — 1983 11 4 3 1 $144,608
Lifetime 12 5 3 1 $187,758

1Jan84-7SA 7f :222 :45 1:221ft *9-5 120 6½ 6¾ 3⁴ 1¹ McCarron C J4 Malibu 89-18 GlacialStrem,TotlDeprture,HulBlze 8
1Jan84—Run in divisions
10Dec83-8Hol 1 1/16 :46 1:102 1:414ft *2½ 114 6⁶ 5½ 4¾ 1½ McCrrCJ6 Affirmed H 86-17 Glacial Stream, Billy Ball, Proof 13
24Nov83-8Hol 1 1/16 :462 1:103 1:421ft 2¾ 113 4½ 4³ 3½ 2nk McCrrCJ4 [S]G S Sires 84-18 SuprDimond,GlcilStrm,ChfCornstlk 6
28Oct83-7SA 1 1/16 :46 1:11 1:434ft 9-5 117 4³ 3½ 1½ 1½ McCarronCJ5 Aw26000 82-19 Glacial Stream,AlKhalifa,Concierge 5
28Oct83—Bobbled start
16Oct83-7SA 1 1/16 :47 1:114 1:432ft *1 116 2² 2½ 1½ 1⁵ McCarronCJ5 Aw22000 84-20 GlcilStrem,Procurer,Lighthwyholm 7
9Oct83-5SA 6f :22 :452 1:103ft *8-5 115 11¹⁵10¹² 7½ 3¹ McCarronCJ7 Aw20000 84-16 PrincOfAsturis,Hnovrin,GlcilStrm 11
9Oct83—Broke in a tangle
3Sep83-5Dmr 6f :222 :451 1:094ft *2-3 114 10¾ 7¹¹ 6⁹ 4½ McCarronCJ2 Aw17000 83-16 PcMni,PrinceOfAsturis,SndDigger 10
3Sep83—Hit side gate, broke in a tangle
12Aug83-7Dmr 6f :223 :454 1:11 ft 3 113 7² 4½ 3² 2² Hawley S10 Aw16000 81-24 Mamaison,GlcilStrem,ChrgerGreg 11
12Aug83—Wide final 3/8
2Apr83-3SA 1 1/16 :47 1:114 1:441ft *4-5 114 3¹ 3½ 4² 5½ McCarronCJ6 Aw21000 77-18 TrickyWillie,StrongDollr,Concierge 6
2Apr83—Lugged in
12Feb83-8SA 7f :223 :452 1:222ft 5¾ 114 5¾ 3½ 6¹¹ 6¾ McCrrCJ6 Sn Vicente 79-14 Shecky Blue, Full Choke, Naevus 7
12Feb83—Lugged in
●Jan 16 SA 7f ft 1:244 h Jan 11 SA 6f ft 1:121 h Dec 27 SA 6f ft 1:114 h ●Dec 22 SA 5f ft :583 h

Interco — Ch. c. 4, by Intrepid Hero—Yale Coed, by Majestic Prince
Br.—Spendthrift Farm (Ky) — 1983 8 5 2 1 $323,258
Own.—Sofro D **123** Tr.—West Ted — 1982 8 1 1 2 $29,944
Lifetime 16 6 3 3 $353,202 — Turf 15 6 2 3 $348,402

17Dec83-8BM a1⅛ Ⓣ 1:493yl *1-2 116 3⁴ 1hd 1hd 1¹ Judice J C5 B M Dby — — Interco,BangBngBng,BronO'Dublin 6
11Dec83-8BM 1⅛ Ⓣ:4821:1411:552sf 4½ 115 7⁵ 3½ 1³ 1¹ Judice J C3 B M H 63-37 Interco, Super Sunrise, Floriano 15
2Dec83-8Hol 1⅛ Ⓣ:47 1:1131:48 fm*2-5 115 6⁷ 6¾ 2½ 1¾ McCrronCJ2 Spnc Bay 90-10 Interco, Anticipative, Pac Mania 8
2Dec83—Wide 3/8 turn
20Nov83-6Hol 1⅛ Ⓣ:4631:1131:481fm*9-5 122 9¹⁵ 7¾ 3½ 2¾ McCarronCJ8 Hol Dby 88-13 Royal Heroine, Interco, PacMania 11
20Nov83—Run in two divisions, 6th & 8th races; Lugged in late
9Nov83-5SA 1⅛ Ⓣ:4731:1211:501fm *1 113 5¾ 4³ 2hd 1¾ McCarronCJ4 Aw26000 76-24 Interco, Debonair Herc, Bosto 12
9Nov83—Bumped 1/8
22Oct83-7SA 6f :22 :451 1:094ft 4½ 116 5½ 5½ 4½ 2½ McCarronCJ6 Aw24000 85-19 Expressmn,Interco,PrinceOfAsturis 6
22Oct83—Broke slowly, wide 3/8 turn
15May83♦5Longchamp(Fra) a1 1:454sf 2½ 128 Ⓣ 3¾ AsmnC PxdeLaJhe(Gr3) Aragon, Ginger Brink, Interco 7
3May83♦5StCloud(Fra) a1 1:46 gd*6-5 123 Ⓣ 1½ AsmnC Px Le Pompon Interco, Bal des Fees, Ginger Brink 6
11Oct82♦6StCloud(Fra) a7½f 1:423sf *9-5 119 Ⓣ 4½ AsnC Px Thms Bryn(Gr3) BaldesFees,NorthernFashion,Alluvi 9
19Sep82♦4Longchamp(Fra) a7f 1:231gd 4½ 123 Ⓣ 4½ KsssJL Px D L Smnd(Gr1) DeepRoots,Maximov,CrystlGlitters 6
Jan 17 SA 7f m 1:301 h Jan 11 SA 4f ft :47 h Jan 6 SA 4f ft :49 h

Desert Wine * — B. c. 4, by Damascus—Anne Campbell, by Never Bend
Br.—Jones & Warnerton (Ky) — 1984 1 0 0 1 $9,000
Own.—Cardiff Stud Fm & T90Ranch **123** Tr.—Fanning Jerry — 1983 9 2 4 0 $344,793
Lifetime 18 5 7 2 $714,793

1Jan84-8SA 7f :221 :443 1:223ft *3-5 123 5³ 6¾ 5⁷ 3½ DelahoussyeE2 Malibu 86-18 Pac Mania, RetsinaRun,DesertWine 8
1Jan84—Bumped in rear quarters at 5/8; run in divisions
15Oct83-8Bel 1½ :48 2:01 2:261ft 37 121 2½ 7²¹ 9²⁶ 9³¹ Toro F10 J C Gld Cp 58-14 SlwO'Gold,HghlndBld,BondngBsq 11
17Sep83-10LaD 1¼ :48 1:371 2:013ft 2½ 126 1¹ 2½ 3⁵ 4¹¹ McCarronCJ1 Spr Dby 89-18 Sunny'sHalo,PlayFellow,MyHbitony 6
27Aug83-8Bel 1 :441 1:083 1:35 ft 2¾ 124 4¾ 4½ 3½ 2¹ McCrrnCJ1 Jerome H 89-12 APhenomenon,DesertWine,Copelan 8
27Aug83—Bore in
21May83-8Pim 1 3/16 :464 1:103 1:542sy 4½ 125 1½ 1½ 2½ 2¾ McCrrnCJ7 Preakness 90-11 DptdTstmony,DsrtWn,HghHonors 12
7May83-8CD 1¼ :471 1:364 2:02ft 16 126 3² 2hd 2½ 2² McCrrnCJ5 Ky Derby 84 10 Sunny's Halo, Desert Wine,Caveat 20
28Apr83-7Kee 1⅛ :464 1:11 1:492sy 4 121 2¾ 2½ 4¾ 3¾ McCrrnCJ5 Blue Grass 82-16 Play Fellow, ‡Marfa, Desert Wine 12
28Apr83—Placed second through disqualification; Steadied 3/16
10Apr83-4SA 1⅛ :46 1:102 1:492ft *4-5 120 6¾ 4½ 7½ 6¾ McCrronCJ10 S A Dby 73-19 Marfa, My Habitony, Naevus 10
10Apr83—Wide 7/8 turn
27Mar83-8SA 1 1/16 :454 1:094 1:413ft *2-3 124 1½ 1hd 2hd 2hd McCrronCJ6 Sn Flp H 93-13 ‡Naevus, Desert Wine,FifthDivision 6
27Mar83—Pl 1st thru disq
5Mar83-8SA 1 :46 1:103 1:353ft *6-5 119 1½ 1½ 1² 1½ McCrrnCJ3 Sn Rafael 90-16 Desert Wine, Naevus, BalboaNative 7
Jan 17 SA 6f m 1:14 h ●Jan 11 SA 6f ft 1:12 h ●Dec 29 SA 4f ft :461 h ●Dec 23 SA 7f ft 1:242 h

understands a good bet from a bad one, as well as an opportunity to score from just another wager.

Having marked Interco to begin, I revisited its past performances. Three points looked suddenly larger: (a) the colt did not finish out of the money in eight starts at three, which included four graded stakes events; (b) Interco had recently won the $200,000-added Gr 2 Bay Meadows Handicap; and (c) Interco had finished second in the Gr. 1 Hollywood Derby. Moreover, Interco had finished second on the dirt track to the top sprinter Expressman on October 22, following a five-month layoff and after breaking poorly and traveling wide.

The final straw came from the most important source of soft information, the tote board, which, after all, is merely the collective array of the public's opinions. Desert Wine was 5 to 2; Glacial Stream was 9 to 5; Interco was 9 to 1. Mrs. Eilken had been persuaded by West's strong judgment, and she cashed a nice bet. Thanks to her, her escort cashed a nicer one. Interco won handily, paid $20.40. As all now remember, West later won the $500,000-added Gr. 1 Santa Anita Handicap with Interco and an amazing seven consecutive stakes, six of them graded, four Gr. 1.

So an information management approach to handicapping invokes the occasional use of soft information. To preclude its possibilities would be a mistake nowadays. With handicapping knowledge and skill so well distributed and so many informed, diligent, and excellent handicappers out there daily, all handicappers can profit occasionally from well-grounded points of view, judgments, and opinions that are not their own. A prerequisite to the profits resulting from soft information is the willingness to change one's mind, or at least to open it, an attitude this book hopes to slam home hard. A second prerequisite that serves well is the benefit of a good price.

In descending order of their usefulness, below is an array of soft information sources sometimes important to making handicapping decisions. Following the list are a few guidelines for making certain the occasional change of mind does not descend into anarchy.

- The totalizator board
- The opinions of talented handicappers
- Trainer judgments

- Professional seminars
- Professional media
- Inside information

So that matters do not get out of hand, handicappers are recommended to stay close to these guidelines:

1. *Soft information should have a rational basis.* Knowledgeable handicappers are not on the prowl for advanced notice about well-disguised stable maneuvers or shifty betting capers. Subjective information must be supported objectively, with facts, statistics, performance patterns, or comparable evidence. West's judgment notwithstanding, had I not marked Interco myself and judged its trainer's confidence realistic, I would not have bet on the horse.
2. *The sources of information must be reliable.* The more frequently soft information is brandished, the less reliable it tends to be. This is particularly so with trainers, acutely so of other insiders. Few trainers have enough time to handicap, and fewer still are talented or sophisticated about it. Many trainers who understand their own horses very well know correspondingly little about the opponents. That is, as much as they might like their chances today, they have not evaluated the competition closely.
3. *Prefer talented handicappers with ideas and methods that differ from your own.* The modern transformation of handicapping from its rudimentary past to the multifaceted intellectual challenge game it is today means the value of solidly grounded handicapping judgments has risen sharply. A good handicapper's judgment can be a valuable resource, especially if that judgment proceeds from valid perspectives and procedures your own do not encompass.
4. *Personal opinions of handicapping friends and associates should be accompanied by persuasive explanations of their merits.* An engaging opinion of another is easily validated. Simply ask its source to explain the reasons for forming the opinion. The explanations should make sense in terms of fundamental handicapping, or of well-known knowledge and evidence.
5. *Inside information should be factual, not judgmental.* In con-

cert with this book's heavy priority on amassing information, stable information is always welcome. But just the facts, please, no strong opinions. To be avoided, for sure, is the old saw that "they're really trying today." So what? But if the stable foreman mentions that a horse has been entered at the racing secretary's special request, that's nice to know. It's also marvelous to be told that a stable's horse bled last out. The bleeders will be medicated with lasix next time, and some will pay boxcars.

On the other hand, handicappers should discount jockey information that intends to explain recent losses. The explanations are too often facile and self-serving. Similarly, when trainers note that horses have been working and training particularly well, discount that. And owners' inside information, even the factual kind, about their horses is almost never helpful or insightful, except as it helps to sustain the owners' personal spirits.

6. *Never ignore the information on the odds board.* Investing on overlays and avoiding underlays can only result from skillful handicapping, but the converse does not hold. Skillful handicapping can result, and too often does, in betting underlays and ignoring overlays. If the most likely winner of the season is not a fair price, it is not a good bet. The Interco tale would not be worth its telling had the odds been less than five to one.

   In this special sense soft information is part and parcel of all final handicapping decisions. The tote board will always be the most important source of soft information. As with Interco, all other considerations being roughly equal, or uncertain, handicappers should take the odds.

FINANCIAL RECORDS

Even recreational handicappers proceed from a profit motive. That arouses a greatly dismaying thought. Very few regular handicappers bother to keep financial records. The cynical explanation that handicappers do not care to face their losses may be true enough, but it misses a large, large point. The incentive to improve and to win depends on knowing exactly where one stands in relation to those goals. Keeping records of their play helps handicappers play better.

As for the administrative inconvenience, information management technology rushes to the rescue here. Simple spreadsheet applications for microcomputers simplify financial record-keeping; they also facilitate individual analyses, projections, and simulations never before feasible. Handicappers can not only track their own money-management practices without performing the arithmetic, they can now simulate the relative effectiveness of their actual play and of various alternative procedures. Simulation becomes an individualized technique for studying money management at the racetrack, now available to all.

By entering the personal win percentage, average odds on winners, number of races to be played, investment capital, betting patterns, and the instructions for manipulating the numbers (money-management rules), handicappers can determine how well they should do and compare potential outcomes to the real results. That kind of feedback can only stimulate the will to win. By simulating money-management outcomes at various levels of handicapping proficiency, many handicappers will learn for the first time how sensitive profit margins are to the win percentage. They will thereby determine to raise their proficiency level a percentage point or two. In his wonderful softcover book, *A Winning Thoroughbred Strategy,* Dick Mitchell shows that a boost of three percentage points in handicapping proficiency, from 30 to 33 percent winners, results in a tripling of profits.

This book proposes a portfolio style of money management that combines straight wagering and exotic wagering. Methods, procedures, sizes of wagers, and types of wagers will vary from person to person and from race to race. Moreover, many handicappers' betting portfolios will change several times a season, perhaps, depending on periodic results. Wagers that are not working well enough will be eliminated, at least temporarily, and wagers working well can be boosted.

Portfolios aside, the following financial data items will be collected continuously.

- Number and type of wagers
- Wins
- Winning odds

- Sizes of wagers
- Chronology of wagers
- Methods of wagering
- Profits
- Losses
- Dollar returns on investments

It's appropriate to repeat for emphasis that the monitoring and studying of handicappers' betting activity is a formerly cumbersome routine now easily controlled by resorting to information technology.

HOW HANDICAPPERS COPE WITH NEW INFORMATION

A reliable guide to the intelligence, maturity, and worldiness of high-level decision-makers is their reaction to new or unfamiliar information that conflicts with the status quo. Quite a few, fearful of the unknown or untrusting of the unfamiliar, instinctively screen it out. Insecure with change, they'd rather not know. Then you have the intellectual conservatives, rational types, numerous in number, who are more comfortable with things the way they are. As problem-solvers and decision-makers, these are usually the last to identify the real problems or issues, the last to recognize the need to change, and the first to retreat to the old and familiar ways when something goes wrong. Few conservatives can be found on the intellectual horizons of change. They prefer instead to lay back, to play it safe. Another group are egocentric types wholly convinced of the profound correctness of their own ideas, experience, and actions. When these people change at all, they normally do so as a consequence of ideas or actions they are very proud to call their own. Ideas and actions that fall outside of their own frames of reference are simply not entertained. Not much can be done, actually, to influence the decision-making behavior of many members of those three groups.

What has this discussion to do with handicapping? In the high-tech, information context, everything. In this book's approach, handicappers are viewed as information managers who control information and use it to make numerous and various decisions about horses and wagers. While not so many can be classified as among the instinctively insecure or the intellectually conservative, many handicappers indeed are card-carrying members of the third group.

Handicappers tend to be egocentric. Perhaps that explains why the most common reaction to new or unfamiliar information among practiced handicappers has been to filter it through existing perceptions, beliefs, and practices. If the new material fits comfortably enough, it might be used; if not, throw it out. The more things change, the more they remain the same.

Not too many years ago class handicappers and speed handicappers perceived each other as sitting atop opposite poles. Even though class and speed as factors in handicapping are entirely interlocking, should a new book emphasize or prioritize either factor, the other camp could be expected to criticize it roundly and perhaps deny its usefulness.

When speed handicapping became fashionable in the mid-seventies, the same conflicts arose between adherents of figures based solely on final times and those who honored fractional and final time figures in combination. Hybrid methods of any kind were likely to be judged at least slightly inferior by both the speed and class authorities of the handicapping spectrum. Form analysis has not yet been fully assigned its fundamental position in handicapping orthodoxy, especially by method players who do not include form analysis in their methods.

Specialized material, such as body language, trips, dosage indexes, foreign race records, or the Z-System of place and show wagering, at first are widely perceived as too far outside the norms of conventional handicapping and thus are shunted into the background of events until the force of their own utility in turn forces handicappers to come to grips with the information, however technical or esoteric.

Fortunately, these misdirections have been changing. In their latest books, speed mentors Andrew Beyer and Bill Quirin each took a more analytical, more interpretive approach to their speed figures. Class handicappers increasingly recognize the key role of speed and early speed in their analyses. William L. Scott published a comprehensive study of form analysis but carefully related form to other fundamentals of handicapping. From all perspectives the message is being disseminated repeatedly that handicapping information cannot be segmented. It is, instead, relational. This book intends to carry the relational character of handicapping information to its logical conclusion. All types of information can be related to each other in multiple ways for analyzing different races. If handicapping information is fully integrative, handicappers benefit from more information, not less, and can help themselves best not by screening out large amounts of information that do not mesh well with existing methods but by incorporating all of it in the broader information management approach described here.

Handicappers generally have employed two other means of assimilating the information glut. One involves selective perception. Handicappers selectively accept some types of information as meaningful and dismiss the other types. The decisive variable is perceived need. If new or unfamiliar information is thought pertinent to a perceived need or area of weakness, handicappers will entertain it. If not, it's judged not relevant.

Other handicappers experience the kinds of analytical problems Mike Mason articulated in the seminar excerpted at the beginning of this chapter. These are informed, progressive players attracted to all types of information as potentially worthwhile, but they eventually tend to get bogged down in the mass. The weight of the information exerts its own kind of pressure on their thinking, and a disorientation sets in. As Mason says, when so much information abounds, some will favor certain horses, some will discount other horses, and still some other types of information might favor different horses and perhaps downgrade the original contenders. To the extent that these thought patterns dominate the handicapping process, there's a portentous kind of stark reality to Mason's dilemma. What to do? How to sort it all out? How to end with reliable handicapping decisions? The answer: Develop an information management approach to handicapping.

As potentially decisive information continues to be upturned and the role of information in modern handicapping intensifies for growing numbers of regular handicappers, this basic problem, of managing vast amounts of information effectively, will increasingly become the practical circumstance. If unalerted, or unprepared, more and more out-of-date handicappers will find themselves figuratively buried under an avalanche of handicapping information, unable to make choices in which they can invest serious money.

THE PROBLEM-SOLUTION STRATEGY

In 1982 when Mike Mason asked me, "What do we do with all this information?" I responded with comments that felt unsatisfactory at the moment and sound foolish now.

The important distinction, I said, was between the fundamental and the incidental. Fundamental information, I explained, reflects the competitive qualities of racehorses; class, speed, form, distance capabilities, bloodlines, or relates to the methods appropriate for assessing those qualities. Incidental information relates to circumstances affecting the running of races: pace, post position, track bias, trainers, jockeys, footing, and trips. Handicappers should identify logical contenders from fundamental considerations, I insisted, and use incidental information in combination with the basics to separate those fundamentally solid contenders.

Handicappers who become information managers will quickly understand what was so wrong with that pat answer. All handicapping information is relational. It's that simple. The relations of greatest value will depend on the particular handicapping problems to be solved or the particular decisions to be made. If class and speed are entirely interlocking, each of those factors will be related at times to trips or track biases in important ways. To conceptualize information as segmented, that is, the fundamental vis-à-vis the incidental, is to misapprehend all those opportunities in handicapping where situational factors predominate—and there are many.

Class might be decisive more frequently than track bias, to be sure, but class can be nullified by track bias as well. To argue that class is fundamental but track bias merely incidental, and therefore subordinate to class, is to miss the cheap but speedy colt breaking from the inside post at six furlongs on a track with a strong positive rail-speed bias. On biased surfaces that cheaper horse often wins wire-to-wire at 12 to 1.

The information management approach to handicapping is therefore entirely relational and integrative. At any time certain types of information might be most important to effective decision-making. At other times the same information types might be meaningless. The notion that multiple types of information can be comparably decisive becomes plausible. To the extent that a relational and integrative view of handicapping information can contribute to making more effective decisions, the information management approach succeeds.

Once the commitment to collecting the many types of handicapping information has been made, the real, practical problem of managing "all that information" becomes the operating condition. Handicappers who understand that information management has less to do with computer technology than with problem-solving and decision-making can simultaneously understand that computerized information systems represent the best solution to the practical problems of managing "all that information."

In conclusion, Table 2 summarizes the information explosion of contemporary handicapping. The handicapper's ideal information system includes all of it.

Table 2
The Handicapper's Information System

Components		
I. Theories and methods		
1. Comprehensive handicapping		
2. Speed handicapping		
2.1 adjusted final times		
2.2 speed figures		
2.3 projected times		
2.4 pace pars		
3. Class appraisal		
3.1 eligibility conditions		
3.2 ability times		
4. Early speed: speed points		
5. Form analysis		
5.1 form defects, form advantages		
5.2 situation handicapping: stage of the meeting		
6. Trip handicapping		
6.1 running position		
6.2 track bias		
6.3 trouble lines		
7. Body language		
7.1 sharp, ready, and dull horses		
7.2 frightened, angry, and hurting horses		
II. Probability studies		
1. Percentages and Probabilties	Fred Davis	1974
2. Computer Discoveries in Thoroughbred Handicapping	Bill Quirin	1979, 1984
3. Master Grass Sires List	Bill Quirin	1975–79–82
4. Dosage—A Practical Approach	Steven Roman	1981, 1984
III. Statistical compilations		
1. Trainer statistics		
2. Trainer patterns analyses		
3. Dosage indexes		

Table 2 *(continued)*

Components
3.1 nonclaiming three-year-olds
3.2 two-year-olds
3.3 selected older horses
4. Stakes races
4.1 grade designations
4.2 purse values
4.3 eligible ages
5. Foreign stakes races
5.1 grade designations
5.2 U.S. equivalent purse values
5.3 eligible ages
6. Sire lists
6.1 sprints
6.2 routes
6.3 slop
6.4 grass
7. Leader lists
7.1 owners
7.2 breeders
7.3 trainers
7.4 jockeys
7.5 horses
IV. Reference materials
1. Results charts
2. *American Racing Manual,* annually
3. Past performance histories
3.1 all horses on grounds
3.2 selected horses on local circuit
3.3 shippers
4. Periodicals
4.1 *The Blood Horse*
4.2 *The Thoroughbred Record*
5. *Trackside*
V. Soft information
1. The totalizator board
2. Opinions of talented handicappers
3. Trainer judgments
4. Professional seminars
5. Professional media
6. Inside information

Table 2 (*continued*)

Components
VI. Financial records
1. Data items
1.1 wagers
1.2 wins
1.3 winning odds
1.4 sizes of wagers
1.5 chronology of wagers
1.6 methods of wagering
1.7 exotic wagers and procedures
1.8 profits
1.9 losses
1.10 returns on investment
1.11 costs
1.12 expenses
2. Portfolios
2.1 straight wagers
2.2 combination wagers
2.3 evaluation data
3. Simulations
3.1 Similar methods
3.2 Other methods
3.3 Kelly criterion

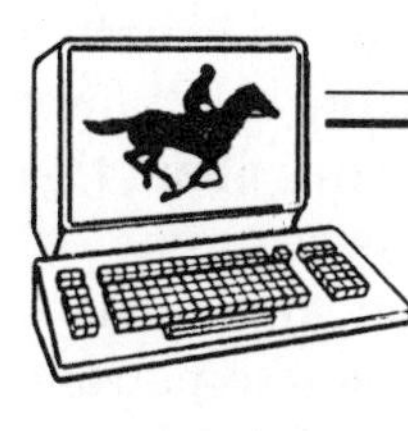

Chapter 3

HOW HIGH-TECH INFORMATION SYSTEMS HAVE CHANGED THE GAME

We have to bring the product to the people.

WILLIAM BORK
President and General Manager
Penn National racetrack, on the *Autovend* System

AT THE OPPOSITE extremes of the country sit two minor racetracks, Longacres, near Seattle, and Penn National, in Grantville, Pennsylvania. Each year on his rounds of the nation's racetracks, *Daily Racing Form* executive columnist Joe Hirsch visits Longacres and reports to all of us that this is still one of the greenest, loveliest, best-managed plants in the nation. Hirsch does not stop in Grantville, which is brown, desolate, and surrounded by fields. By this omission the columnist has no way of comparing the marketing information systems of the two racetracks.

MARKETING AND HIGH-TECH INFORMATION SYSTEMS

Morrie Alhedeff, who runs Longacres, is also the former acting president of the Thoroughbred Racing Associations (TRA), an organization that claims of itself to be deeply into marketing. As the national attendance dwindles, racetracks sputter and close, competition for the gaming dollar intensifies, and the young adults they have never learned how to attract continue to stay away, racing associations have turned their full attention to the acknowledged leader of the modern corporation in the service economy

—marketing. To racing executives, marketing means promotion and sales. In media interviews and executive sessions the shiny new directors of marketing at the nation's tracks proclaim their united belief that racing has simply got to be promoted better. No doubt Morrie Alhedeff totally agrees. So does anyone who cares about this sport. The divisive issue concerns what it is that gets promoted. Theoretically, in the marketing texts, it's the products and services that satisfy the identified needs of the market, that is, the patrons and customers of racetracks. This is where executives like Alhedeff and William Bork, who runs the show at Penn National, must part company. Bork faces intense competition in the marketplace; Alhedeff has none.

Thus Alhedeff promises his customers horseshoe-pitching contests and all-weather vinyl jackets if they will come to his racetrack, but once inside he will not give them projected payoffs on Exacta combinations. Alhedeff remains comfortable in the conviction that racing's customers need only to be entertained between races, the prevailing marketing strategy in relatively uncompetitive markets, but he does not think bettors care to be informed as to what the Daily Double combinations will return for each $2 they bet. Customers cannot get that basic information at Longacres, which apparently does not rattle Hirsch, the TRA, Longacres' marketing management, or many national racing officials, but upsets numerous regular racing patrons in the far northwest.

If asked what he thinks of *Autovend,* Bork's latest marketing system at Penn National, would the acting president of the TRA stare blankly at the issue, or might he be prepared to respond knowingly, based on his up-to-the-minute marketing information?

In general, are the nation's racetracks poised to enter the high-tech age of information? Bork is; Alhedeff is not. Will marketing be the savior of modern racing's attendance problems? Will marketing improve customer services; give repeat customers what they need and want? Will marketing attract the young adult market and convert large numbers of them into loyal fans? Will marketing transform the occasional racetrack visitors into repeat customers? Based on its first ten years—devoted to entertainment activities and giveaways, and pitched to the occasional customer who does not care about racing as a sport or about picking winners as a stimulating intellectual participation game at which knowledgeable, skillful players can win—very probably not. The prevailing marketing strategy deserves attention here, as high technology and the information systems of today afford racing unparalleled opportunity to capture and hold the educated, affluent, professional and managerial young adult upscale markets that have historically eluded the tracks.

High-tech information systems can impact the marketing of racing in essential ways never before imagined, but not without first identifying the information needs of racetrack customers and supplementing the prevailing entertainment promotions with information services. Market research would surely find that those information needs include instruction on handicapping knowledgeably and wagering intelligently. In short, information about playing the races more effectively. Conditions for marketing racetrack information services are finally advantageous. They are, in fact, exciting. First, tremendous knowledge and multiple sources of vital information about handicapping and pari-mutuel wagering now exist, where formerly they did not. I do not refer to the baseless, ill-tested systems and methods of racetrack hustlers and fast-buck artists that have historically retarded all progress in the information services arena, but to solid facts, statistics, information, and well-documented methodologies, the latter supported by evidence of their utility, produced by serious researchers who follow the game, and all of it of real value to racing's loyal fans and customers. Second, cost-effective computer technology now exists to encode, store, process, retrieve, and distribute the information, not only at the track but to networks of terminals and communications devices outside of the track, even into the home.

The first high-tech communications systems have already been installed by racetracks in the most competitive markets. The systems have been designed to distribute wagering opportunities and the telecasts of races, but not, sorrowfully, handicapping information. The most interesting of them all has been developed and implemented by tiny Penn National Racetrack.

William Bork cannot lure patrons to Penn National by offering them free tote bags, jackets, or jewelry. Bork has been forced to come to grips with the real reasons people flock to racetracks. They go there to handicap and to bet; in other words, to play the races. Racing is a participative sport for repeat patrons. Perceptive marketing managers will realize soon enough how information systems can enhance that participation motive. Not surprisingly, Bork has fastened on the betting purpose of racetrack attendance and has extended those opportunities to people in numerous Pennsylvania locales who cannot get to Penn National frequently. Bork's innovations include tele-bet, inter-track wagering, and simulcasting, but the most intriguing by far is *Autovend,* a high-tech system of off-site betting machines similar to the ready-teller machines installed and promoted by the banking industry.

Off-site betting, importantly, should not be confused with off-track betting (OTB), the New York model. Off-track betting is operated by third parties, and revenues are shared three ways: among the state, the tracks, and OTB. Off-site betting is controlled solely by the host racetrack. In the Penn National system the Autovend machines plug into the telephone lines used by its tele-bet service, but Autovend avoids the administrative hassles and personnel costs tele-bet entails, and from the customer's point of view provides the privacy telephone betting does not. The system is also much more flexible to market. The machines can be installed anywhere: bars, restaurants, supermarkets, sports arenas, shopping malls, office buildings, gymnasiums, street corners.

At first Bork installed Autovend machines in bars and restaurants, but his vision preferred freestanding facilities owned and operated by Penn National. Such facilities would afford racing patrons all the niceties—simulcasting, comfortable seating, work space, information services, food and beverage service, rest rooms. Simulcasting is deemed essential. Tele-Bet systems, for instance, the forerunner of self-betting machines, add numerous accounts whenever host tracks pick up another cable system.

When Autovend was scheduled to begin, in the summer of 1984, Penn National reached 420,000 homes on cable systems and serviced 6500 telephone accounts. The daily Tele-Bet handle was $65,000, but numerous accounts lie dormant. Bork's master plan intends building ten Autovend "locations" at $100,000 each. If each handled $30,000 daily, the nightly Autovend gross would equal what is bet at the track. Autovend bets are tied directly to the track's computer. The pari-mutuel odds are impacted, the payoffs are the same and so is the takeout, thus no opposition is heard from special-interest groups, such as horsemen. If Pittsburgh is added to the cable system, Penn National will reach another 250,000 homes, Autovend systems will proliferate, and the track's handle will double again.

The larger scenario is still more tantalizing. High-tech communications systems make feasible the marketing of Penn National racing throughout central and western Pennsylvania, which has no thoroughbred racing besides Penn National. Bork understands that merely be beaming his television signals to a communications satellite, the track's future can be secured. At any site Bork chooses to beam his races, he simply sets up a downlink to the satellite and a dish. By satellite communications, cities and towns without cable can watch the races on local independent stations. Handicappers throughout central and western Pennsylvania can make a convenient trip to the local Autovend machine, even as they might stop at the local bank's ready-teller outlet, make various bets on the evening's program, and later enjoy the races on television.

Autovend is quick and easy for the bettor.

At each off-site location the bettor first opens an account with a mutuel clerk who staffs a machine called a J-25 Autotote Limited. In exchange for the bettor's capital, say $100, the clerk supplies a voucher showing the balance and an account number. The voucher looks like a mutuel ticket sold at the track's windows. The bettor takes the voucher to an Autovend machine and fills out a betting slip similar to a Pick-Six ticket. The bettor inserts the voucher and the betting slip into the machine. The display on a seven-inch screen flashes the account balance and the current bets. The bettor verifies that information by pressing a bet "Approval" button. The computer processes the transactions. Betting tickets are issued by the machines. A new voucher is returned with the new balance calculated. If $20 has been bet, a voucher that showed a $100 balance will now show $80. If the bettor wins, he takes the voucher and winning tickets to Autovend and a new voucher is issued, showing the new balance. Vouchers can be cashed by Autovend clerks or at the track whenever the bettor chooses.

Autovend has been marketed by Penn National precisely as an off-site betting system, which when combined with simulcasting, represents to Bork the high-tech instrument for increasing racing's fan base. I think Bork overlooks an essential point here. As do Tele-Bet accounts, Autovend machines can lie dormant. Any form of racetrack betting that does not result from informed decision-making about horses and racing, from knowledge and skill in handicapping, if you will, cannot possibly increase either the revenue base or attendance base of racing. Gimmick betting, as Bork has imagined it—lottery-style jackpots won by holding winning horse numbers, serial bets modeled on the Pick-Six, multiple parlays for a buck investment, exotic numbers games dependent on winning horses—is not the answer, and is not a tenable marketing technique among sensible racetrack operators. The impulse bettors, the gamblers, should be strictly a tertiary market for racing in the information age. Informed bettors, or handicappers, and racing enthusiasts, the loyal fans are the two primary markets; the casual racegoers represent an enormous healthy secondary market. The only plausible strategy for increasing racing's attendance base and revenue base is to convert casual racegoers into informed handicappers and bettors, and racing enthusiasts. Autovend-type systems are marketed best not as betting systems but as information systems.

By this strategy Autovend provides not only a system for off-site betting but a means for the off-site accessing of basic information about handicapping, horses, track conditions, and odds lines.

It's imperative, for example, that Autovend systems display up-to-date betting odds at the push of a button for the next race on the program. Any racetrack that does not or will not do that does not understand its product well enough to market it effectively. *Horseplayers do not bet on race outcomes.* They bet on the public's proposition—odds—that a particular outcome will occur. If those propositions are unknown, intelligent betting cannot proceed; only gambling (guessing) can occur. If betting has not begun, morning lines can be displayed. Once betting has begun, however, current odds are greatly preferred.

The Autovend screens must also display jockey changes, overnight scratches and late declarations, track conditions, and the results of previous races. The results of previous races display more information than the first three finishers and payoffs. Results information for each race should include final time and fractional times, the par time for the race and race variant, complete order of finish, the winning jockey and trainer, and the description of the running from the official result chart. Similar information for previous days' races could also be available to the customer. To the extent that useful information were provided, betting would increase. That assertion could be tested by installing dual systems in comparable environments. One would be an information system, the other strictly a betting system.

Bork has been quoted, "Other sports have increased their fan base with TV, so why not us?"

Because other sports are essentially spectator sports, and racing is essentially a participation sport. Marketing the spectacle of racing is not enough; never has been. Too many spectators have no rooting interest in the outcomes and are not emotionally or intellectually involved, unless bets have been placed. The bet represents a participative interest. But as consecutive bets are lost, this level of interest quickly subsides. Young adults, to refer to the population racing courts but does not win, usually have growing but limited bankrolls and normally will avoid playing a game in which they take a repeated financial thrashing.

But a bet that represents an informed choice and deliberate investment strategy is quite different. That kind of bet represents the use of considerable knowledge and skill. It reflects a rational decision and the ability to play a complicated game intelligently. This kind of participative racetrack experience makes all the difference, especially as more and more of the participants come to understand an attainable level of knowledge and skill will be rewarded with profitable results. This—and nothing less—converts the casual racegoer into a loyal fan or racing enthusiast who attends and bets repeatedly. Only by converting casual racegoers into informed handicappers and bettors will racing's attendance base increase.

High-tech information systems now offer racing's marketing experts the greatest potential of all time for increasing the sport's attendance base. Bork's market research informs him that only 10 percent of all the folks living in Pennsylvania have ever attended the races. "We can tap into the other 90 percent, but we have to take the product to the people," he says, referring to the marketing potential of Autovend and simulcasting.

Yes, indeed, but the marketing possibilities of the high-tech information age hold out a much more meaningful message to racetrack executives everywhere. In taking the product to the people, with Autovend, tele-bet, inter-track wagering, off-track betting, and simulcasting, racing must also take the participative process to the people. The process of handicapping.

RACETRACK DESIGN AND LAYOUT

The Meadowlands, in Secaucus, New Jersey, is said by its design engineers to be a second-generation racetrack. This features vertical integration, or horizontal separation, the concept of building the

Figure 1
The new Garden State Race Track, a multi-purpose facility.

New tote and computer technology, moneyhandling procedures, economic and aesthetic considerations are changing the layout of the racing plant. In diagram 1, the grandstand and clubhouse are separated. This served as a simple crowd control device to separate clubhouse patrons from grandstand masses. But today it is no longer economical to construct what amounts to two buildings. Consequently, track planners are laying the clubhouse right on top of the grandstand, as shown in diagram 2. Tomorrow, as racetracks strive to appeal to the racing fan of the 21st century, a true multi-purpose structure may be connected to the main building, as shown in diagram 3.

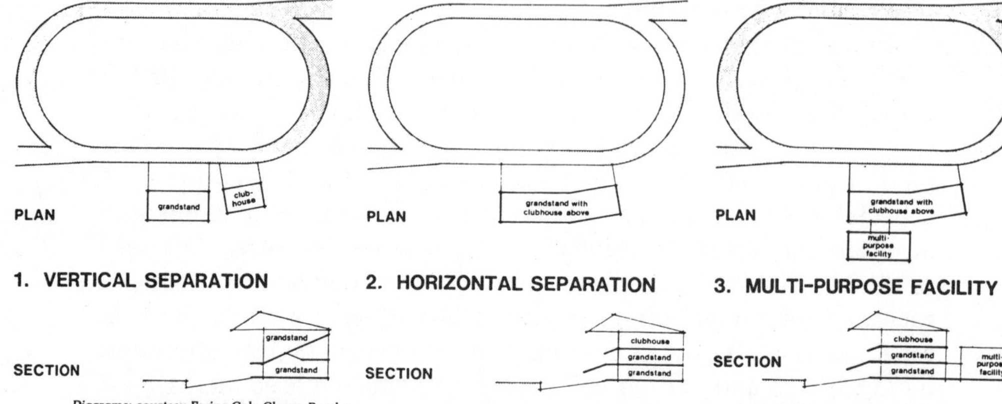

Diagrams: courtesy Ewing Cole Cherry Parsky

Photo: courtesy Ewing Cole Cherry Parsky.

I've seen the future and it is the past. *The rebuilding of Garden State Race Track (rendering shown above) may herald an end to the rectangular-shaped racing plant. A glass-covered paddock and clock tower-like structures will give the building an entirely new appearance: "a look of days gone by," said architect Parsky.*

clubhouse above the grandstand and not beside the grandstand. Building the clubhouse beside the grandstand represented first-generation design. That resulted in two separate buildings, with separate entrances, escalators, and partitions for different groups of customers. Vertical separation of buildings raised construction costs astronomically and complicated the movement of people.

Garden State Racetrack, in Cherry Hill, New Jersey, was reconstructed in 1984 as a third-generation racetrack, and this is the most fascinating design of all, greatly beneficial to handicappers. As seen in Figure 1, Garden State is now a multi-purpose facility, the chief characteristic of third-generation design. The track features a glass-enclosed paddock at the far left of the vertically integrated building, which will be visible from all levels of the grandstand and clubhouse, from the rear viewing areas adjoining the clock tower-shaped structures. Paddock inspections will be feasible using binoculars from open areas near the seats. A short walk replaces a long trek, and shelter from rain and cold is provided.

What truly distinguishes third-generation design, however, are the "multi-purpose" facilities and self-betting systems. A multi-purpose center is a large amphitheater sectioned off from the track's central corridors, without a trackside view, and intended in part for the simulcasting of races from other tracks. The multi-purpose center might be operated during the off-season or concurrently with a day's local card, perhaps to handicap, bet, and watch important stakes scheduled nationally. A tremendous advance for fans and handicappers would be an arrangement among racing associations for the simulcasting to all operating tracks of all graded stakes. This would stimulate as nothing else could the level of interest in racing as a national sport. Simulcasting and the resulting inter-track revenues would grow, with little or no depreciation of local pools. Regular attendance would be stimulated. If multi-purpose centers at racetracks became the only sources of daily out-of-state simulcasting, handicappers and racing fans would have even more incentive to attend the races and not do their shopping at off-site systems. High-tech handicappers can readily imagine a large carousel of computer terminals in these theaters, the terminals plugged into computers that contained regional or national databases of information useful to handicappers and bettors. The daily customers could query the database at the track. Would that be an incentive to go to the races?

Third-generation design also involves heavy electrical systems, data communications ducts, and physical layouts that can facilitate the use of self-betting machines at trackside, in the infield, at the paddock, at dining tables, along the walls, at Autovend carousels on each level, and even in the parking lots and stable areas. The potential for eliminating long betting lines is a loaded issue for architects, racetrack designers, and plant managers, as unions would fight that, and racing fans are thought to enjoy waiting in lines to do their betting. Few of them enjoy being shut out, however, which happens far more frequently than is now necessary.

If third-generation design features become popular with racing's customers, today's plants will find themselves building multi-purpose centers for simulcasting and datacom (data communications) networks for installing self-betting machines. At Garden State the planners and designers have prepared for the electronic developments of the twenty-first century, including whatever might emerge from the tote companies' research and development arms. Modern office buildings face the same problems. The designers need to prepare for the electronic desks and computerized work stations of the near future. Office automation is an eventuality. Existing office buildings will be forced to adapt to electronics by changing to accommodate information systems. So will existing racetracks.

HIGH-TECH HORSEMEN

Southern California trainer Darrell Vienna for several seasons now has used a microcomputer to manage pedigree and performance information about thoroughbreds his clients might want to buy, breed, or sell. The information also guides the horseman in claiming horses from others. Vienna has been active in the import markets, so it's reasonable to expect that he has access to computerized files of information about the horses of Ireland, England, France, and South America as well.

Other trainers and numerous owners subscribe to pedigree information services. With a receiver, a printer, and a telephone link to the communications system, owners can dial a computer and within minutes read a printout of a horse's extended pedigree, including the updated performances of the family members at all American racetracks. This informs owners who are breeders as to whether their equine families have appreciated commercially of late and generally provides valuable information about potential matings, purchases, and sales.

Other obvious applications of high-tech information systems for horsemen have been slow to materialize.

With a desktop computer and a few basic business applications programs, trainers could automate the stable's bookkeeping and accounting functions; keep completely accurate records on personnel, payroll, and inventory; schedule all routine stable operations and daily work tasks; and file every imaginable piece of information about each of their horses, including bloodlines, training patterns, workouts, past performances, nutrition, and veterinary histories. Trainers would enjoy instant access to this information and lots more. No programming or other computer skills are demanded beyond the ability to handle the keyboard, a typing skill, and the wherewithal to follow the input, processing, and retrieval instructions that appear on the terminal screen.

With a second microcomputer and a few floppy disks, trainers would have the same information available in the quiet of their residential dens. Most of the routine business operations associated with managing a large stable, in fact, could readily be done off-site. For trainers such as Lazaro Barrera, whose main strings are bolstered by large numbers of horses at the off-track on the circuit, by more horses and two-year-olds at faraway training centers, and even by the upscale breeding operations of key clients as far away as Kentucky farms, high-tech information systems and communications networks represent the advances of a lifetime. If leading horsemen do not care to operate the system themselves, they can hire computer technicians and information specialists to do the work and to present the information to them.

But horsemen will probably be among the rear guard of those moving slowly into the age of information. Trainers especially are creatures of long-forming habits, and basic changes or adjustments usually come piecemeal, if at all. The sources of change that are trusted are usually personal experiences. Not many trainers on the scene at select yearling sales, for example, know what a dosage index is, how to obtain dosage information, or how to interpret it, but the few who do have leaped far ahead of their colleagues. They will select better classic prospects on the whole and eventually make more money for their clients and for themselves as a result of obtaining the information.

Information systems can help horsemen make better management decisions about horses, and high-tech developments can assist in training and caring for horses, so progressive trainers should begin to feel more incentive than before to come to grips with several contemporary trends. Lasers, for example, can now accelerate the healing of a two-year-old's bucked shins without the painful blistering or pin-firing of the flesh on the cannon bones. As the great majority of two-year-olds buck their shins, and by traditional treatments would miss the chance to dip into what has become a wonderfully rich pot of gold in two-year-old racing, many trainers can be expected to get better acquainted with this high-tech veterinary advance.

Computers and high-speed photography can now inform horsemen how efficiently young untested thoroughbreds use their limited energies while running. The technique is called gait analysis. A camera cranking away at 500 frames per second captures a horse's strides in minute detail. A subsequent computer analysis reveals the overlap time the legs remain on the ground. The measurement is taken in fractions of seconds, and the longer that two legs are on the ground at once, the poorer is the horse's running stride, or extension, as horsemen call it.

Gait analysis has been developed by Massachusetts Institute of Technology (MIT) professor of electrical engineering and computer science George Pratt, who got the brainstorm a decade ago after analyzing a high-speed camera sequence of Secretariat catching and passing stablemate Riva Ridge during the 1973 Marlboro Cup at Belmont Park. At the instant the two horses were nose to nose, the left forelegs of each had been fully extended, but Secretariat's *right* foreleg was about to leave the ground while Riva Ridge's right foreleg was not yet midstride. Gait analysis indicated Secretariat's footfalls were more evenly spaced than his rival's. Each of Secretariat's front legs touched the ground only as the other was pushing off. A computer analysis of the two champions' gaits revealed, as only computers can, that no two of Secretariat's legs were on the ground at the same time for more than 81 thousandths of a second (approximately 18 percent of each complete stride), compared to 117 thousandths of a second for Riva Ridge (about 27 percent of a stride).

Subsequent gait analyses of numerous champions demonstrated that Secretariat probably had the most efficient stride ever. The all-time champion covered more distance per stride, 23.8 feet, than other horses. Riva Ridge covered 23.2 feet. Pratt has noted that Secretariat was so smooth-gaited you could have balanced a glass of water on his rear. By analyzing the strides of top horses, the efficiency of their gaits can be evaluated and converted to a set of standard scores, zero to one hundred, with the highest scores representing the best strides. By this technique all horses can be compared to the champions and ranked on a similar scale.

MIT insisted Pratt obtain a copyright for gait analysis in the university's name. In consequence, since 1980, equine-biomechanics has been offered to horsemen as a service that evaluates racing potential. The most extensive application so far has been at major select auctions of yearlings and two-year-olds. Research on gait analysis has shown that strides do not change as horses grow older, excepting the lengthening due to maturation.

At the auctions gait analysts film the horses in a gallop and then project the films, frame by frame, onto a device known as a digitizer. The digitizer enables technicians to plot the exact positions of the horse's hoofs, knees, hips, and other moving parts, from stride to stride. The information is fed into a computer, and the results of a gait analysis are compared to the comparable profiles of past champions. The computer scores a sales prospect from zero to one hundred, and the evaluation is relayed to clients, the prospective horse buyers.

Research has shown that good racehorses use little up-and-down motion in their strides, but rather exhibit instead a powerful rear-quarter thrust that results in an elongated extension that represents the most efficient gait. Proponents of gait analysis concede that the technique cannot predict which horses will be winners or champions. But they argue strongly the method is reliable in identifying which horses will not run well. To be sure, gait analysis often discourages buyers from making expensive purchases of well-bred horses having inefficient strides.

Can gait analysis be helpful to trainers on the backstretches of racetracks? One who thinks so is the intelligent, progressive, and much respected New York trainer of Triple Crown champion Seattle Slew, Billy Turner. Turner uses the technique routinely at his barn. "If gait analysis says a horse won't run," Turner has been quoted, "it just won't run, no matter how long you try."

Turner realizes an efficient gait cannot make a racehorse a winner, but he is also convinced no horse can become great without a favorable stride. Pratt and other promoters of gait analysis agree that simply because a horse can run does not mean it will. As Turner puts it, "A horse that has the gait but not the heart will not run very long." On the other hand, Turner, Pratt, and other scientists believe that gait analysis is a valuable tool for distinguishing young horses as superior prospects or not, and for making some basic training decisions about a young horse's career potential.

High-tech backstretch services that trainers feel far more comfortable with have emerged from orthopedics. Veterinarians now are routinely saving racehorses from injuries that not long ago were incurable. Sesamoid bone fractures are a highly visible case in point. Ohio State veterinarian Lawrence Bramlage has developed surgical techniques for sesamoid fractures utilizing metal plates, steel wires, bone drilling, and a relaxingly graduated administration of tranquilizers and anesthesia. As long as injured horses have not lost bone, as long as it's a clean snap, the sesamoids can be reconstructed using the stress-reduction principles of physics, and horses can sometimes race again. Previously, 90 percent of sesamoid victims were destroyed. Approximately 60 percent of Bramlage's patients have been saved for breeding purposes, and he's had three rejuvenated winners as well.

Surgical instruments and surgical routines for horses have been changed radically by high-tech advances, such that today the equine operating room looks like those equipped for humans, including anesthesiologists and scrub nurses who use tranquilizers, painkillers, and anesthetics to calm injured animals. On another revised front, the postoperative objective usually requires getting injured horses on their feet as soon as possible.

To allow leg fractures to heal, horses traditionally had to be immobilized following surgery, lest frightened and threatened by the surgical experiences, they kick themselves to death once they awakened. Ruffian died in that tragic manner, kicking herself after awakening from extensive surgery. Now horse patients are slowly tranquilized before surgery and not anesthetized until fully secure and relaxed. "If they go into surgery relaxed," says Bramlage, "they will not be so scared when they come out." Immediately following the rub-off of anesthesia, horses are stood on all fours again. The stakes horse Star Gallant, for instance, was on its feet immediately following leg surgery, with only a lightweight cast and bandages on the injured left foreleg.

As students of equine body language will remember, Bonnie Ledbetter, the foremost authority, has labeled the fall or supine position the most threatening of experiences for horses. Being able to stand upright, says Ledbetter, is a matter of survival for thoroughbreds. As veterinarians have long known, if a limb has been operated on and continues to cause pain, horses will overload the other limbs, usually overcompensating for the pain and usually causing themselves additional serious injury. Those problems have been confronted by numerous high-tech refinements in equine orthopedics, affecting preoperative treatment, surgical technique, instrumentation, monitoring, and postoperative treatment too. It's been veritably a revolution, and trainers have little psychological difficulty accommodating these high-tech veterinary advances.

The quintessential high-tech horseman, a man already operating on the cutting edge of the information age, is D. Wayne Lukas. He is the new leader of the pack, poised as this is constructed to set standards for money won, stakes won, and stakes-money won that have never before been approached.

How does Lukas do it? The horsemanship is obvious, and it includes as keen an eye for perfectly conformed yearlings as has ever been witnessed at public auctions. Lukas has become in seven short seasons the most talented trainer of two- and three-year-olds as this sport has ever boasted, dominating the stakes in both divisions and winning consecutive Eclipse awards with Landaluce, Althea, and Life's Magic. In these arenas, Lukas has been a man for his times, the game having drifted toward younger horses incessantly since the colorful trainer's debut in 1978.

Beyond the basics, Lukas does it with computers. The trainer employs an astonishing fifty-seven people and has major strings competing simultaneously in southern California (home base), New York, the South, and the Midwest. He manages the daily activities of this national network of horses and employees with microcomputers. The Lukas computers talk to one another, probably the only datacom network on the backsides of racetracks.

Once the star filly Althea needed a race, nothing severe, just a tightener, but none at the home base suited her. Lukas consulted his computer. It informed the trainer of the upcoming Barbara Fritchie Handicap, a Gr. 3 $100,000-added stakes for 3up in Maryland. Lukas shipped Althea to Bowie the next day. She won easily.

Computers. Datacom networks. Information systems. They are basic equipment for horsemen in the age of information.

HIGH-TECH HANDICAPPERS

Handicappers face the same challenges and choices as do racetrack executives and horsemen. In short, they can shift into the high-tech age of information deliberately, confidently, purposefully, and quickly, or they can be dragged along kicking and screaming, unready, unable, and unwilling to change conventional practices until it has become socially unfashionable or personally impractical to resist any longer. The most desirable and useful changes, however, are rather drastic, involving not only new concepts, methods, and ways of thinking and making decisions but also new technology, which itself is costly, psychologically disorienting, and demanding. In the normal case, considerable time and commitment to the skill training and practice are required to use the technology well.

It can be fairly assumed that the general lag between technological change and the cultural change, social change, and personal change it presses will be just as prevalent among handicappers as among racing officials and horsemen.

The ultimate challenge for handicappers who move into the high-tech information age will be the careful design and development of handicapping information systems. Such systems are often denoted by the acronym MIS. The acronym means Management Information Systems, a concept borrowed for handicappers from the managers of modern organizations, who also depend on information they cannot organize, process, and manage by themselves to make their key decisions. The handicapper's MIS consists of manual files, microcomputers, peripherals, applications software, database software communications networks, databases, off-line information services, human information teams, and most important, new ways of using information to analyze past performances and make handicapping and wagering decisions.

A worst-case scenario for handicappers of the high-tech information age is to fail to change at all or to avoid the issues. The motivation in part can be reduced to expediency. Two major recent book releases on handicapping, by Andrew Beyer in 1983 and by William

Quirin in 1984, both alluded to the kinds of information needs now central to success in handicapping, namely, the contemporary handicapper's need to go far beyond the information in the past performance tables and results charts.

Thousands of well-schooled, talented handicappers are playing the game now, and practically all of them can interpret the past performance and chart data intelligently. That has resulted in too many underlays and low-priced favorites. The best way to get ahead today is to obtain solid information that not as many other handicappers will access or use.

Another weak-case scenario relates to high-tech handicappers who move into the future by acquiring the technology but not the new ideas and methods for utilizing the technology and information most effectively; that is, handicappers who do the same things as always but now do them with computers. The purpose and promise of this book is not so much to promote the use of computer technology in handicapping, as to indicate how computers might be used to implement a new approach to handicapping: an information management approach.

The role of the microcomputer in the full-blown approach is dual: to store and retrieve large quantities of handicapping data and information quickly, and to process the data and information accurately, as instructed. As presented here, importantly, the processing function does not entail the making of handicapping selections or betting decisions. The computer is a tool, an instrument, a means toward greater ends. Human handicappers make the decisions. It is a very large point.

Handicapping software is best, too, when merchandised as a means of supplying management information to handicappers for the greater end of making informed decisions about races, horses, and wagers. By that standard the high-tech handicapping software merchants have dashed off to a bad start, which will be taken up with some displeasure in the next chapter.

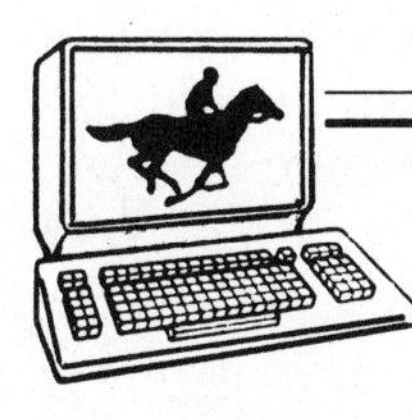

Chapter 4

STATE OF THE ART: A BLAZING NEGATIVE CRITIQUE

For every good handicapping program we found in researching this article, there were about three that were absolutely worthless.

You'd be better off throwing darts to pick a winner—or playing pin the tail on a favorite—than you would be using one of these systems.

One program we reviewed picked only two winners in 150 races we tested.

ROGER E. MOTT
Software Supermarket, a guide to personal computer software, May, 1984

A MELANCHOLY ASPECT of the high-tech movement is the back door it has held wide open for the hucksters of handicapping. To be sure, the rip-off artists and quick-fix merchants have not been bashful about streaming in. This is something of a setback.

For the past several seasons the literature of handicapping and its accompanying instruction in most major markets have been suffused with an accumulation of knowledge, evidence, and scholarly intelligence that has stood as counterpoint to the traditional stereotypical merchandising of threadbare, ill-tested systems and methods for "beating the races." Even the scientific method has edged its way into the study of handicapping, lending its rigor and helping to assure that future claims of success unsupported by facts cannot be held as tenable. With each succeeding season the hucksters and rip-off artists were losing considerable ground to solid, serious efforts to illuminate the theory and practice of handicapping.

But now the hucksters are back. They have returned full force to the advertising pages of *Daily Racing Form* and on fanciful direct-mail flyers, armed now with computerized handicapping programs that are offered as eternal salvation to racegoers who also happen to own personal computers. Beating the races is possible and easy at least, the ads proclaim; simply follow the instructions on your terminal screen, type in a handful of data items, and the computer will do all the rest.

A particularly insidious feature of the new-fangled ripoffs is the selling prices of the programs. The cost of touting has gone up—way up. Prices are often outrageous. As this is written, a demonstrably worthless piece of software is being brandished to racegoers for $1,300. Some hucksters will sell computer-programmed selections to those who don't own computers for the same kind of inflated prices. And in a fiendishly criminal component of the merchandising of high-tech handicapping, a few high-tech hucksters have attempted to sell hardware that costs approximately fifty bucks from Radio Shack for a thousand or even two thousand dollars. An uneducated racing public had better beware!

The temptation to dismiss the current market of high-tech handicapping software as entirely pointless must be resisted, however, because in the context of generally unsubstantial, overexpensive programs and cheap, sensationalistic advertising, some very brainy people are also developing handicapping software that is not only providing excellent high-tech models of handicapping but has been unusually powerful in getting results.

What's what? Who's who? This chapter is intended to clear the air, to help.

THE HIGH-TECH MARKETPLACE

One of the few qualified individuals to complete an in-depth review of the available software in handicapping is George Kaywood, of Omaha, Nebraska, who prepared the high-tech activities and presentations for the Second National Conference on Thoroughbred Handicapping at the Meadowlands Hilton Hotel in October 1984. Kaywood is not very sanguine about the high-tech marketplace. In reviewing applications programs for inclusion at the national exposition, he accepted a few and turned down dozens. Kaywood reported a steady submission of computer programs characterized by poor documentation, unsatisfactory tests of results, and unreasonably high costs. Several of the programs were just badly written, unsubstantial. They did not reflect a knowledge and understanding of handicapping consistent with today's generally accepted principles and practices.

At the same time, as proprietor of *Com-Cap*, a small midwestern business that markets various computer racing programs, Kaywood has collected evidence that what the high-tech handicapping audience wants are prepackaged applications programs that produce selections or rankings. These handicappers are thought to prefer to stick a ready-made applications program into a disk drive and in a few effortless minutes be visited by a potential winner on the terminal screen, notwithstanding the substance or effectiveness of the handicapping model inside the computer. Kaywood suggests that perhaps not many regular handicappers are sophisticated as yet about the possibilities of high-tech handicapping.

The author believes a reason for that is an understandable failure among high-tech handicappers to distinguish between the logical representation of data, which handicappers are qualified to do, and the physical representation of data, which few of us are qualified to do. Assuming—wrongly—that handicapping by computer requires special high-tech skills in programming and computer technology, technically unskilled racegoers choose instead the only available alternative, purchasing the user-friendly applications programs of others. This precondition is seriously flawed and must be remedied.

This book argues that handicappers do not need computer skills, notably programming skills, to practice high-tech handicapping in the information age. Further, it asserts that handicappers should be in no hurry to develop those skills. In many cases trying to program a computer to process complicated handicapping methods would likely be the worst step handicappers could take. Computer programming requires formal training and practice.

Handicappers who own microcomputers and prefer to do their own handicapping by utilizing the machines can definitely do so—beyond their high-tech imaginations. High-tech handicappers should be primarily concerned with the logical representation of data but not with the physical representation of data. Later chapters in this book spell out the differences. Computer technicians can convert the handicapper's logic to language the computer can understand. High-tech handicapping becomes the collaborative effort

Figure 2
The Racetrack Robot I, a hypothetical composite of a high-tech handicapping advertisement in *Daily Racing Form*.

The National All-Purpose High-Tech
Handicapping Selection Program

* THE RACETRACK ROBOT I *

A Guaranteed Short Cut to High-Tech Handicapping Success

Several prominent computer scientists from a highly respected firm and a well-known veteran handicapping professional of impeccable credentials have been working for five years to produce the ultimate computer handicapping program. It was hard work all the way—hours and hours of testing. Now they proudly present—THE RACETRACK ROBOT I. It does just about everything for you.

* Evaluates more than 20 handicapping factors—more than twice the number of most programs.
* Averages 5 to 1 and up on its winners. Stays clear of phony favorites.
* Picks Daily Doubles and Exactas with amazing consistency—a big bonus the program developers did not anticipate.
* Works especially well at minor tracks.
* Comes with a special instructional manual, chock-full of handicapping tips from the veteran professional.
* No personal judgment required of users—ROBOT I eliminates the guesswork.

WHY SO LONG IN PRODUCTION? It takes years to condense an old pro's vast knowledge and experience into a couple of dozen variables that can be computerized in a single winning package. Now you can get the same winning information within minutes—easily computed!

Make checks or money orders payable to: The R.R.I., Inc.

The Racetrack Robot: $425.00

(Send your name and address. Be sure to tell us your computer make and model.)

NOTE: If you do not own a personal computer, we will gladly sell you one especially useful for running THE RACETRACK ROBOT I.

it should be, between handicappers and computer specialists. It is likely that the technicians will themselves be handicappers or at least racegoers.

Racegoers or handicappers who own microcomputers and prefer to buy ready-made programs they can slide into a disk drive will be susceptible to advertisements such as the hypothetical composite of a typical *Daily Racing Form* ad presented in Figure 2.

What is wrong with this advertisement? Aside from its patently fraudulent claims of success, thinly veiled references to reputable professionals (who are they?), and the slick, sensational tone and content, the specific claims for the Racetrack Robot I are so diametrically opposed to what a reasonably satisfactory computer program would offer that handicappers can use the copy to alert themselves to similar high-tech schemes.

For technical reasons to be touched upon later in the chapter, well-designed computer models of handicapping incorporate a half dozen factors or less, not more. By promoting "more than twenty handicapping factors . . . more than twice the number of other programs," the Racetrack Robot I marks itself down as pointless. I cannot even imagine more than twenty factors of handicapping, let alone believe that more than a handful of them operate independently of the others. That is, the handicapping factors overlap; they measure the same thing. Moreover, although more than twenty variables might be easily computed by computers, they are not easily entered by users. A nine-horse field requires at least 180 inputs. A nine-race card of ninety horses would require 1800 inputs. In computer handicapping, twice as many is usually much, much less.

No full-blown handicapping methodology in history has delivered horses that average 5 to 1 or greater on winning. The upper limit on success is approximately 7 to 2, and that is seldom reached. Truncated systems consisting of two or three factors occasionally return 4 to 1 or better in the short run, but none of them endure. Approximately 70 percent of all races are won by horses sent off at odds of 3 to 1 or less.

How convenient that the Racetrack Robot I not only picks 5 to 1 shots to win, but also has the uncanny sixth sense to pick Daily Doubles and Exactas. As this was not predicated on any feature of the system design, naturally the program developers should not be expected to explain the achievement in handicapping terms.

Another characteristic of computer models is a sensitivity to the track surfaces and horse populations on which they are modeled. Programs that work well at major tracks are unlikely to do as well at minor tracks. The horses and track surfaces are vastly different. By claiming a significant edge at minor tracks, the Racetrack Robot I programmers intimate a corresponding lack of success at the majors. Does this magical program work well for all horses at all racetracks?

The advertisement is more conspicuous for the information it fails to present than for the information it does. What hardware models is the software compatible with? All?

What kind of documentation is available for users? Does the instructional manual "chock-full of tips from the veteran professional" also show customers how to run the computer program?

Do customers get a diskette or a cassette tape? Where is the effectiveness data and workouts? How do prospective consumers obtain that evaluation information?

If handicappers buy the program and can't run it on their personal computers, do they get their money back? If the Racetrack Robot I proves ineffective at the local track are there any guarantees or provisions for refunds?

Finally, the price tag of $425 is far too steep. Handicapping applications programs should never exceed $250 and usually should cost $150 or less. Any programs priced above $250 are simply not priced competitively.

On the matter of price, in my opinion the greatest high-tech rip-off to date is a computer program marketed as the *Handicapper's Guide to Fiscal Fitness,* developed by a gentleman billed as Lee Lawrence. Lawrence has managed to get on television and into the print media with reports of his handicapping prowess utilizing computers, and the result has been an almost scandalous blitz of multiple-media and direct-mail advertising pitched at unaware horseplayers. Figure 3 shows the prices Lawrence wants for his products. Systems software costs less than Lawrence gets for a handicapping application. I find the prices offensively shocking.

Evidence has long since begun to mount as well that not only is Lawrence's computer program a flimsy model of marketing, but also his product guarantee is a sham. A responsible consumer protection resource in racing that has locked horns with Lawrence and his merchandising techniques is *Phillips Racing Newsletter,* which has evaluated products and services in the handicapping market for its

Figure 3

Five Computer Applications Programs of Handicapping, As Evaluated by *Software Supermarket*, May 1984.

COMPUTER ANALYSIS FOR THOROUGHBRED RACING ($200), for the Apple II, II Plus and IIe, the Commodore 64 and VIC 20, the IBM PC, the Timex/Sinclair 1000, TRS-80 Models I, II, III and 4, and virtually all CP/M machines, from the Cannella Corp., 420 E. Genesee St., Syracuse, NY 13202, 315-457-8804.

Class—good. Produced a fair percentage of winners based on class alone.

Speed—poor. Does not rate.

Consistency—good. Appears to be on target.

Pace—poor. Does not rate

Recency—good. Uses date of last race.

Jockey—poor. Does not rate.

Trainer—poor. Does not rate.

HANDICAPPER ($29.95–$36.45), for the Apple II, II Plus and IIe, the Atari 400 and 800, the Commodore 64, the IBM PC, TRS-80 Models I, II, III and 4, and virtually all CP/M machines, from Dynacomp, 1427 Monroe Ave., Rochester, NY 14618, 716-442-8960.

Class—fair. Could be misleading.

Speed—poor. Does not rate.

Consistency—good. Looks at finishes.

Pace—good. Rates style of runner.

Recency—poor. Does not rate.

Jockey—poor. Does not rate.

Trainer—poor. Does not rate.

The Leader (s)

THE SPORTS JUDGE—THOROUGHBRED SYSTEM ($129.95), for the Apple II, II Plus and IIe, the IBM PC and TRS-80 Models I and II, from The Sports Judge, 310-320 The Village, Redondo Beach, CA 90277, 213-516-2044.

Class—excellent. Produced a good percentage of winners based on class alone.

Speed—good. Appears to be on target.

Consistency—good. Appears to be on target.

Pace—good. Uses the right calls.

Recency—good. Uses date of last race.

Jockey—good. A must in handicapping as top jockeys usually get top horses.

Trainer—good. Definitely a difference between the best and the worst trainer at the track.

Win %—37.
Average Mutuel—$7.41.
Dollar Return—30%.

IN$STRIDE THOROUGHBRED SYSTEM ($200), for the Apple II, II Plus and IIe, the Commodore 64 and VIC 20, the IBM PC, the Timex-Sinclair 1000, TRS-80 Models I, II, III and 4, and virtually all CP/M machines, from Racing Data Systems Inc., P.O. Box 2214 Station D, Ottawa, Ont., Canada KIP5W4, 416-242-3775.

Class—good. Produced a fair percentage of winners based on class alone.

Speed—poor. Does not rate.

Consistency—good. Appears to be on target.

Pace—poor. Does not rate.

Recency—good. Uses date of last race.

Jockey—poor. Does not rate.

Trainer—poor. Does not rate.

THOROUGHBRED SYSTEM ($99.95), for the Apple II, II Plus and IIe, the IBM PC, and TRS-80 Models I, II and III, from Computer Research Tools, 725 S. Evanwood Dr., West Covina, CA 91780, 213-962-1688.

Class—fair. Could be misleading.

Speed—fair. Does not rate.

Consistency—good. Looks at finishes.

Pace—good. Uses all calls.

Recency—good. Uses date of last race.

Jockey—poor. Does not rate.

Trainer—poor. Does not rate.

Note: Each program was tested on the same 150-race sample.

regular subscribers since 1974. In August 1984 Phillips commented on the Lawrence program. I present the comments verbatim:

COMMENTS ON LAWRENCE'S HANDICAPPER'S GUIDE TO FISCAL FITNESS

This is one of those costly computer methods you read about but seldom can afford to buy. Lee Lawrence is the principal and the method program costs $1,300, or with a 24K TRS-80 RS100 computer, $2,100, and a printer is an extra $295. All from Lee Lawrence, Inc. Bosley Bldg., 210 Allegheny Ave., Baltimore, MD, 21204. (Wonder if that closeness to Mike Warren tends to rub off?)

Lawrence sent us his method and a lot of literature but not the computer. As Stephen Duma reported last issue, Steve's method and figures did better in a test on a certain day at Aqu recently. Less winners with Lawrence's material and it was mostly chalk. Steve couldn't figure out the system sent—same one we had—as some figures just didn't work out or fit. We had the same problem but we thought, maybe you needed the computer. We also kept in mind the saying, "garbage in, garbage out."

A long-time reader of ours who is also a veteran in horse race handicapping, bought the whole works from Lawrence to give it a thorough test. If anyone knows about handicapping, he does. He gave it a thorough test and here are some of his thoughts. . . .

"To say the least, Lee Lawrence's Handicapper's Guide to Fiscal Fitness is distinctly without merit! The basis of the method is:

1. Last 4 speed ratings and variants (Racing Form) at required distance. In sprints, 5–7f races (last 4). In routes, 1 mile to 1⅛ (last 4). Turf races, last 4 turf spd + var. Pure rubbish!
2. In sprints you take best ½ mile fraction of horse's best race, whatever it is, win, place or show or any other position. That means if he has 10 races in his record, you take, say, a race with a ½ mile fraction of 44.1 run at AC, which might have been run a year ago, but it's his best race. Pure junk! A half-mile fraction run 6 months or a year ago means nothing.

"Instructions say Bet Top Horse. Very seldom bet 2nd choice and then perhaps place or show." He then gave us the computer printouts for five consecutive days at Hol recently. Computer is supposed to pick the horses in order. Out of 38 races, he caught just 4 winners at prices of $3.20, $6, $6.40 and $4.40, all chalk or favorites. Also had nine 2nds . . . even lower prices. He said during the previous week, he had 37 races and 6 winners or 16%. His total for 10 days was 75 races, 10 winners or 13% and top mutuel (not average) was $8.

He sent the whole works back along with a detailed workout and, of course, they said he was doing it wrong. If they knew this fellow, they'd know that it was not done wrong.

He told us that he acquired the name of another person in the area who had one of the computers and who had the same experience finding it worthless. She sent it back and was also told "you are doing it wrong. The old ploy!" Our friend finally got them on the phone and they said they would send some of the money back. He said, "I'll believe it when I get the check. And I would put old Lawrence in a class with Warren."—Phillips Racing Newsletter, August, 1984

A general trade consumer protection magazine that has examined a wide array of the handicapping software that has spilled onto the market in recent years is *Software Supermarket*, which reported its findings in May 1984. The report was bleak. It characterized three of every four programs tested as "absolutely worthless." Each program was tested on the same 150-race sample. The worst picked two winners, but the best picked 55 (37 percent), leading the magazine to conclude that a few computer models of handicapping can actually throw profits.

The consumer magazine eventually analyzed thousands of races by computer and identified the handicapping factors common to the great majority of winners. Seven were selected as most important and recurring: class, speed, consistency, pace, recency, jockey, and trainer. Next the magazine used the seven factors to eliminate unsubstantial programs and to identify five they wanted to evaluate in-depth. The five programs were assessed in four ways: (1) coverage of the seven key factors, (2) win percentage, (3) average mutuel on winners, and (4) rate of return on the dollar invested.

In reporting its findings, *Software Supermarket* compared each program on the seven handicapping factors (see Figure 3), but presented its performance data as ranges. The win percentages ranged from 22 to 37 percent; average mutuels ranged from $5.26 to $7.41; ROIs ranged from −17 percent to +30 percent.

The most effective program was *The Sports Judge,* which not only was rated good to excellent on all the important factors and picked 37 percent winners at average payoffs of $7.41 and a .17 dollar profit but also had 26 of its second-highest rated horses win. Its one-two winning percentage was a powerful 63 percent.

More interesting perhaps to handicappers everywhere are three substantive guidelines the magazine presented to help consumers evaluate computer programs in handicapping. The rules make sense, suggesting that the author of the report knew more than a little about racing and handicapping:

1. The program should be designed so that in sprint races it gives more importance to speed than class, whereas in longer races (a mile or farther) it shifts the emphasis to favor class over speed.
2. The program should evaluate pace or early speed. (The author points out that it is a fact that frontrunning horses with good early speed win more than a fair share of their races, which is true.)
3. The program should also evaluate jockeys and trainers.

Computer programs that were tested by *Software Supermarket* could not yield profits if they ignored jockeys and trainers, notwithstanding their performance on other variables.

In a final section of the chapter the author presents a more comprehensive set of standards by which high-tech handicappers can evaluate various computer programs in handicapping.

THE SPORTS JUDGE

The Sports Judge is a regression* model of handicapping that demonstrates how powerful a computer program can actually be-

* Regression is a form of statistical correlation by which predictor variables, such as handicapping factors, are related to a predicted variable, such as race outcomes.

come. The program has been tested variously at southern California tracks and never yet found wanting. At Del Mar 1984 it beat the entire meeting (43 days), picking practically every race and ending with 31 percent winners, an average win mutuel of $6.60 and an 8 percent dollar return. In key races it outperformed a group of 11 professionals, 31.4 percent to 23 percent winners, earning a 29 percent dollar profit while the handicappers were losing 18 percent. Prior to that it handled Hollywood Park, in the three weeks from June 27 through July 15, selecting an astonishing 62 winners of 135 races, or 46 percent proficiency. The weekly dollar profits were 45 percent, 35 percent, and 14.5 percent. Best bets won 25 of 41 races, or 61 percent.

How does a computer model of handicapping become that proficient and that powerful?

It happens gradually, statistically, scientifically, and invariably at the hands of intelligent computer scientists who also know a lot about the art of handicapping. Handicapping entrepreneurs Steve Arthur and John Erickson fit that mold, and each will tell you *The Sports Judge* has been under development for years. The research and development process has been painstaking, involving thousands of races and repeated modifications of the program. Each modification demands another round of testing. The computer is the necessary handmaiden.

Essentially, combinations of handicapping factors judged to have predictive value are combined, weighted, and tested across large random samples of races. The combinations of factors and relative weights are changed incessantly, striving to improve the predictive power of the program. Eventually, long after the initial enthusiasm, energy, and commitment has waned, an acceptable level of proficiency is established, or it is not. The computer permits the amount of mathematical processing without mistakes that would be literally impossible to accomplish manually. Handicappers without computers will never piece together a powerful regression model of handicapping, of that they can be certain.

The Sports Judge as a model consists of seven factors in combination: class, speed (pace), distance, recency, track category, jockey, and trainer. The weightings among the factors remain the privileged property of Arthur and Erickson. High-tech handicappers who buy *The Sports Judge* ($129) input seven-plus data items, standard formulas are applied, and ratings for each horse appear on the terminal screen. An unusual and potent dimension of *The Sports Judge* insists that no horse warrants support unless it earns a distinct edge in the ratings. The mere top rating is not enough. That steers high-tech handicappers away from overly competitive races, a nice service.

High-tech handicappers of the information age can expect to hear more from Arthur and Erickson, who operate as Post Data Sports (PDS), a division of Post Data Services, which is allied with Texas Instruments, a genuine leader in computer technology. A visit to the PDS offices reveals the outfit has been financed as few racing businesses before it have. Several technicians are busy continuously at program development and testing, conferences are held in oak-paneled rooms, and there is an extensive library.

Arthur envisions a regional and national network of low-cost selection services, whereby subscribers can dial the PDS computer daily from residential telephones for the picks at local tracks. The computer responds to voice recognition, perhaps the names of racetracks, and provides the appropriate selections. In a futuristic scenario, subscribers will dial the PDS host computer from telephones connected to personal computers and the PDS host will download its ratings for the local tracks across telephone wires and directly into the personal computers of subscribers. Low-cost datacom networks for handicappers will be part of racing's near future.

To fly closer to reality, high-tech handicappers should not get the impression that regression models of handicapping represent a high-tech panacea for clobbering the races. They do not. *The Sports Judge* is an exception to the rule. Its power may persist only temporarily. To appreciate that, handicappers need to entertain a brief nontechnical discussion of regression and its applications to handicapping.

MULTIPLE REGRESSION MODELS

The microcomputer plays a vital role in two basic approaches to handicapping. One concerns any systematic methodology that necessitates a heavy amount of data processing, such as speed handicapping or pace handicapping. The methods normally concentrate on a single quantifiable factor, with adjustments attached to basic ratings due to the influences of other strongly related factors. The second approach is multiple regression.

Regression is a statistical technique for predicting the outcomes of unknown events from events that are well known. In the simplest case, an unknown variable (beaten lengths), called the predicted variable, is estimated from a known variable (class ratings), called the predictor variable. When two or more known variables are used as predictors (class ratings, speed figures, consistency, etc.), the technique is known as multiple regression. Thus multiple regression is a correlation test by which the strength of the relationships between multiple predictors and a predicted variable is established. The classic case occurs in higher education, where college grade-point averages are routinely predicted from intelligence test scores, high-school grade points, and perhaps a measure of motivation or interest. The correlations can be relatively strong or weak, or nonexistent.

If a positive correlation is actually found, the regression technique produces an equation that indicates how the values of the predictors are associated with the predicted variable. A set of *best values* on the predictors can be identified to predict which horses are likely to score best on the predicted variable or finish first. The best bets will be horses whose scores on the predictors come closest to the *best values* revealed by the regression equation.

In order to improve the strength of the correlations, researchers vary the combinations of predictors that show some initial promise and tinker with the relative influences (weights) of each factor, as indicated by the regression equations. Each predictor variable is treated independently by the regression formula, and the influences of each are added together. The idea is to produce a set of predictors and weights that when added together will give the strongest prediction of the outcome.

The author has always taken a rather smug, bemused stance in reaction to the use of multiple regression formulas to predict race outcomes. Because the technique treats the influences of the several predictors *independently,* researchers must be careful to select predictors that contribute influences that are more dissimilar than alike. The strength of the predictions should improve only to the degree that they result from factors that are in some basic way different. Worse, if influences from predictors that are essentially the same are added together, the apparent strength of the prediction will have been artificially inflated.

In the higher education example, if the predictors consist only of scores on a number of intelligence tests, when the influences of each score are added together, essentially the same ingredient (intelligence) is being added again and again. Thus the strength of the prediction of college grades, based on intelligence scores, will be artificially inflated. To the extent success in college is predicated on knowledge, motivation, interest, perseverance, or time management, the predictions based on intelligence alone will be inaccurate and misleading. But if intelligence as a predictor is combined with knowledge (course grades), motivation, and time management skills—nonintellectual factors—the resulting strength of the prediction will be far more accurate.

It can be similarly argued that the effectiveness of multiple regression formulas in handicapping can be thwarted or minimized by a lack of independence among the factors of handicapping. Are class and speed essentially independent factors, or are they strongly related to each other? What about class and consistency? Speed and pace? Class, speed, and form? How independent are the factors of handicapping? If they are interrelated but treated independently, how contrived will the resulting correlations be? An apparently strong prediction might actually be quite weak. It might be based not on a combination of different handicapping factors that add together strongly but on a combination of essentially overlapping factors instead.

These doubts enlarged when mathematician Bill Quirin published the first computer-generated multiple regression models of handicapping back in 1979. Quirin produced a sprint formula and a route formula. The sprint formula was derived from a study of 5800 horses in 646 sprints; the route formula was based on 2419 horses in 300 routes. The sprint formula combined seven weighted factors; the route formula, five weighted factors.

When Quirin used the regression formulas to make predictions in hundreds of sprints and routes, each formula yielded an 8 percent dollar profit. Knowing that Bill Quirin is perhaps the best, most rigorous scientist studying handicapping, the profit margin of his multiple regression formulas did not strike me as meaningful.

First of all, in producing his regression formulas, Quirin could be expected to control for the problem of overlapping factors by taking intercorrelations of the predictors he combined and eliminating those factors that were too strongly related to other predictors. Future, less rigorous computer-program developers could be expected to omit this crucial step, thereby inflating the strength of their predictions. Quirin's formulas were surely clean, yet they yielded just 8 percent.

On the matter of overlapping factors, I confess to a change of heart. The power of *The Sports Judge* cannot be ignored. Let's examine the combination of factors its formula relies upon:

- Class
- Speed
- Consistency
- Recency
- Track category
- Jockey
- Trainer

Now consider a subset of those factors:

- Class
- Speed
- Consistency
- Recency

Are those four factors relatively independent, or are they interdependent?

In classical handicapping terms, speed is an expression of class, consistency is an expression of class, and class itself is a function of form. Conceptually, the four factors are greatly interdependent. Yet obviously they each contribute to the multiple regression formula employed by *The Sports Judge*. Statistically, they are relatively independent.

How to reconcile?

The answer is that thoroughbred racing has changed. Since the roaring inflation of the seventies and the consequent extensions of the racing calendars to their end points, the overall quality of the competition, especially overnight competition, has declined drastically. The interrelatedness of the handicapping factors has been altered just as dramatically. Class and speed are more disparate than ever, particularly in sprints. Early speed especially has rushed to the forefront as an independent factor in handicapping. Standards of form have been so liberalized that expressions of class are now less a function of sharp form than ever.

If key factors of handicapping are more independent than before, multiple regression models of handicapping make more sense than previously. They are less susceptible to strong intercorrelations among the predictors, just in time for the information age and the widespread use of personal computers. Multiple regression models of handicapping as powerful as *The Sports Judge* are therefore more plausible than ever.

That consideration aside, there is a second technical problem much more practical in its consequences. Multiple regression equations are built on samples of horses and racetracks that have characteristics peculiar to them. How well do the predictions generalize to other horses at other racetracks? Not so well.

In technical terms, the problem is called external validity. The application of the multiple regression formulas is limited to the kinds of horses and racetracks from which the regression samples were taken. Any real change in the characteristics of the horse population (predictors) or the racetracks where the finishes have occurred (predicted variable), and the computer model cannot be expected to work as well.

For this reason the 8 percent profit margin of the Quirin formulas could not be expected to travel far. If the formulas were applied at Arlington Park, but the original samples did not tap Arlington, the effectiveness will be eroded to the extent that the horses in the original study are not represented at Arlington. Racetracks do differ. Even major tracks on different circuits house varying kinds of horses. Generalizations from major tracks to minor tracks are specious. If demonstrably effective regression formulas do not work well locally, it's fair to conclude they do not apply. New formulas are needed. An 8 percent profit margin can evaporate quickly.

Quirin himself advised handicappers that his formulas may not work as well at local plants. Not surprisingly, various serious handicappers who applied Quirin's multiple regression models at different tracks reported varying results. Some reported that the formulas worked well. Others said they did not work well at all. Quirin would not be surprised by that.

In the same way, the 17 percent profit margin obtained by *The Sports Judge* may be eroded whenever racing conditions differ from the program's regression formulas. Shrewdly, the program's authors have attempted to control for that by including a "track category" factor in their formulas, but events can get complicated. In southern California, for example, Hollywood Park has just changed its circumference to a mile and one-eighth (from a mile) and its surface to include more sand. Those kinds of influences on the predicted variable (finish position) could alter the effectiveness of *The Sports Judge* at Hollywood Park radically. The researchers would need to do more tinkering with their factors and weights, and a new multiple regression formula would be the end result. Unless PDS supplied them with the new computer programs, consumers of *The Sports Judge* at Hollywood would be out of luck.

The preceding discussion is germane to high-tech handicappers everywhere, as the spread of personal computers among handicappers assures that several multiple regression models of handicapping will soon be forthcoming. All varieties of applications programs will continually enter the marketplace. *Software Supermarket* has estimated that of the 200,000 personal computers sold monthly, approximately 10 percent wind up in the hands of racegoers. That's quite a market for designers of computer-generated multiple regression programs. As handicappers already know, most of the selection programs will be poorly designed, and will not work effectively.

The best of consumer protections includes an awareness of the program developers' professional credentials, information about the regression samples, and workouts of results that can be fairly generalized to the local situation. If local workouts are not available, handicappers should consult the consumer protection magazines and *Phillip's Racing Newsletter* for the information there. If no information about the program's development and evaluation is available, do not buy the computer software.

Perhaps the most unfortunate tendency among high-tech handicappers today is their stated preference for prepackaged computer programs that provide ready-made selections. Consistent with the approach recommended by this book, another kind of prepackaged computer software is far more suitable for high-tech handicappers who prefer to do their own handicapping.

INFORMATION PROGRAMS

Far better than applications programs that provide selections are computer programs that supply information. This book asks high-tech handicappers to make decisions, not selections. Information guides decision-making. It also informs the basic analysis of horses' records and helps solve the stickiest problems of handicapping. The more information, the better. The personal computer replaces the time-consuming drudgery inherent in manual record-keeping files. It provides storage, updating, and retrieval. If processing is required, the computer does it instantly and without error.

Five types of handicapping information are especially troublesome for manual systems but especially conducive to computer storage and processing: trainer and jockey statistics; speed figures; pace ratings; class ratings; and records (data items) of races and horses not contained in the past performances, such as biases, trip notes, and body language. All of the above demand large amounts of data collection or data processing, with continual updating of records. It's hard work for humans but child's play for computers. Note that speed figures, pace ratings, and class ratings are treated as information, not as equivalents to selections. When combined with other kinds of handicapping information, these information indexes are strong determinants of numerous handicapping decisions.

Unfortunately, the high-tech marketplace does not yet offer handicappers much to choose from among information programs. The leading source is the aforementioned *Com-Cap,* owned by Kaywood, but the small midwestern company's product line until lately has been programmed primarily for user groups possessing either the Commodore 64 or VIC 20 microcomputers. A look at the Com-Cap programs demonstrates the possibilities for high-tech handicappers to increase their information supply at low cost and small bother:

Money Program #1	$5
The Speed Pars/Ratings Program	$25
The Beyer Speed Program	$50
The Com-Cap Racing Records Program	$50
The Com-Cap Pace Program	$65

The money program takes this year's and last's earnings and provides a weighted average-earnings figure for each horse in a field. Kaywood believes the figures are handy for a number of class applications, notably at smaller tracks.

Described as an incredible time-saver by Kaywood, the Speed Pars/Ratings Program converts par times to speed ratings using the concept of proportional time promoted by Beyer. The result is a speed chart. Users simply punch in a final time (1:12.1) and the computer presents the corresponding speed rating (100). Users can adjust raw times for variants and beaten lengths before or after they consult the computer.

The Beyer Speed Program computes speed figures according to the method promoted in *Picking Winners*. The program stores and retrieves daily variants automatically and also makes adjustments for sex and age automatically. Up to four races back, users can tell the computer how many speed figures they want to check. Beaten-lengths adjustments are also computed automatically.

The Racing Records Program is unique; handicappers can keep account of as many as 130 trainers and jockeys at once in *each* of 41 different performance categories. The program permits separate records for *fast* and *off* tracks. The idea is to make available instantly to high-tech handicappers the strengths and weaknesses of every jockey and trainer on the circuit. Moreover, the input demands are few, as the computer completes all the necessary categorizations and math. Updating the records takes a few minutes a day.

Figure 4 shows a partial display of the Racing Records Program for a hypothetical jockey.

The Pace Program computes nine variations of pace ratings from fractional times and final times, and converts the rates of speed to corresponding pace-class ratings (a single numerical rating). The program requires a working knowledge of velocity ratings (feet per second). Figure 5 shows a sample printout of the Pace Program.

When considering information programs that supply speed figures, pace ratings, or class ratings, high-tech handicappers need to know whether the processing tasks reflect their personal handicapping beliefs and methods. They must therefore review the program documentation. The Com-Cap programs are accompanied by extensive documentation and clear explanations of the input, processing, and output operations, as well as personal discussions of the programs by Kaywood, if that is desired. In high-tech merchandising, a service orientation is always appropriate. That is, vendors should be selling both products and services, not just the computer programs.

If information programs do not provide the *precise* kinds of handicapping information users prefer, the alternative is to write

Figure 4
Sample Printout (Partial) Com-Cap Racing Records Program

RACING RECORDS FOR JOHN A. JOCKEY

SPRINTS

	M	W	P	S
ALL	4	3	0	1
MDN	1	1	0	0
CL	0	0	0	0
ALW	1	0	0	1
ST AL	1	1	0	0
DD	1	1	0	0
EX	2	2	0	0

ROUTES

	M	W	P	S
ALL	2	2	0	0
MDN	0	0	0	0
CL	2	2	0	0
ALW	0	0	0	0
ST AL	0	0	0	0
DD	1	1	0	0
EX	1	1	0	0

MAIDEN

	M	W	P	S
ALL	1	1	0	0
SPR	1	1	0	0
RT	0	0	0	0
DD	1	1	0	0
DD SP	1	1	0	0
DD RT	0	0	0	0
EX	0	0	0	0
EX SP	0	0	0	0
EX RT	0	0	0	0

CLAIMING

	M	W	P	S
ALL	2	2	0	0
SPR	0	0	0	0
RT	2	2	0	0
DD	1	1	0	0
DD SP	0	0	0	0
DD RT	1	1	0	0
EX	1	1	0	0
EX SP	0	0	0	0
EX RT	1	1	0	0

ALLOWANCE

	M	W	P	S
ALL	1	0	0	1
SPR	1	0	0	1
RT	0	0	0	0
EX	0	0	0	0
EX SP	0	0	0	0
EX RT	0	0	0	0

STR ALW

	M	W	P	S
SPR	1	1	0	0
RT	0	0	0	0
EX	1	1	0	0
EX SP	1	1	0	0
EX RT	0	0	0	0

DOUBLES

	M	W	P	S
ALL	2	2	0	0
SPR	1	1	0	0
RT	1	1	0	0
MDN	1	1	0	0
MD SP	1	1	0	0
MD RT	0	0	0	0
CL	1	1	0	0
CL SP	0	0	0	0
CL RT	1	1	0	0

EXACTAS

	M	W	P	S
ALL	3	3	0	0
SPR	2	2	0	0
RT	1	1	0	0
MDN	0	0	0	0
MD SP	0	0	0	0
MD RT	0	0	0	0
CL	1	1	0	0
CL SP	0	0	0	0
CL RT	1	1	0	0
ST AL	1	1	0	0
SA SP	1	1	0	0
SA RT	0	0	0	0

MUTUELS

AVG WINNING MUTUEL= $ 18.88
TOTAL: 5
AWM $10 OR MORE= $ 27.07
TOTAL: 3
AV DD WINNING MUTUEL= $ 16.50
TOTAL: 2
AV EX WINNING MUTUEL= $ 20.47
TOTAL: 3

AGE GROUPS

	M	W	P	S
2 YO	1	1	0	0
3 YO	2	1	0	1
4 YO	2	2	0	0

DISTANCE SWITCHES

	M	W	P	S
S-R	0	0	0	0
R-S	0	0	0	0

FIRST-TIME STARTERS

	M	W	P	S
ALL	0	0	0	0

LAYOFFS

	M	W	P	S
ALL	0	0	0	0

RECENT CLAIMS

	M	W	P	S
ALL	0	0	0	0

TOTAL RACES

	M	W	P	S
ALL	6	5	0	1

STAKES

	M	W	P	S
ALL	1	1	0	0
SPR	1	1	0	0
RT	0	0	0	0

CLASS MOVES

	M	W	P	S
DOWN	0	0	0	0
UP	0	0	0	0

Figure 5
Sample Printout. The Com-Cap Pace Program

COMPARATIVE ANALYSIS FOR RACE # 2 5/1/84

HORSE	ST-1	1-2	2-END	ST-2	AOP	AFP	ES	LS	LPS	AVG$	ITM
1 FAVIU	59.23	55.72	50.98	57.64	55.25	55.31	58.44	54.31	5.5	933	.33
2 TIMMI	59.99	55.45	51.56	57.64	55.46	55.67	58.82	54.60	5.6	1403	.31
3 HAYN	59.46	56.41	51.96	57.89	55.77	55.94	58.68	54.93	5.6	1160	.42
4 DANDY	58.78	54.24	51.80	57.10	55.93	54.94	57.94	54.45	5.5	529	.30
5 L. FI	57.39	55.00	49.42	56.17	53.77	53.94	56.78	52.80	5.4	889	.58
6 DANNY	59.46	57.39	50.72	58.41	55.60	55.86	58.94	54.57	5.6	826	.45
7 SOMEG	61.67	54.91	50.78	59.23	56.41	55.79	60.45	55.01	5.6	556	.33

PACE & CLASS RATINGS

5/1/84 RACE # 2

HORSE	S-1	1-2	2-E	S-2	AOP	AFP	ES	LS
1 FAVIU	689	654	606	673	649	650	681	640
2 TIMMI	743	698	659	720	698	700	732	689
3 HAYN	715	684	640	699	678	680	707	670
4 DANDY	644	598	574	627	615	605	635	600
5 L. FI	669	645	589	656	632	634	663	623
6 DANNY	682	661	594	671	643	646	677	633
7 SOMEG	676	608	567	651	623	617	663	609

COMPARATIVE ANALYSIS FOR RACE # 8 5/5/84

HORSE	ST-1	1-2	2-END	ST-2	AOP	AFP	ES	LS	LPS	AVG$	ITM
1 WAR F	60.29	57.08	51.72	59.51	58.29	58.89	59.90	55.62	5.9	1200	.67
2 CINTO	59.87	55.45	51.32	58.37	56.87	56.53	59.12	54.85	5.7	1967	.57
3 SAVES	59.20	55.21	45.66	58.49	55.65	54.97	58.85	52.08	5.5	883	.58
4 BEAFI	53.83	46.12	43.63	53.88	53.32	50.24	53.86	48.76	5.0	1861	.46
5 INEBR	59.08	53.17	49.56	57.63	57.26	57.66	58.36	53.60	5.8	1725	.40
6 RUN F	57.83	55.10	51.76	58.17	56.74	55.07	58.00	54.97	5.5	815	.50
7 ZEPPY	58.61	51.86	47.60	58.12	56.64	54.46	58.37	52.86	5.4	719	.45

PACE & CLASS RATINGS

5/5/84 RACE # 8

HORSE	S-1	1-2	2-E	S-2	AOP	AFP	ES	LS
1 WAR F	730	698	644	722	710	716	726	683
2 CINTO	801	757	716	786	771	768	794	751
3 SAVES	686	646	551	679	651	644	683	615
4 BEAFI	729	652	627	730	724	693	729	678
5 INEBR	767	708	672	753	749	753	760	713
6 RUN F	665	638	604	668	654	637	667	636
7 ZEPPY	662	595	552	658	643	621	660	605

personal programs. As will be seen, high-tech program development by individual handicappers does not require computer programming skills, a circumstance that opens many doors to computerized handicapping information until now thought shut.

Will information programs proliferate in the near future? Absolutely. They are usually easier to develop, do not require extensive tests of effectiveness, and will be perceived increasingly as more useful than selections programs by high-tech handicappers.

STANDARDS FOR EVALUATING COMPUTER PROGRAMS

The first line of defense against a high-tech marketplace that is already out of control is to develop a set of standards for evaluating computer programs. Whether they deal in selections or information, all computer programs should satisfy the following criteria.

1. Consistent with the known probabilities and generally accepted principles and practices of handicapping.

   The lead guidepost is something of a catchall but can be interpreted specifically. High-tech handicappers, for instance, are encouraged to apply the guidelines found by *Software Supermarket* (see page 99) as common to the great majority of effective selection programs. Moreover, computer programs should be based on the up-to-date knowledge base of handicapping. Programs are not intended to develop or extend that knowledge base themselves; that's the province of hard-nosed research. Computer programs are designed to make the working world of handicapping lighter and more accommodating. Claims of original know-how or methods, unavailable except by computer programming, can be discounted as probably unsubstantial, ill-tested, and ineffective.
2. Fully documented.

   In clear, exact, stepwise manner the documentation tells users what the computer will do, what users must do, and what the output will look like. In the face of poor or unavailable documentation, computer applications should be avoided.
3. Facilitate complex processing or cumbersome storage requirements.

Computer programs are not intended to rescue users from mental or manual operations that are quick and easy. Tabulations of speed points or form defects hardly require a micro's special talents. But obtaining velocity ratings for pace analyses begs for computer processing. So do calculations of handicapping probabilities for fields of horses and the creation of handicappers' morning lines.

In *A Winning Thoroughbred Strategy* computer scientist-handicapper-author Dick Mitchell invokes a pocket-size computer to produce handicapping ratings based primarily on speed and class, converts the ratings of probabilities, translates the probabilities into bettable odds, and finally specifies the sizes of Kelly-type wagers on selected overlays that guarantee a minimum 15 percent advantage.

Once the basic ratings have been obtained, the computer, not the handicapper, uses special built-in formulas to do all the dirty spade work. It's a near-perfect exercise in putting the computer to its best advantages in handicapping. The high-tech output is not feasible by mental or manual labor alone.

4. Easy daily input requirements.

If this standard subverts full-dress handicapping by computers, so be it. Applications programs best consume just a thick slice of the handicapping process, not the whole. The most powerful, most technically sophisticated computer model of handicapping on the market today demands no more than six inputs per horse. Let that be the standard. A race should be put to rest within fifteen minutes, closer to ten. Otherwise, technical complications arise, human impatience interferes with other aspects of the handicapping, and the inefficiency subverts the very purpose of computer programming.

The periodic updating of databases is entirely different. That represents record keeping, not handicapping.

5. Adequately tested.

If computer programs provide selections, they must also provide workouts of results and a clear summarization of all the performance data, especially percentage of winners, average odds on winners, and rate of return on the dollar invested (ROI). Vendors should also supply descriptions of the horses and racetracks on which the programs were tested and evaluated.

6. Compatibility information.

   All applications programs must advise prospective customers of the operating systems or hardware models on which they can be run. Do not buy computer programs for which this information is missing.

7. Reasonable cost.

   Depending on storage requirements and processing complexity, computer applications programs in handicapping should cost from $250 downward. Most will retail between $50 and $150. Ignore obviously overpriced programs. As a pointed testimonial, Com-Cap also sells the original Quirin multiple regression programs for sprints and routes. Each costs $10. Both cost $15. Both programs, together with a program that converts the final ratings to odds lines, cost $25.

8. Clearly defined service provisions.

   Computerized products must come with well-defined services. One is provision for a complete rebate if the program cannot be run on the user's personal computer. A second provides for any necessary debugging of the software. A third provides for program updates at low prices. A fourth should stipulate consumer privacy, so that customer names are not willfully sold or distributed to every high-tech mailing list in the nation.

The following two high-tech applications programs meet all of the above standards. They are advanced, with complex processing, but also unusually powerful. Each is presented in outline form only.

THE SARTIN METHODOLOGY

A full-blown handicapping methodology that satisfies each of those performance standards in the extreme has begun to be referred to throughout the country by the name of its developer, Howard Sartin, a clinical psychologist at the Inland Empire Institute, Beaumont, California, whose fascination with handicapping began years ago in concert with therapy groups he led for gamblers. Sartin decided to restrict the gamblers' betting to horse races but allowed no gambling at all until the group could elaborate a method that produced 45 percent winners at average odds of 3 to 1.

Sartin chose pace as the factor on which to build a winning system, and he has been refining first calculations ever since. The breakthrough occurred when Sartin discovered that velocity ratings—feet per second—were more instrumental than final times in finding winners. Velocity ratings dispel the notion that a thoroughbred's rate of speed is five lengths a second, or one length every one-fifth second. Instead, horses accelerate and decelerate individually, and Sartin's method emphasizes the relationships among the accelerating-declining rates of speed. The method is centered on what Sartin has called "the dynamics of incremental velocity." It can show how horses with faster final times have actually run considerably slower. The method focuses on three race intervals: the first quarter, the half mile, and the final quarter. Numerous pace ratings are calculated, and final ratings are modified by considerations of class, distance, and racetrack.

In recent months the power of the Sartin Methodology at several major tracks had become so renowned that rumors of 50 to 60 percent winners had spread to the far west shores from multiple sources. This was mind-blowing, even unsettling, until review of the method as described by Sartin's personal and impressive technical papers revealed that it supported two horses a race. Sartin encourages consistent profit-making by betting 60 percent of a single-race investment on a lower-priced selection and 40 percent on a higher-priced selection. The top selection wins only 40 to 45 percent of the bettable races!

The Sartin Methodology is available in computer software form, complete with documentation plus a technical paper entitled "The Dynamics of Incremental Velocity and Energy Distribution," as well as personal consultation by Sartin himself, if necessary. It is highly recommended for high-tech handicappers interested in obtaining pace ratings of the highest caliber.

THE Z-SYSTEM

The Z-System is a highly mathematical fail-safe approach to place and show wagering. The system picks place and show overlays that also have strong probabilities of finishing in the money. It gets two to four plays a day. The system is explained in technical detail by its developers, William Ziemba, professor of management science at the University of British Columbia in Vancouver, Canada, and Donald Hausch, doctoral candidate at Northwestern University, in *Beat the Racetrack,* published in October, 1984. Among the mathematicians who have promoted the Z-System is the renowned Dr. Edward Thorp, author of *Beat the Dealer* and creator of the original card-counting blackjack systems. Thorp wrote the brief foreword to *Beat the Racetrack.*

The Z-System depends on a securities concept of investing that Ziemba calls the inefficiency of market. No knowledge of handicapping is needed. The system relies on a trend analysis of the tote. The basic premise holds that while the public estimates the winning probabilities of horses exceedingly well (efficiency), it sometimes makes glaring mistakes when betting to place and show (inefficiency). A discrepancy or inefficiency occurs when a much smaller percentage of the place or show pool is bet on a particular horse than this horse's proportion of the win pool.

When an inefficiency is spotted, Ziemba advises handicappers to pose three questions.

1. How good is the bet?
2. Is the bet really good enough?
3. How much should you bet?

The Z-System, which can be implemented by using a programmed calculator or a hand-held computer or even by consulting removable cards in the book version, provides the three answers. As it turns out, the first must be equal to a 14 percent expectation or advantage at twenty-two major tracks, an 18 percent expectation at other tracks. The single-race expectation and the Kelly method of wagering are combined to determine the optimal size of each bet. As technically advanced and challenging as the Z-System is, its actual application can be simplified for handicappers who use the high-tech instruments. A software version is available and costs $140.

On the inaugural Breeder's Cup Event Day, November 10, 1984, at Hollywood Park, Dr. Thorp joined Ziemba, the author, and others for a rare day at the races. Thorp wanted to try the Z-System. He did not buy a *Racing Form.* He did not buy a track program. He knew nothing about the horses. He watched the tote board devotedly and wrote numbers using a pencil and pad. He found six bets; won five. Thorp's profit was $1,851.

THE POSITIVE SIDE

Earlier I mentioned that the general pollution of today's high-tech marketplace was susceptible to a countervailing force: that a number of very brainy people were also toiling away in this specialized sector of handicapping. It's highly appropriate to end this survey on that upbeat promise of better things to come. Once the primitive stage of high-tech handicapping has passed, and the hucksters, quick-fix merchants, and ripoff artists have been put to pasture, as they will be, a number of computer scientists, mathematicians, design engineers, statisticians, and information scientists who are also handicappers will take over with an array of newfangled products and services that will boggle the minds and stretch the imaginations of all of us. In a sport and pastime that depends absolutely on data processing and information management, the possibilities are not only real but endless.

The present is bleak, but the future is bright.

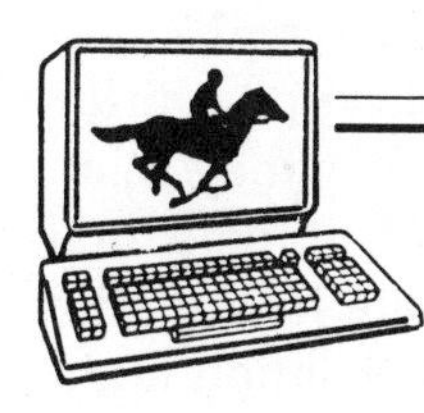

Chapter 5

HANDICAPPING THEORY AND METHOD IN THE INFORMATION AGE: Information Management and Intuitive Thinking

THE REDOUBTABLE ANDREW BEYER won $95,000 when Swale, a 3-to-2 shot, won the 1984 Belmont Stakes. Lee Rousso, age 25, of Los Angeles, won $63,000 by picking six overlays in 27 bets at a three-day handicapping contest in Reno, Nevada, in May 1984. Rousso won $80,000 in the same contest in November 1982. Jeff Siegel, a syndicated newspaper selector based in southern California, led the handicapping for a multimember syndicate that cashed the only winning Pick-Six ticket in a 1984 Golden Gate Fields pool worth $210,000 to the winners. By unverified account, Siegel's syndicate cashes several munificent Pick-Six tickets a season. It's true, too, that a losing day pinches where it hurts. The ticket on the huge Golden Gate payoff reportedly cost $3,700.

What Beyer, Rousso, and Siegel have in common, beyond the capacity to win huge amounts of money at the races through the painstaking application of considerable knowledge and skill in handicapping, is an approach to the game that is not easily classified. Beyer is no longer a speed handicapper, and to categorize the man by that narrow definition is to miss him by a wide margin. Rousso has read every decent book ever published on handicapping, and after playing with him on numerous occasions I can attest that he has become one of the sharpest handicappers on the scene simply by applying everythng he knows to each race on the card. Siegel's game is not so familiar to me, but I can testify at least that he would not be properly defined as a class handicapper, speed handicapper, trip handicapper, trainer analyst, or any hybrid definition of those specialties. Beyer, Rousso, and Siegel are generalists, not specialists.

What Beyer, Rousso, and Siegel do, in some essential and fundamentally valid way, is relate diverse items of handicapping information to one another. You might say, as I like to, that they relate information that is fundamental to information that is incidental. Or, to put it differently, you might say that they relate information about the abilities and preferences of horses to information about the circumstances of races.

The relations among the numerous information items are themselves numerous. Different races are analyzed in different ways. What is incidental, or circumstantial, can be and often is more decisive than what is fundamental, horses' abilities. Moreover, and more important, the relationships that will become significant in handicapping any particular race are not known until after the race has been analyzed for a time and after other circumstances attending to the running have been evaluated—closely. That is, the key relationships are race-related and emerge; they are not superimposed. There is a method to this madness, but it is not a systematic method in the classical sense. It is not a stepwise method that applies race to race, again and again.

The general procedure resembles what this book refers to as an information management approach to handicapping.

Beyer, Rousso, and Siegel do something else as well. They not only let the total context of the particular race dictate the handicapping of that race, but they also let the relative odds on more than one horse dictate the bet. That is to say, they do not bet on selections that figure to win, they bet on overlays. Most of all, they prefer to bet on horses that have a handicapper's reasonably good chance to win but are offered at unreasonably high odds. In exotic pools, they often bet on method selections and less obvious overlays in combination, but only when the combination bets either (a) figure strongly at satisfactory prices or (b) represent generous overlays themselves. In all cases, and without violating the tenets of fundamentally sound handicapping, the opportunity to make either a quick, generous return on a normal investment, or a big score, influences the final decisions.

The final decisions do not result from adherence to the precepts of logical, deductive reasoning. The decisions are far more intuitive than logical, resulting as they do from the adding up of numerous information items and relationships that make the most sense under the circumstances.

For the past three seasons my personal style has been evolving in similar directions, and in this book my aim is to draw upon those experiences to reach a set of conclusions. My purpose is to translate the use of multiple types and souces of information into a coherent theory and method of handicapping.

I believe it represents the most powerful way nonprofessionals, that is, recreational handicappers, can beat the races for important money. The approach promises both consistent seasonal profits for the most knowledgeable, skillful handicappers and repeated opportunities for large payoffs in the short run for just about everyone. It does put a heavy educational burden on handicappers. The trick, if there is one, is to learn just about everything there is to know, keep the knowledge base current, and make that information accessible at the appropriate times.

THE INFORMATIONAL, INTUITIVE APPROACH

Systematic method players have an extraordinarily good chance of selecting horses that fit their methods well but of missing just about everything else. The informational management approach seizes upon all the possibilities, including method selections. Whereas method players proceed by eliminating horses that do not satisfy their basic standards, identifying contenders, separating the contenders, making a selection, and determining finally whether the selection warrants a play, the informational approach involves none of that and embraces a very different sequence of steps or operations. The process begins before the past performances are analyzed, by understanding the class demands of the eligibility conditions and the peculiar characteristics of today's racetrack, including footing and track biases.

Next is an analysis of each horse's record. Starting wherever handicappers please, the intention is to recognize components of the record that can be considered positive or favorable under today's conditions, notably any positive aspect that is frequently decisive, including high early speed, a class edge, impressive final time, and trainer maneuvers and credentials on the fundamentals of class-speed-form-distance-pace in combination. All information that is

7th Hollywood

6 FURLONGS. (1.07⅖) CLAIMING. Purse $14,000. 3-year-olds. Weight, 122 lbs. Non-winners of two races since April 15 allowed 3 lbs.; a race since then, 6 lbs. Claiming price $32,000; for each $2,000 to $28,000, allowed 2 lbs. (Races when entered for $25,000 or less not considered.)

Mister Gennaro

Ch. c. 3, by Sonic Shuttle—Pleasure Royale, by Pleasure Seeker
Br.—Montanino G (Cal) 1984 5 1 0 1 $10,775
Own.—Alter H & Kahala 116 Tr.—Spawr William $32,000 1983 2 M 0 0
Lifetime 7 1 0 1 $10,775
14Mar84-7SA 6½f :22 :452 1:183gd *2¼ 116 8⁶½ 77½ 65½ 44 Shoemaker W5 c40000 73-26 Afcondo,KnghtSkng,KyToThCondo 9
14Mar84—Veered in sharply, bumped hard at start
26Feb84-7SA 1 :463 1:111 1:364ft 11 115 96 812 817 817¾ ShoemkrW8 Ⓢ Aw23000 66-15 SarosSaros,Fortune'sKingdom,Tbre 8
26Feb84—Wide 7/8 turn
15Feb84-3SA 6f :214 :444 1:101ft 4¼ 116 2½ 2½ 2hd 3hd Shoemaker W6 50000 87-19 IrishS'gtti,DustToRichs,MistrGnnro 6
27Jan84-4SA 6f :214 :452 1:104ft 7¼ 118 1hd 1½ 14 12 ShoemakerW5 Ⓢ M32000 84-18 MisterGennro,NuclrAttck,MontM. 12
18Jan84-4SA 6f :22 :453 1:122ft 31 118 3nk 2½ 2hd 51½ Shoemaker W8 Ⓢ Mdn 74-23 Hrdtohndl,Dd'sQust,Orphn'sNight 12
18Jan84—Bumped start
12Nov83-6SA 6f :22 :452 1:121sy 16 117 3¾ 45½ 612 615¾ Hawley S8 Ⓢ Mdn 62-24 Contequos,Carrizzo,OnaLuckyStrek 8
2Nov83-6SA 6f :214 :453 1:121ft 9¾ 117 1½ 31 44½ 712 Pierce D7 Mdn 65-23 KyToThCondo,OnLckyStrk,Contqs 7
2Nov83—Broke slowly
Jun 10 Hol 5f ft 1:00 h May 23 Hol 5f ft 1:031 h May 19 Hol 5f ft 1:014 h Apr 14 SA 6f ft 1:133 h

In Natural Form

Ch. c. 3, by L'Natural—In Rare Form, by Run for Nurse
Br.—Hardcastle Linda L (Cal) 1984 8 0 1 1 $7,750
Own.—De Spenza C V & Dorothy M 116 Tr.—Brooks L J $32,000 1983 2 1 0 1 $11,850
Lifetime 10 1 1 2 $19,600
27May84-1Hol 6f :221 :452 1:112ft 14 116 4¾ 66¾ 711 714 Meza R Q6 32000 66-18 PtternMtch,BrgingAhed,OfficeSekr 7
11May84-3Hol 1 :463 1:101 1:361ft 6 114 21 43½ 613 618¼ Meza R Q5 37500 66-18 PlntyConscos,MksClco,KnghtSkng 6
11May84—Lugged out final 1/2
21Apr84-7SA 6f :22 :45 1:104ft 2¾ 120 3½ 43½ 511 510¾ Lipham T6 40000 73-19 Mingash, KingsJester,SpeedyTalker 6
21Apr84—Veered in, bumped start
25Mar84-4SA 6½f :221 :451 1:161ft 12 120 21 31½ 44½ 611¾ McCrrnCJ2 Ⓢ Aw26000 77-17 LordOfTheWind,Blips,Hardtohndle 6
25Mar84—Broke slowly
15Mar84-9SA 1⅟₁₆ :472 1:114 1:44 ft 7 116 32½ 3½ 1hd 32¼ Toro F7 50000 79-22 Andress, Quati Pink,InNaturalForm 8
17Feb84-3SA 1 :461 1:112 1:38 ft 9½ 116 11½ 12 11 2¾ Lipham T6 40000 77-17 TonophLow,InNturlForm,JollyJosh 6
1Feb84-1SA 6½f :212 :443 1:172ft 15 119 88¼ 76¾ 56 55 Lipham T7 50000 78-17 HilTheEgle,TostMster,TonophLow 10
1Feb84—Bumped start
13Jan84-4SA 6½f :214 :444 1:172ft 19 120 54½ 55½ 610 612 Lipham T5 Ⓢ Aw21000 71-17 LuckyBuccneer,StrMtril,DistntRydr 6
31Dec83-2SA 6f :22 :452 1:111ft 6¾ 118 31 31½ 2½ 1nk Lipham T4 Ⓢ Mdn 82-18 InNturlForm,UnfrComptton,DggD 12
14Dec83-2Hol 6f :214 :444 1:113ft 32 118 65½ 26 34½ 32 Lipham T7 M50000 77-25 CopyMster,NoblFury,InNturlForm 11
May 23 Hol 4f ft :471 h ● May 1 Hol 5f ft :59 h Apr 18 SA 3f ft :362 h

Speedy

B. g. 3, by Diplomatic—Charissa, by Field Master
Br.—Seley J H (Cal) 1984 4 1 0 2 $10,325
Own.—Seley J H 119 Tr.—Brinson Clay $32,000 1983 0 M 0 0
Lifetime 4 1 0 2 $10,325
24May84-2Hol 6f :22 :451 1:11 ft *9-5 116 42½ 33 3nk 13½ DelahoussyeE6 M32000 82-21 Speedy, Net Points,SirEdgarAllan 12
10May84-4Hol 6f :222 :451 1:102ft 3 116 3½ 11½ 1½ 33½ DelhoussyE8 Ⓢ M32000 81-16 Nuclear Attack,SolidSpirit Speedy 11
10May84—Bumped start
23Apr84-4SA 5½f :222 :454 1:044ft 12 118 41½ 44½ 54 58 Sibille R1 Ⓢ Mdn 79-18 QuieroDinero,SobrPrinc,Tddy'sLov 7
23Apr84—Crowded, rank to place on turn
11Apr84-2SA 6½f :214 :45 1:18 ft 5½ 118 1½ 1½ 21½ 36¾ CastanedM6 Ⓢ M32000 73-20 Sparkling Rose,Princeville,Speedy 12
Jun 10 Hol 5f ft 1:022 h Jun 3 Hol 5f ft 1:012 h May 20 Hol 5f ft :594 h May 9 Hol 3f ft :354 h

Benedict Canyon

Ch. c. 3, by Sun Canyon—Balcony's Delight, by King's Balcony
Br.—Mole-Richardson Farm (Cal) 1984 2 2 0 0 $4,620
Own.—Mole-Richardson Farm 1145 Tr.—Bradshaw Randy K $32,000 1983 0 M 0 0
Lifetime 2 2 0 0 $4,620
3Jun84-7Fno 5½f :213 :45 1:034ft 4½ 1125 33 43 21 11½ DominguezRL3 Aw5680 91-10 BndctCnyon,LgsBNmbl,CndyInCort 7
19May84-4Fno 5½f :222 :452 1:042ft *2¼ 118 21 2hd 1½ 12¼ Aquino C9 Ⓢ M12500 88-15 BndictCnyon,FriscoFlt,LgsBNimbl 10
May 25 Fno 4f ft :513 h May 16 Fno 2f ft :241 h May 11 Fno 6f ft 1:14 h ● May 7 Fno 6f ft 1:13 h

Kings Jester

B. g. 3, by Distant Land—Quick Jest, by Jester
Br.—Pinkcastle Inc (Cal) 1984 5 1 2 1 $17,850
Own.—Maven Stables 116 Tr.—Rowan Mary $32,000 1983 2 M 0 0 $1,700
Lifetime 7 1 2 1 $19,550
19May84-1Hol 7f :22 :452 1:23 ft 5½ 116 3nk 73½ 77¾ 78¾ DelahoussayeE4 50000 73-17 Purloin,NuclearAttck,AmzingSport 9
9May84-7Hol 6½f :22 :45 1:162ft 3¾ 116 2hd 1hd 1hd 33¼ DelahoussayeE2 50000 85-19 SpeedyTlkr,AmzingSport,KingsJstr 8
21Apr84-7SA 6f :22 :45 1:104ft *2 120 42½ 33 33½ 22 Sibille R4 40000 82-19 Mingash, KingsJester,SpeedyTalker 6
21Apr84—Bumped start
24Feb84-7SA 6f :214 :444 1:101ft 3¼ 120 53 32½ 33 21¼ DelahoussayeE8 40000 86-19 PrincePeninsul,KingsJstr,TostMstr 8
10Feb84-2SA 6f :22 :45 1:104ft 4¼ 118 32 32½ 21½ 1¾ DelhoussyE6 Ⓢ M40000 84-21 KingsJstr,InAMssg,UnfrComptton 12
10Feb84—Veered in, bumped after start
24Jly83-6Hol 5½f :223 :453 1:034ft 13 116 1hd 33 38½ 512¼ DelahoussyeE7 Ⓢ Mdn 80-14 LrdOfThWnd,FftySxInRw,BldT.Jy 11
14Jly83-6Hol 6f :222 :453 1:114ft 29 116 32½ 42¼ 43¼ 47½ DelahoussyeE1 Ⓢ Mdn 70-19 Benji King, Vencimiento, Bozina 9
Jun 8 Hol 5f ft 1:012 hg May 3 Hol 5f ft 1:004 h Apr 14 SA 6f ft 1:133 h

Galaxy Ruler — Ch. g. 3, by Orbit Ruler—Pretty Swish, by Windy Sea

Br.—Parsons Mary (Cal) — 1984 3 1 0 0 $1,705

Own.—Parsons Mary — 119 — Tr.—Valenzuela Martin — $32,000 — 1983 0 M 0 0

Lifetime 3 1 0 0 $1,705

7Jun84-7Hol 6f :221 :451 1:101ft 13 119 84 911 1018 1014¼ Drexler H4 40000 71-16 ThreeForTwo,OfficeSkr,Af'ciondo 10

7Jun84—Bumped start, lugged out 3/8 turn

20May84-7AC 6f :223 :442 1:082ft *9-5e114 12 11 12 15 Olguin M1 Mdn 97-11 GlxyRuler,OurKplu,NumbersRunner 9

18Jan84-4SA 6f :22 :453 1:122ft 15 118 42½ 55½ 910 1211¼ ValenzuelPA12 [S]Mdn 65-23 Hrdtohndl,Dd'sQust,Orphn'sNight 12

18Jan84—Lugged out

Jun 5 Hol 3f ft :36 h ●May 29 AC 5f ft :594 h May 15 AC 5f ft 1:014 h May 9 AC 5f ft :593 h

Jolly Writer — B. g. 3, by Staff Writer—Jolly Midget, by Aforethought

Br.—McFadden Judy (Ida) — 1984 6 1 2 1 $5,095

Own.—Rimrock Stable — 116 — Tr.—Hutchinson Kathy — $32,000 — 1983 1 M 0 0

Lifetime 7 1 2 1 $5,095

24May84-7Lga 6f :214 :453 1:111ft 6 117 31½ 1hd 12½ 2hd Baze M B4 25000 80-23 RstlssRuful,JollyWritr,Sooty'sPrinc 9

17May84-8Lga 6f :214 :451 1:104ft 2¾ 116 64½ 64 44 55 Baze M B4 25000 77-22 LongTrmIntrst,UrsrLcky,LvrsLstng 7

27Apr84-5Lga 1 :464 1:123 1:394ft *3 114 64½ 22½ 23 33½ Frazier B3 25000 66-26 Grnny'sGrits,FlyingRegl,JollyWritr 10

18Apr84-9Lga 6f :22 :45 1:094ft 3¾ 117 31 32½ 33½ 55½ Frazier B1 Aw8400 81-25 Irunforten, Daegon, Literaki 6

7Apr84-8Lga 5½f :213 :444 1:04 sy 13 117 67½ 34 23½ 22½ Frazier B2 Aw8400 89-15 Magnatice, Jolly Writer, Cracovyak 9

11Mar84-3YM 5f :222 :46 :582ft 5¼ 120 43 22 2hd 1¾ Cooper B4 Mdn 88-11 JollyWriter,SilverLegend,DiscoTble 9

30Jly83-4Lga 5½f :222 :464 1:051ft 21 120 65¾ 88½ 815 913½ Zubieta F M8 Mdn 72-20 SharperOne,SonarBem,BnjoMusic 10

●Jun 11 Hol 3f ft :352 h Jun 5 Hol 4f ft :483 h May 6 Lga 5f ft 1:05 b

Bower Song — B. g. 3, by Messenger of Song—Lady's Bower, by Royal Tower

Br.—Warwick G M (Cal) — 1984 5 0 1 2 $6,450

Own.—Four Four Forty Farms — 116 — Tr.—Borick Robert — $32,000 — 1983 8 1 3 1 $12,438

Lifetime 13 1 4 3 $18,888

6Jun84-1Hol 6f :222 :46 1:112ft 11 115 31 53½ 33 23 Hawley S2 25000 77-20 RoylCycl,BowrSong,TrumphntBnnr 8

6Jun84—Bumped, jostled at start

13May84-1Hol 6f :221 :444 1:104ft 16 115 52¾ 73¾ 47 34½ Hawley S9 25000 78-18 OvrlndJrny,TrmphntBnnr,BwrSng 12

6May84-3Hol 6f :221 :45 1:11 ft 33 114 3nk 43½ 67½ 89¾ Fell J6 28000 73-17 JohnJrry,PttrnMtch,TrmphntBnnr 10

6May84—Lugged in badly down backstretch

26Apr84-1Hol 6f :22 :444 1:103ft 7½ 115 31 34 36 35¾ Meza R Q3 [S] 25000 78-19 PtternMtch,TostMster,BowerSong 9

26Apr84—Bumped start

13Apr84-3SA 6f :213 :45 1:103ft 12 116 31 31½ 55 59¼ Meza R Q2 [S] 32000 75-20 Tonopah Low, Nosfos, John Jerry 7

13Apr84—Bumped at 5 1/2

4Nov83-3SA 6f :22 :453 1:121ft *4-5 117 1hd 1½ 2½ 21¼ Pincay L Jr5 c25000 75-23 Aficionado,BowerSong,SonOfChief 6

21Oct83-3SA 6f :214 :45 1:103ft 4¼ 117 44 36 26 37¾ Pincay L Jr1 32000 77-20 DistntRyder,ProudYnke,BowrSong 7

21Oct83—Stmbld, lugged in

8Oct83-6BM 6f :22 :441 1:084ft 9½e114 41½ 43¾ 48¼ 713¾ DillnbckBD10 Aw15000 82-05 Bold T. Jay, Nak Ack, B. In Time 10

8Sep83-11Bmf 6f :221 :451 1:11 ft 21 114 76¼ 65¾ 34½ 2¾ DillenbeckBD6 Midpen 84-14 JstTwoBlocks,BowrSong,Mstr'sTch 9

8Sep83—Rough trip

19Aug83-11Stk 6f :22 :451 1:104ft 21 113 32 32 3½ 71¾ DillnbckBD7 John Peri 94-05 Larrocca, Foreign Legion,MeNoLie 8

May 27 Hol 5f ft 1:012 h May 4 Hol 4f ft :483 h Apr 24 Hol 4f ft :501 h Apr 20 SA 6f ft 1:15 h

relevant to the racing situation at hand is assumed to be useful and potentially decisive.

At this early analytical point, no particular information items should be too heavily weighted or accepted as decisive. Systematic method analysis can be part and parcel of the information analysis, but its selection, if found, is held as tentative and inconclusive.

Counterbalancing the emphasis on identifying the positive aspects of the horses' records is a simultaneous recognition of any fundamental information that is negative and will likely hinder the horses' chances to win.

Few horses are eliminated at this point. The only eliminations will be horses with no interesting information in their favor.

This analytical step of utilizing knowledge, information, and systematic methods to break a race into its components parts most often ends with alternating views of the probable outcome. Analyzing the alternatives further usually entails the clarification of a few specific handicapping problems peculiar to this particular race. If the race is to be understood in depth, those problems must be solved, and information management can often make the crucial difference in problem-solving.

To illustrate, we examine a six-furlong claiming sprint for $32,000 three-year-olds, run at Hollywood Park, June 14, 1984. Eight have been entered.

The race conditions alert handicappers to prefer the kinds of horses that most often win open claiming races, namely horses that combine good early speed, improving or peaking form, and a drop in class or in a claiming situation limited to three-year-olds, that proverbial improving three-year-old moving up in class after winning or performing powerfully at a slightly lower level. Speed handicappers would prefer the high-figure horse but would also prefer any improving three-year-old to have run last out two ticks faster than par at today's higher class level.

At Hollywood Park on June 14, 1984, the track was fast, and a speed bias that had dominated the dirt course earlier in the season had lately disappeared. No track biases were apparent. The information analysis might proceed as follows:

Mister Gennaro. Eliminated at the start last out and parked wide from the outside post February 26, two bad trips, the colt otherwise shows consistently fast early speed. It gets seven speed points and should be an early leader. It has also finished a head behind $50,000 three-year-olds, after being hooked on the front all the way. Dopes out nicely on early speed and class. On the downside, it was claimed by a weak trainer two months back, has not run since, and shows an irregular workout pattern, including a bothersome break of five weeks from April 14 until May 19. Form is problematic, but the colt does not have a "form defect," as identified by William L. Scott's research—it had a saving five-furlong workout four days ago.

By traditional methods, handicappers should note, Mister Gennaro would be eliminated at this first step, as lacking acceptable current form. Yet the colt has some positive points in its favor. The information management approach deals more kindly and effectively with this horse than most systematic methods normally would.

In Natural Form. Some early lick, but not that fast and not much else in its favor. Weak trainer, terrible recent finishes. Never beaten winners; usually not close.

Speedy. Just beat maiden claimers at this level when favored to do so. Handicappers know these kind normally need a drop in claiming price against winners, often as great as 50 percent, unless the adjusted final time beats today's class par. This gelding's does not. Nothing else looks positive or favorable; no advantages.

Benedict Canyon. Multiple impressive winner at Fresno, a minor, fair circuit oval in northern California. Allowance time better than maiden claiming time, an impressive improvement. Handicappers realize three-year-olds can improve dramatically. Is this colt impressive enough in this spot? Do handicappers know? Or not yet? Might handicappers need additional information about the trainer? Works are sharp enough and the colt was favored first time, but the horse that finished third that day also finished second June 3 in the allowance heat. Not much of a class rise there, after all. The improvement pattern suddenly is not so impressive, but what's the bottom line here? Handicappers simply do not yet know. That's important to admit.

Kings Jester. This is the typical claiming class-dropper and one with good early speed. Form is acceptable; no defects, if not sharp. The gelding finished in the money previously in three of four races versus better and will probably be competitive against $32,000 horses, unless form is dulling. Trainer is minor and largely an unknown. Figures and pace ratings look ordinary. Is this horse the class of the field? Maybe. Maybe not.

Galaxy Ruler. Wire-to-wire maiden win at Caliente in blazing time surrounding two awful efforts at the major tracks. Favored at the border, no betting support in the big leagues. Gets a form defect for last race. First quarter time at Caliente slow. Cannot take the lead here from Mister Gennaro.

Jolly Writer. Shipper from Longacres is up in class after a last-out disappointing effort there, relinquishing a big lead from eighth pole to the wire. Good trainer. To be ridden today by leading jockey Laffit Pincay, Jr., usually competitive, but final times poor since April 7 allowance effort in the slop. Doesn't figure on class or speed here, but owner-trainer-jockey connections cannot be lightly dismissed.

Bower Song. Up in class after finishing second eight days ago, following a bad start from the inside posts. Several rough trips, with repetitive trouble at the start. Minor trainer, major jockey. Hawley stays on after two troubled trips, but solid finishes and sharp current form is the attraction. Nice style to benefit from a tiring pace. Not outclassed here. Adds up nicely as a traditional kind of class-form-pace contender and has trouble lines to boot. Handicappers know the statistical advantages held by last out, second-place finishers.

The information analysis recognizes five horses with positive, favorable characteristics in their records. Handicappers should note that the analysis is characterized by a kind of stream-of-consciousness information flow, with the positives and clear advantages gathering a kind of additive force as the analysis continues. Mister Gennaro holds the early speed advantage, Kings Jester might have a slight class edge, and Bower Song figures strongly on class-form-pace-trouble in combination. Similarly, Benedict Canyon and Jolly Writer cannot be dismissed out of hand. Each has a real, predictable chance.

What to do? The next step is called problem-solving. The intent is to apply additional information sources and a more rigorous analysis of the specific handicapping factors identified during the preceding step. The problems still to be solved should be clearly formulated. The problems in our illustration might be listed as follows.

1. Will Mister Gennaro control the early pace and retain sufficiently sharp form to maintain its lead to the wire?
2. Will Kings Jester's class edge tell and be sufficient to overcome a tiring Mister Gennaro?
3. Is Bower Song approaching a peak effort; if so, will it be decisive regardless of the early speed and class advantages of the other two horses?
4. Is Benedict Canyon in fact improving, and is the shipper talented enough to handle this class at Hollywood Park?
5. Will its new trainer and jockey Pincay help the consistent Jolly Writer sufficiently if the pace develops favorably today?

As stated, the task at this point is to bring as much handicapping knowledge and information to bear on the problems as is feasible, hopefully, to find convincing solutions. Regarding Mister Gennaro, for example, handicappers know (if they have collected the information) that the trainer is weak, a low-percentage man. Moreover, the workout schedule is not properly patterned nor impressive in the least. The frontrunner may steal the race, but it may tire and lose just as easily.

Of Bower Song, handicappers possessing trip information would know how troublesome the last race had actually been, rougher by far than appears in the running lines, yet the finish had been strong. Improvement seemed probable. Jockey Hawley not only knows the horse but is also the second best gate-jockey in southern California and has returned on the horse eight days later to try for the win he feels he probably deserved last time.

To advance the problem-solving stage further, handicappers would need to consult a database on trainers. They could query the database for the win percentages of trainers Bradshaw and Rowan. If one existed, they might also ask a specialized database on shippers how many fair circuit shippers have won at the southern California major tracks during the past five years. If the number looked intriguing, they might ask the computer to compute an impact value which would inform them of the probability that shipper moving ahead in track class can win at Hollywood Park. Handicappers would value that information greatly but normally do not have it available or accessible.

After specific race-related handicapping problems have been solved to the extent feasible with the available information, the next step is to elaborate the likeliest outcome scenarios or to determine the alternative ways the race will likely be run and decided.

Scenario 1 sees Mister Gennaro winning wire-to-wire on an easy early lead.

Scenario 2 finds Mister Gennaro tiring and the class dropper Kings Jester beating it to the wire.

Scenario 3 has Bower Song tracking the pace of Mister Gennaro, and whatever else might pace forwardly, and beating either the frontrunner or Kings Jester to the wire.

Scenario 4 has either the early speed (Mister Gennaro), the class horse (Kings Jester), or the classic choice (Bower Song) winning and either Benedict Canyon or Jolly Writer finishing second.

The final step is making decisions. At this point the handicapping becomes more intuitive, balancing the probable outcomes against the odds the public is offering on each. This race offered the Exacta besides straight betting. The odds on the five horses near post time follow.

Mister Gennaro	9-2
Kings Jester	8-5
Bower Song	7-1
Benedict Canyon	30-1
Jolly Writer	5-1

In the final analysis, the outcome scenarios were judged to be of roughly equal probability. Kings Jester could not be tolerated therefore at 8-5. Bower Song was an appealing overlay in the straight pool. Intuitively, as the odds were juxtaposed to the probable outcomes, handicappers would become more strongly attracted to Bower Song, somewhat less strongly directed toward Mister Gennaro, and disinterested in the overbet Kings Jester. At this decision stage price becomes the potent factor it deserves to be. It is a primary intuitive force, and rightly so.

Bower Song becomes a straight bet, $50 to win.

A perfectly plausible decision on exotics would be to box an Exacta with Mister Gennaro and Bower Song, and to play each of these horses on top of Exacta combinations having Kings Jester, Benedict Canyon, and Jolly Writer as the second horse. The unit cost of the eight combinations would be $40 ($5 Exactas at Hollywood). At the low price, Kings Jester would not be sensible as a top horse. The projected return would be too low to justify the added investment. If Bower Song won and Kings Jester ran second, however, the projected payoff was $160, and this would be a fair return.

As events proceeded, Mister Gennaro led by three at the first two calls, by two at the eighth pole, but tired noticeably after that. Bower Song and Kings Jester fought closely to the wire, and Bower Song won by a long neck.

The win wager netted $350, the Exacta investment netted $153, and a race unbeatable by conventional handicapping guidelines earned handicappers almost $500.

Handicappers who are now insisting that all would have been lost if Kings Jester had won by a neck instead have missed the boat. Before the race Kings Jester did not figure strongly to win by any account and was offered as a decided underlay. Classical handicapping method likely would have selected Kings Jester, but handicappers would have chosen to pass because the price was unacceptable. Fair enough. An information management approach, alternatively, opens the race to its real, alternative possibilities. Handicappers who become information managers find opportunities to wager shrewdly when presented with race situations that reflect the full range of handicapping know-how and up-to-date information.

Handicappers should understand that the information management, intuitive decision-making approach is not a stepwise logical deductive method of handicapping but rather a kind of inductive analytical thinking process that moves continuously from relatively

SEVENTH RACE **6 FURLONGS. (1.07⅗) CLAIMING. Purse $14,000. 3-year-olds. Weight, 122 lbs. Non-winners of two races since April 15 allowed 3 lbs.; a race since then, 6 lbs. Claiming price $32,000; for each $2,000 to $28,000, allowed 2 lbs. (Races when entered for $25,000 or less not considered.)**

Hollywood

JUNE 14, 1984

Value of race $14,000, value to winner $7,700, second $2,800, third $2,100, fourth $1,050, fifth $350. Mutuel pool $144,558. Exacta Pool $214,481.

Last Raced	Horse	Eqt.A.Wt	PP	St	¼	½	Str	Fin	Jockey	Cl'g Pr	Odds $1
6Jun84 ^{1}Hol^{2}	Bower Song	3 116	8	1	$3^{1\frac{1}{2}}$	2^{1}	2^{hd}	1^{nk}	Hawley S	32000	7.00
19May84 ^{1}Hol^{7}	Kings Jester	3 116	5	3	5^{2}	4^{2}	$3^{2\frac{1}{2}}$	$2^{1\frac{3}{4}}$	Delahoussaye E	32000	1.70
14Mar84 ^{7}SA^{4}	Mister Gennaro	b 3 116	1	4	1^{3}	1^{3}	$1^{2\frac{1}{2}}$	$3^{\frac{3}{4}}$	Olivares F	32000	5.60
27May84 ^{1}Hol^{7}	In Natural Form	3 116	2	5	8	7^{3}	5^{3}	4^{5}	Valenzuela P A	32000	23.50
24May84 ^{2}Hol^{1}	Speedy	b 3 119	3	7	4^{1}	$3^{\frac{1}{2}}$	$4^{2\frac{1}{2}}$	5^{2}	Meza R Q	32000	3.50
24May84 ^{7}Lga^{2}	Jolly Writer	b 3 117	7	2	$6^{\frac{1}{2}}$	6^{1}	6^{2}	$6^{\frac{1}{2}}$	Pincay L Jr	32000	5.30
3Jun84 ^{7}Fno^{1}	Benedict Canyon	3 114	4	8	7^{1}	8	7^{3}	7	Dominguez RL^{5}	32000	15.20
7Jun84 ^{7}Hol^{10}	Galaxy Ruler	b 3 119	6	6	$2^{\frac{1}{2}}$	5^{1}	8	—	Black K	32000	14.10

Galaxy Ruler, Eased.

OFF AT 4:45. Start good. Won driving. Time, :21⅘, :45, :57⅘, 1:10⅘ Track fast.

$2 Mutuel Prices:			
8–BOWER SONG	16.00	5.80	4.20
5–KINGS JESTER		3.20	2.80
1–MISTER GENNARO			5.40

$5 EXACTA 8–5 PAID $153.00.

B. g, by Messenger of Song—Lady's Bower, by Royal Tower. Trainer Borick Robert. Bred by Warwick G M (Cal).

BOWER SONG, in contention outside horses from the outset, was brushed outside KINGS JESTER while rallying in the stretch but outfinished that one in the final sixteenth. The latter, forwardly placed early, rallied between horses entering the stretch, lugged out under left handed whipping in the stretch but finished gamely. MISTER GENNARO sprinted to a long early lead but tired in the final furlong. IN NATURAL FORM saved ground around the turn, angled out to the middle of the track entering the stretch and finished strongly. SPEEDY saved ground on the turn but tired in the drive. JOLLY WRITER was never dangerous. GALAXY RULER, in contention to the middle of the turn, appeared to clip the winner's heels and checked sharply to the outside on the stretch turn, then was eased in the final sixteenth.

Owners— 1, Four Four Forty Farms; 2, Maven Stables; 3, Alter H & Kahala; 4, De Spenza C V & Dorothy N; 5, Seley J H; 6, Rimrock Stable; 7, Mole-Richardson Farm; 8, Parsons Mary.

Trainers— 1, Borick Robert; 2, Rowan Mary; 3, Spawr William; 4, Brooks L J; 5, Brinson Clay; 6, Hutchinson Kathy; 7, Bradshaw Randy K; 8, Valenzuela Martin.

Overweight: Jolly Writer 1 pound.

specific data items about horses and humans to information relationships to more global considerations of the race as a whole. The use of information is generally *additive,* and as more information is added to the brew, opinions and decisions stir, change, and alternate.

The handicapping process perseveres until all the relevant information items have been identified, all the key relationships have been crystallized, and the alternative outcomes begin to gather a relative weight of their own.

While the handicapping process is not stepwise, it does tend to pass sequentially through phases or stages that can be characterized as follows.

- Understanding the prerace information clues, notably eligibility conditions, track footing, and track biases. (The prerace information set)
- Analyzing the past performances by identifying the positive and negative information within each horse's record. (Information search and race analysis)

- Solving the specific handicapping problems peculiar to the race by examing types and sources of information neither obvious nor accessible in the past performance tables. (Problem-solving)
- Elaborating alternative scenarios as to how the race might be contested and what the probable outcomes might be in each case. (Outcome scenarios)
- Making decisions, by balancing the various probable outcomes against the relative odds offered by the public. (Making decisions)

Here are the considerations crucial to each phase of the handicapping.

PRERACE INFORMATION

The prerace information alerts handicappers to the class demands of the race and to the relative importance today of track bias, footing, trips, and even post position and bloodlines. If races will be run on turf, handicappers will need to consult Bill Quirin's Master Grass Sires lists. The lists will be part of the database. If the surface will be deep or cuppy, handicappers will discount previous impressive races on hard, glib strips. These conditions can change within days or weeks.

If a rail bias exists, inside posts and frontrunners will be at advantage. If an outside bias exists, inside posts and lanes will be disadvantaged. If a come-from-behind bias looms large today, recent races where frontrunners coasted wire-to-wire will be discounted. Horses that finished strongly from behind against a strong inside speed bias will be expected to run even bigger when the come-from-behind bias is present to help. Bias information too belongs in the data base. If no bias has been noticeable, post position should be less meaningful.

INFORMATION SEARCH AND RACE ANALYSIS

The second phase, of past performance analysis utilizing information of all kinds, is lengthy, dependent absolutely on comprehensive knowledge, skill, and multiple information resources, and far different from the practice that has been traditional. The critical question asks, what are the positive aspects of this horse's record for today; its advantages today? Eliminations are not actively pursued. The only horses eliminated will be those with little or no information support. The horses virtually eliminate themselves.

This early analytical phase also involves the use of systematic methods, but not to identify contenders, eliminate noncontenders, and make a selection. The methods are simply embedded in the larger information search. The author, for instance, relies on an understanding of eligibility conditions to make class appraisals of the entrants, but within the grander context of identifying the positive, favorable aspects of each horse's record. Horses judged unsuited to the class demands or the conditions are not thrown out at this point. Speed handicappers would produce figures at this phase but would not forsake all but the high-figure horses.

Reliance on systematic methods is not unimportant, just less decisive than customarily practiced. Systematic methods, after all, provide a rigorous rational framework for analyzing the races. Other kinds of information obtain a kind of relative value when considered in relation to methodical selections. The handicapping resembles a balancing act, with systematic methods and other information abutting and abetting each other throughout the race analysis.

When used in the information search, systematic methods do count, and sometimes count decisively, as when the method's selection sticks out against all the other considerations and is being offered at a proper or improperly high price. This happens regularly, perhaps twice a program, and nothing said about systematic method play here is meant as an argument for dropping them. As will be seen, a first decision handicappers should expect to make at all times is whether a single selection figures to win at a fair or generous price.

PROBLEM-SOLVING

The third phase, of solving the problems identified during the preceding search and analysis, usually requires sources of information not available in the past performances. At this point the usefulness of databases and computer technology cannot be denied. The ideal database answers any question handicappers want to ask, and a pragmatic database answers questions that are known to be frequently problematic to a comprehensive race analysis.

The typical problem situations necessitate several kinds of information, as outlined below.

- Trip information, and for several previous races.
- Trainer patterns information, of all kinds.
- Pedigree information, including dosage indexes, turf and mud sires, and sprint and distance sires.
- Speed information, either adjusted final times, speed figures for previous races, or the calculations of pace-speed relationships.
- Class information, preferably qualitative ratings for previous races (manner of performance) and notably for three-year-olds and lightly raced four-year-olds.
- Historical distance information, for older horses, telling whether horses have won before at today's or related distances.
- Statistical information, to include the impact values of characteristics noted in the past performances.
- Lists, of leaders in several categories and of stakes races by purses, grades, and eligible ages.
- Specialized information, to include information about foreign racing, about common shippers (stables), vet information, and other information of special interest.

Beyond these information types for problem-solving, an important feature of the information system will consist of applications programs that the computer can use to process the various kinds of data items in the database. Solving problems often requires arithmetical calculations. Computers handle the processing. Pace ratings can be calculated quickly, for instance, whenever pace figures will shed light on the race, which is frequently the case. Numerous ability times can be determined in a few minutes and "double advantage" horses flagged unerringly. Dosage indexes can be calculated swiftly. Regarding quantitative information, computers can be programmed to process the relevant data stored in different parts of the database in numerous ways, resulting in information productivity. The processing capability of the individual handicapper goes up—fantastically up.

High-tech handicappers will enjoy the problem-solving techniques now at their disposal. They will have almost everything they need to know.

OUTCOME SCENARIOS

Handicappers will become familiar with the common outcome scenarios that emerge from the preceding analyses. Scenarios are hypothetical descriptions of the running and probable outcomes of races. The three in the illustration below are found frequently and are statistically significant.

1. The early speed horse alone on the pace.

   From Quirin's probability studies handicappers have learned that early speed wins approximately 60 percent of all races and that bets on lonely frontrunners can return a sizable profit.
2. The class drop.

   Horses dropping in class win twice their fair share of the races, and the 30 percent drop or greater represents the single strongest probability factor in handicapping.
3. The classic selection.

   This is the horse that figures strongly on fundamental handicapping, including considerations of class, final speed, early speed, form, distance, and pace in combination.

Other common scenarios are less obvious, but only slightly less significant.

1. The early pace survivor.

   Handicappers now appreciate that early pace duels do not necessarily defeat the horses engaged in them. As frequently as not, one of the pace-pressing horses survives the fight and wins the race. Pace analysis always entails an evaluation of the early pace and its likeliest survivor.
2. The impressively improving three-year-old moving ahead in class.

   A most attractive class rise occurs among three-year-olds that have been maturing, improving, and simultaneously benefitting by experience in competition. Because the horses are moving up in class, many bettors shy from them. Yet they often stand out, notably if speed figures, pace ratings, or class evaluations look superior to par for today's higher class. When this kind of improving horse is juxtaposed with the three-year-old dropping down from indifferent or dull races, the improving horse is almost always the handicapper's preference.

3. The older claiming horse that has beaten better previously and is now showing sharply improved or peaking form.

   The older horse most likely to move up in class successfully is the horse that has been there before. Handicappers notice the signs of improving form, plus the back class. If the trainer succeeds with claiming horses as well, or the jockey is a national or local leader, this scenario is even more influential.

4. The special statistical selection.

Handicappers stay on alert for special situations where considerations of form, class, speed, and so forth may be less than telling, but the percentages or probabilities are firmly on their side.

Here are four statistical situations that occur frequently enough to afford us substantial seasonal profits.

- Horses having unacceptable dirt form switch to turf for the first or second time and have outstanding grass sires.
- Trainers repeat favorite maneuvers that have thrown seasonal profits in the past. The maneuver often combines the fundamental and the incidental. The class drop-jockey switch maneuver among a preselected group of "powerful" trainers has thrown tremendous profits on various circuits for years.
- Better, younger horses, juveniles and three-year-olds notably, stretch out to longer distances while up in class and possess dosage indexes that either support the combined moves or do not.
- Maidens drop into maiden claiming conditions after showing satisfactory early speed or an acceptable finish against nonclaiming maidens.

Regarding problem-solving and outcome scenarios, not much has been mentioned so far about two fundamental factors of handicapping: form and distance. That has been deliberate. The information management approach treats the two factors quite differently from traditional systematic methods.

Traditionally, the two factors, form especially, have been used to eliminate unacceptable horses early in the handicapping process. Research from various sources has shown this practice to be mistaken. Form guidelines have been amazingly liberalized in recent years. A horse returning from a lengthy layoff and showing a recent five-furlong workout can be accepted. Handicappers now appreciate, too, that many horses win at distances that are not their best or even particularly comfortable. Sprinters win their fair share of routes all the time.

Thus, the factors of form and distance best come into focus later in the handicapping process, to distinguish horses that have looked otherwise contentious. At the scenario phase, sharp form can bolster a horse's basic advantages. Dulling form can lower expectations. In the example, form defects prevented Mister Gennaro from adding up as the highly probable winner. Had the horse approached the race in sharper form, indeed, it would have won.

Thus, an important point. In the information management approach it's crucial to use *form* to evaluate closer the expected advantages of horses favored by the alternative scenarios and *not* as a fundamental elimination factor. Moreover, the form factor remains at times so vague, complicated, and elusive, it's best considered intuitively in combination with other fundamental considerations and not in isolation. To eliminate horses too early, based on form, is to eliminate too many winners and to miss too many real possibilities. At the last stages of the handicapping, horses having form defects can be discounted, especially at lower odds, and horses having form advantages can be preferred.

Distance is also now known to be an unreliable elimination factor. Sprinters stretching out to routes, the most common distance change, win their fair share of the longer races. Young horses stretch out convincingly at times, when basic abilities and bloodlines support distance racing and when the trainers involved are effective at the maneuver.

More basic, while most horses have preferred, more comfortable distances, most horses also win frequently at related distances. If not always, six-furlong winners can be expected to win often at six-and-one-half and seven furlongs. Middle-distance winners can win at all middle distances, usually. And many horses are versatile enough to perform in sprints and middle-distance routes as well.

My handicapping friend Robert Irwin, the remarkable visual artist, has long specialized in picking reliable hard-knocking claiming sprinters when they are occasionally stretched out to a mile. The "bulldogs," as Irwin refers to them, can take most claiming fields for a mile's run. Irwin has been correct often enough to collect substantial profits on the specialty, a function of trip handicapping skills, both in the straight pools and in Exactas.

As a general guideline, whenever form and distance appear problematic, handicappers should become strict when the price is relatively low, liberal when the price is relatively high.

In inspecting scenarios, finally, handicappers should appreciate that opposite logics do attract. The same race can be analyzed effectively in different ways, from different handicapping perspectives. Tenable outcomes can be predicted for opposite reasons. Thus, the same race often should be played in seemingly contradictory ways. Exotic wagering supports this point of view and the wagering practices it entails.

In evaluating the likelihood that any of the scenarios will actually occur, handicappers will be influenced first of all by the weight of handicapping information supporting each alternative. As information items are added and subtracted variously, the respective outcome will begin to look more or less probable. As a keener sense of the probabilities is developing, the matter of price, or the relative odds, comes strongly into consideration

Final decisions will be influenced by probabilities of events occurring in relation to the odds offered on each. The best decisions combine the likeliest outcomes and most advantageous odds.

MAKING DECISIONS

Final handicapping decisions are dependent on the relational nature of information. The process is *additive*. This is the essential point. As information items accumulate, are processed, and are converted into more meaningful management information, opinions are strengthened or weakened, and decisive issues begin to come more clearly into view. When the relevant information has been exhausted, the decision alternatives are balanced against the odds, and final decisions result.

Typically, the decision-making process among conventional handicappers differs from this description in a crucial way. Most handicappers rely on certain general principles or systematic methods to sort the available information in accord with the general guidelines. Decision-making proceeds from the general to the particular, involves deductive reasoning, and is reductive, largely ignoring or discounting the possibilities of decisions inconsistent with the methods. Typically, that is, handicapping decisions have been framed, or prepackaged, beforehand by the principles and procedural guidelines inherent in methods of most interest. In my personal play, for instance, I confess a strictly methodical approach to maiden claiming events. I know before beginning exactly what kinds of horses I will select and bet in those awful races. My obstinance is arguable but sustaining. I cannot stand to handicap those races seriously. In the same manner conventional handicapping methods make selections, not decisions. The information management approach begs to differ.

The information management approach is thus presented as a decisions model of handicapping, whereby large amounts of information, much of it ambiguous and contradictory, will lend itself less to making selections and more to making decisions. Handicappers benefit if they become sensitized to the final decisions that will typically have to be informed. The decisions represent frames of reference and points of departure for the handicapping process as a whole.

Below is a taxonomy of handicapping decisions in the information management context.

1. Is there a single solid selection at a favorable price?
2. What are the alternative outcome scenarios? Of those, which horses can be bet advantageously in the straight pools?
3. Which are the key horses in the outcome scenarios and combination bets?
4. Which horses must be included in the combination wagers?
5. Which horses and combinations are overlays and underlays, and to what degrees?
6. Which horses and combinations represent both good values and strong probabilities of winning?

7. How must the race be played to optimize the value of the bets? What will be the size of each bet?

Excepting the first decision, the taxonomy indicates multiple bets on a single race will be a norm. So it is. A constraint on that norm will be a lack of exotic betting opportunities. Where that condition persists, at times two selections can be covered advantageously in the straight pools. At other times the most likely winner in relation to the best price must preside as the most sensible final decision.

Here are some specific recommendations regarding decision-making.

When it exists, the single solid selection at a favorable price should *always* be played in the straight pool. The reason should be immediately obvious. Any series of wagers on a corresponding series of true overlays guarantees profits. By a single solid selection is meant a fundamentally sound horse that figures to beat the others clearly. It is distinguished from the reasonable choice at a reasonable price among several horses that are judged relatively close, as in this chapter's illustration race. The latter can sometimes be a good bet to win, but the price must be better than favorable; it must be generous.

A caution about the single solid selection regards the size of the odds. Thoroughbred racing is characterized by a sizable error factor. Errors can be human or circumstantial, and both occur all the time. Horses figure to win but lose nonetheless for a multitude of unpredictable, uncontrollable reasons, as practiced handicappers need not so much to be advised but reminded. Thus at low odds, below 8–5 perhaps, the single solid selection should be neglected in straight wagering regardless of the likelihood it will win, and certainly so if the horse has been trouble-prone, has a come-from-behind style, is an erratic or inconsistent three-year-old, or will be handled by low-percentage trainers and jockeys. On this matter as well, a mark of the expert is the ability to make substantial profits on relatively low-priced horses. Handicappers as a group may be spinning wheels on straight bets below 3 to 1, as research has indicated, but experts should be prepared to distinquish the attractive overlays at low odds from the hundreds of relatively poor risks.

When making decisions about outcome scenarios of roughly comparable strength, any attractive overlays can be bet both in straight pools and in exotic combinations.

The key horses in exotic combinations are normally the hypothetical winners favored by the alternative outcome scenarios. *All* key horses must be coupled with the other real possibilities, or the race has been misplayed. Handicappers should not leave key horses out, unless there would be no opportunity for a fair return. Key horses are linked as a rule to *all* the real possibilities having good value.

Low-priced key horses that are underlays in straight pools can often be coupled in exotic combinations that yield fair to generous profits. There is a more favorable expectation when low-priced horses represent only the bottom side of exotics, as did Kings Jester in the illustration. These are often overbet as potential winners but underbet as second-place finishers.

Aspects of the information management approach to handicapping will be treated again in more detailed, illustrative discussions in the following chapters. What must be clear by now is the overarching role of information. The more information, the better the final decision. That is the golden rule. It follows that the most informed handicappers will be the biggest winners of the information age. For the great majority of practicing handicappers the first implication is a need for much more information. The second is a corresponding need to improve one's knowledge and skill in several diverse facets of handicapping: class appraisal, speed handicapping, trip handicapping, pace analysis, and all the rest. And the third is to put it all together, as this book recommends: to become information managers.

Chapter 6

INTRODUCTION TO INFORMATION MANAGEMENT

Knowledge is power
FRANCIS BACON

THE VIEW OF HANDICAPPERS as information managers, as this book promotes, departs radically from the conventional definition of handicappers as functional specialists who apply systematic methods to race analyses. Speed handicappers, class handicappers, trip handicappers, form analysts, pace analysts, trainer experts, and handicappers relying on hybrid methods emphasizing combinations of key handicapping factors are functional specialists one and all. They apply their methodologies to the races in patterned and oftentimes stepwise manner. The outcomes they achieve are selections keyed to the dynamics of their methods. If the functions and methods are well comprehended, the resulting selections are routine and predictable.

Handicappers who are information managers behave quite differently. Their activities resemble the behavior, no less, of top managers and executives in the large organization. The management function involves planning, goal-setting, the organization and allocation of resources, problem-solving, decision-making, and an evaluation of results. While functional specialists are concerned primarily with processes, managers are concerned primarily with results. While the behavior of functional specialists is method-driven, that of managers is outcomes-based.

Managers in the organization, of course, must concern themselves with the effective management of numerous resources—human, material, and financial—in large-scale systems characterized by an increasing complexity of technology and technological change. Handicappers who are information managers can concern themselves solely with the management of topical information in a small, individualized system whose other prime resource is the handicapping bankroll.

An important distinction between information management and information technology should be emphasized at the outset. Information technology, which consists of computers, peripherals, communications networks, and software, exists to support management policy, information system goals, and problem-solving and decision-making needs. Without the policies, goals, and identified needs, the information technology remains unconnected, operates largely for its own sake, and often will be superfluous and wasteful. Before investing in a computerized information system and commencing to store information electronically, to put it differently, handicappers have a considerable amount of logical thinking and planning to complete. They will first need a logical system design that reflects real, purposeful information needs. Without the identified information needs and a system design, the technology makes little sense and usually makes little difference in the end. Handicappers who prefer to buy a ready-made information system, including prepackaged handicapping application programs, may be called high-tech handicappers in a narrow sense, but they are not information managers. They are not the audience this book seeks.

Handicappers who become information managers should understand that the first phase of the changeover will be entirely conceptual and logical. The first question is not which comes first, the computer or the program. The first question is which comes first, the system design or the system. The answer is the system design.

INFORMATION AS A CAPITAL RESOURCE

Unlike handicappers, modern corporate managers have never lacked for information resources. Abundant information has always been available. The organizational problem has been accessing the information efficiently and within a time frame that supports managerial problem-solving and decision-making. Computers and datacom networks solved the access problem. Once that progress was achieved, information soon became a primary resource of the firm.

It was judged comparably important to people and money, such that larger organizations that now lack management information systems and distributed data processing have already fallen far behind the competition.

Until recent years the handicapper's problem proved far more fundamental. It was not technological, but the mere existence of meaningful information. The information needed to help handicappers solve problems and make decisions was simply not available. Thus, the dependence on a few select systematic methods for analyzing the past performances. In the past twenty years, however, the information frontiers of handicapping have greatly receded, and today handicappers can find out almost anything they need to know, from methods of analysis to probability data to statistical compilations to descriptive information of all varieties.

Access to handicapping information is no longer a practical problem either. A virtual library shelf of handicapping books that serve as standard texts are now available to provide a knowledge base that was nonexistent two decades ago. Professional figures can be purchased for every major track in North America. Local information services on various circuits normally provide subscribers with daily variants, adjusted times, speed figures, trip descriptions, bias notes, eligibility conditions, purse values, track conditions, workouts, and trainer and jockey data.

Other specialized products supply trainer statistics, reference lists, stakes information, pedigree information, and on and on.

With handicapping information now available in abundance, and access to it convenient and affordable, regular handicappers can rightly decide to view information as a primary resource. All that is needed is a vision of the game that believes that race analyses, handicapping problems, and final racetrack decisions can be supported variously by all sorts of valid information. Once that view has been adopted, information of all varieties becomes a capital resource. It becomes just as important as money, more important than systematic methods of analysis.

If handicapping information is viewed as a capital resource, the implication is that it can be managed—must be managed—to enhance its effectiveness. That's the problem situation handicappers now confront in regard to the multiplicity of information resources. Not availability, not accessibility, but rather effective information management.

In this regard handicappers have a tremendous advantage over the corporate executives and managers. As events have proceeded, the same information technology that solved the access problem in the large organization has contributed to costly, messy management problems of one kind or another there, as executives, managers, administrators, technicians, and users fight about how information will be stored, processed, distributed, and used. Gratefully, handicappers do not operate within large-scale organizations having numerous subsystems and interdepartmental conflicts. They operate alone or in small teams. To a very considerable extent, therefore, individual handicappers can use information technology to solve the problems of information management. They can rely on personal computers and set up computerized work stations that are tailored to individual need and taste.

HOW INFORMATION CAN BE MANAGED

The information management considerations of high-tech handicapping can be grouped under three headings.

1. Identifying the handicapper's information needs
2. Designing the information system
3. Developing and implementing the system

The first two considerations represent logical purposes, requiring neither computers nor technical skills nor the physical collection and storage of information. The first precedes from a philosophical base.

What types of information are important to a handicapper? What are the handicapper's assumptions about the role of information resources in the handicapping process? Do you want to have it all? Do you want only the "good stuff"? How about just having the qualitative but not anything quantitative or vice versa? Maybe you want only procedural information (systems) but not the substantive part (knowledge)? Whatever.

These questions relate to personal beliefs, values, and attitudes. The answers to considerable degree will therefore be subjective, but they presumably will be informed by an experience and understanding of handicapping that has come to grips with the issues, realities, and priorities of the contemporary sport and pastime.

My colleague Andrew Beyer, for example, once upon a time believed that speed handicapping was the beginning and the end. So he produced speed figures. Later Beyer began to value the influence of trainers on race outcomes. So he produced speed figures and trainer data in combination. Lately Beyer has come to believe that trip handicappers are finding the overlays, speed handicappers too many underlays. So Beyer now produces speed figures, trainer data, and trip notes. He interrelates three information resources. Beyer essentially ignores class, form, pace, distance, dosage, and body language. Within his handicapping value system, Beyer probably produces more information and manages it more professionally than just about anybody playing the game on a grand scale. Notably, his methods have broadened through the years, not narrowed. Perhaps most important, Beyer has been a handicapper open to change and revision. Some might say growth.

Another colleague, William L. Scott, rather opposite of Beyer, believes that class, form, and late speed are the prime considerations in handicapping. He produces ability times, a class-speed construct, or figure, modified by considerations of early speed and current form. Lately Scott has placed added emphasis on the form factor and now focuses more on information that reflects form defects and form advantages.

Bill Quirin is another proponent of speed handicapping, including early speed indexes, but his value system broadened too in the 1980s, and he now combines speed figures and pace figures. Recently Quirin has added new information to his methods for evaluating shippers. In analyzing turf races, moreover, Quirin values pedigree information more than speed figures. He therefore produces a master grass sires list, chock full of significant impact values, and updates it periodically. Finally, Quirin values statistical significance and verification, so he produces probability values for just about every handicapping characteristic or relationship that has ever grabbed his attention.

Bonnie Ledbetter believes strongly in the ultimate communication value of body language, or how horses look and how they behave, in the paddock, post parade, and prerace warm-ups, and she keys evaluations of past performance data to body language profiles.

Another colleague, Paul Braseth, who plays professionally at Longacres, believes that trainer patterns are indispensable links to race outcomes at minor tracks. Braseth collects speed data, bias notes, and trip descriptions, but his trainer information is far more extensive relative to the other informational resources he values and much more extensively employed in decision-making.

Each handicapper named above surely collects and uses more handicapping information than the kinds attributed to them here, but information outside of their basic beliefs, values, and attitudes about handicapping and racing is not held in particularly high regard and is not used as frequently in making decisions. It is rarely used when deciding to place serious, sizeable wagers.

My own handicapping values at first emphasized class appraisal, form analysis, and pace analysis to identify contenders and make selections, and I used other handicapping factors negatively to eliminate horses from deeper investigation. I ignored speed figures, trainer patterns, track bias, early speed, body language, and trips (as opposed to trouble lines). Eventually I discounted form, intensified my preoccupation with class appraisal, and supplemented pace analysis with indicators of early speed. In the past few years I have found myself relying more and more at various times on adjusted final times, probability statistics, trainer patterns, dosage, turf breeding, form defects, and the trips, without discounting my fundamental belief in the importance of class or the advantages I previously attributed to early speed and pace analysis. Track bias and body language have remained largely outside of my repertoire, mainly because my aptitude for the observation skills is weak.

As my beliefs, values, and attitudes about the relative importance of diverse kinds of handicapping information changed and broadened, my play evolved naturally into an approach best described as a predisposition to apply a wide array of knowledge, information, and techniques to the analysis of every race. Information supplements methods. Making decisions supplants making selections. That philosophical stance, the notion that all kinds of dissimilar handicapping information, interrelated in numerous, sometimes unpredictable ways, can be important at varying times to predicting the outcomes of races, forms the underpinning for what has here been termed the information management approach.

Below are five assumptions of the information management approach as I like to conceptualize it. All are crucial to the design of an information system. They also imply that the handicapper's information needs will be all-inclusive.

1. All kinds of handicapping information are relevant and variously important to the successful prediction of race outcomes.
2. Various systematic methods of handicapping each work more effectively with certain kinds of races, but none work effectively at all times with all kinds of races.
3. Different systematic methods at times can be useful in analyzing the same races.
4. Similar kinds of races (claiming, allowance, stakes, maiden) can actually be highly diverse, susceptible to different kinds of logical analysis, both inductive and deductive, for which different kinds of information will be required.
5. Handicapping information will be relevant or decisive variously, depending on numerous situational factors; thus several types of information are potentially relevant and decisive in analyzing any race.

These assumptions demand that handicappers who are information managers concern themselves with various methods of analysis, as well as the production, collection, and management of the several diverse kinds of handicapping information that have been demonstrably effective in analyzing the races or any subset of races. As all kinds of information will be relevant and decisive at various times, handicappers need a complete array of potentially useful information. Those who do not have the information will be unable to solve sticky problems and unprepared to make completely informed decisions numerous times a season.

Once assumptions about racing and handicapping have been crystallized, specifying the information requirements that follow from those beliefs often can be done swiftly. Below, for example, are the hypothetical information requirements of Beyer's speed-trip-trainer approach. In each case, we begin with the desired output (speed figures), and list in order the information requirements for achieving the outcome.

Information requirements: Speed

1. Speed figures
   1.1 Par times (projected times)
   1.2 Final times
   1.3 Daily variants

1.4 Beaten lengths
1.5 Beaten-lengths chart
1.6 Speed chart
1.7 Inter-track variants
1.8 Calculation rules and procedures

Information requirements: Trips

2. Trip notes
   2.1 Notation codes
   2.2 Observation schedule (what to look for at which calls and how to observe)
   2.3 Views of races (at track, replays, television)
   2.4 Position notes (at gate, 1st turn, backstretch, 2nd turn, entering stretch, in stretch, at finish)
   2.5 Trouble lines (codes)
   2.6 Track biases
   2.7 Track conditions

Information requirements: Trainers

3. Trainer performance data
   3.1 List of trainers
   3.2 Trainer statistics (previous season and 5-year baseline)

.1 4 up	starts, wins, percentages, impact values, $2 net
.2 3 yo	
.3 Males	
.4 Females	
.5 Sprinters	
.6 Routers	
.7 Layoffs	
.8 Class drops	
.9 Distance changes	
.10 Jockey switches	
.11 Shippers	
.12 Turf races	
.13 Odds levels	
.14 Exotics	
.15 Claims	
.16 Maidens	
.17 Stakes	

3.3 Trainer-jockey statistics—starts, wins, percents
3.4 Trainer-owner statistics—starts, wins, percents
3.5 Descriptions of positive patterns
3.6 Descriptions of negative patterns
3.7 Ratings (various, of effectiveness)
3.8 Rules for calculating any effectiveness ratings

Beyer's approach obviously demands a large amount of information, even though arrayed for only three handicapping factors. Interestingly, handicappers should notice that practically all of Beyer's information requirements fall outside of the standard information items contained in the past performance tables. From Beyer's list, only final times and beaten lengths can be found in the past performances.

High-tech handicappers will need to identify their own information requirements. For practice, refer to the above model and try now to identify your current information needs. Be specific.

After the information requirements have been identified, the next step involves designing the *logical* information system. At this point, knowledge of personal computer systems helps but is not essential. The microcomputer will be at center stage of the system design, for example, but the actual computer to be used is a concern of the physical system and not relevant until the implementation stage. For now, handicappers are preoccupied solely with the design characteristics of the *Logical Information System*. It is enough to know that computers encode-store-process-retrieve-display data, and that they engage in input-processing-output operations, according to sets of instructions.

By examining Beyer's information requirements, his storage and processing needs can be estimated, and an information system well-suited to his handicapping needs will begin to take *logical* shape.

To obtain speed figures, for example, Beyer will prefer a heavy reliance on the computer's processing capability. Par time charts (or projected times), beaten-lengths charts, and speed charts can be stored permanently, and an applications program can be developed to manipulate daily inputs of final times, variants, and beaten lengths. The computer would produce the speed figures quickly, exactly as Beyer instructed. Beyer himself would not have to write the computer program, but merely submit his sequence of instructions to a programmer. Eventually, a small speed database could be constructed. It might contain for each horse the latest ten races and dates, race conditions, and speed figures. The informtion retrieval would be splendidly convenient. When a horse's name has been entered, the computer searches the database and retrieves its speed information, including the figures.

The trainer data would be stored in a large database. Now the storage requirements would exceed the processing requirements, which would involve calculating only percentages and impact values. Trainer statistics for entire racing circuits could be stored in a single database and handicappers could search for information about a particular trainer or for a subset of trainers. They could also relate the trainer data items to one another as well as to other kinds of handicapping information. For the most part, handicappers would prefer merely to retrieve trainer information quickly. For instance, list the trainers who have not won a race this season. The trainer database could be updated periodically, monthly perhaps, or weekly, or daily, as handicappers prefer. Although database design technicians would be needed to develop the physical system, the *logical design of databases* is strictly the handicapper's province. Happily, logical database design is not complicated, as will be shown in chapters 8 and 9.

Trip information is largely notational and can be recorded on daily programs. The notation is simple, short. Trip information could be added in notation form to the speed database, such that speed figures and trip notes for a horse's previous races would be juxtaposed on a terminal screen at the push of buttons. A speed-trip database keyed to each horse active on the circuit would be a handy information guide for handicappers of all persuasions. The design and development tasks are neither overly complex nor unmanageable. The daily input takes only minutes. In the not-too-distant future high-tech handicappers on every circuit will be able to subscribe to regional and national database services that produce that kind of information for a yearly fee plus access charges.

The logical system design also specifies the *alternative* hardware and software components that best satisfy the handicapper's information requirements. The design specification for systematic method players will be vastly different from the system required by information managers. Method players can rely on application programs exclusively, but information managers will need databases.

When a logical system design has been specified, the development and implementation phases can begin. These are highly technical phases, involving programming and physical storage, and will require technical assistance. Most handicappers will need help. They must consult with information systems analysts, or programmers, and design technicians. Information specialists and computer technicians who are handicappers will be the preferred consultants. Advice will be free, labor costs low. Development timelines can be lengthy, and development can proceed slowly. The overall database especially should be constructed slowly, piecemeal, both conceptually and physically. Special software will be required. Any applications programs should be developed according to the instructions of the individual handicapper and not purchased from vendors. Essentially, throughout the collaboration handicappers provide the logical representation of the data and the information system, and technicians translate those specifications into their physical counterparts. In this way handicappers maintain control. Handicappers take responsibility for the system design and system development. Handicappers are managers, not technicians.

DATA ARE NOT INFORMATION

The terms data and information are often used interchangeably, much to the dismay of information scientists. The distinction is concerned with *meaning*. Information forms a *basis for decision-making*, data do not, and those distinctions matter in handicapping.

Data refer to elements of fact or items of interest, or several facts and items in combination. Examples would be final times, or the array of fractional times, racing positions, and beaten lengths that comprise the running lines of the past performance tables. Final times are data. Data are useful as a basis for discussion, calculation, or assessment.

Information is processed data, or meaningful data. Information tells people something they did not previously know and when related to a goal or problem represents a basis for making a decision.

Speed figures are information: processed final times. Final times and speed figures look alike numerically, but final times remain raw data, unrefined, unprocessed, unrelated to making handicapping decisions. They are converted to speed figures by a calculation that involves other data items, including pars, daily variants, and beaten lengths. Final times are converted into speed figures; data are converted into information.

Qualitative material has similar properties. The conditions of eligibility under which a horse last competed are factual elements. They have little meaning. Race conditions become meaningful, however, when discussed in relation to other elements of fact, or data items, such as clockings, manner of performance, beaten opponents, or full race records. In making those connections, the conditions of eligibility can be processed mentally against performance patterns. Factual elements are processed intellectually into information. The information forms a basis for making handicapping decisions.

Conversely, data are unprocessed information. It is raw material. Data are stored in computer memory as bits, bytes, and data items, but can be displayed on a terminal screen as information, after an application program has processed the bits, bytes, and data items according to a set of instructions. The process by which crude oil becomes refined gasoline is a familiar, concrete analogy.

Handicappers should not confuse data with information. The distinction can get fuzzy at times, yet can be held clearly in mind if two characteristics of information are applied to any material at hand. The material must have meaning. That is, handicappers must be able to interpret the factual elements or data items before them in some meaningful way.

If the factual elements or data items have no meaning—cannot be valued or interpreted unless discussed intelligently, measured by a standard, or calculated for a formula—the factual elements or data items are data. But not information.

Second, information represents a basis for making some kind of decision. If factual elements or data items cannot be used to make a handicapping decision of some import, they represent unprocessed data, not information. When the stable foreman tells you that "they" are going for it in the seventh race today, he is supplying data, not information. It's nice to learn the stable's horse is well intended, but there's no basis for interpreting the intention or for making a handicapping decision. If nothing else comes of this discussion ever, I shall consider it a breakthrough of information science that handicappers now understand that the inside dope dished out routinely at racetracks is merely data, and not information. It cannot be used to make decisions.

When this book uses the term information, it refers to factual elements or data items that have been processed, discussed, calculated, or summarized and can be interpreted or evaluated. We are referring to processed data. The information can be used to make a handicapping decision.

DATA PROCESSING AND INFORMATION PROCESSING

A data processing operation converts factual elements and data items into information. The process can take several forms. Mental processes, such as analytical reasoning or logical deduction, are one form. Arithmetical calculation, or the use of mathematical formulae, is another. Computer programming is a multifaceted form of data processing, particularly well suited to processing large amounts of factual elements and data items and to multiple processing operations.

Information processing can involve the same kinds of operations as does data processing but refers to factual elements or data items that already have been processed and have obtained meaning. Information is processed further in order to create a higher order of information having greater decision-making value. When speed figures are analyzed in relation to trip notes, that represents information processing. The new information is more meaningful and decisive. When class valuations are interpreted in terms of form defects and form advantages, and that information is again interpreted in a context of early speed indexes and pace analyses, a synthesis has occurred, and that represents information processing. The new information will usually be more decisive.

Making decisions in handicapping usually proceeds from data processing to information processing. Data processing operations might recur several times during the race analysis. As final decisions are approached, however, information processing takes over. Numerous data items have been processed in multiple ways, and several types of information are processed in relation to one another.

Whether performed manually, with a calculator, or with a computer, data processing consists of one or more of the following operations.

1. *Recording* data to create records or files of information.
2. *Sorting* or sequencing or arranging records or files.
3. *Merging* the contents of two or more records.

4. *Calculating* amounts by performing mathematical operations.
5. *Accumulating* calculated amounts to obtain summary totals.
6. *Storing* data or information for future use.
7. *Retrieving* stored data or information when it is needed.
8. *Reproducing* or duplicating data or information for multiple uses.
9. *Displaying* or printing the output of the processing (the information) for intended users.

Experienced members of data processing units within organizations and institutions will recognize the list as the data processing functions of the computer. Information processing includes those operations at times, but goes beyond the mental capacities reflected by computerized data processing operations to include analysis, synthesis, and evaluation. That is, information processing that contributes directly to effective decision-making normally requires higher-order thinking skills to which the computer is not easily adapted.

Handicappers should perceive clearly that making meaningful decisions in handicapping usually depends on mental reasoning skills that computers do not possess. When computers spit out speed figures, they complete data processing operations, not information processing. Handicappers who depend on computers to make handicapping decisions are therefore settling for data-processing techniques.

To the extent that successful handicapping will be associated with decisions that demand analytical reasoning, or breaking a race down into its logically related parts; synthesis, or the creation of new wholes from the constituent parts of a race; and evaluation, or the use of criteria to form valid judgments about probable outcomes and appropriate wagers, handicappers who are the real experts will always outperform the computers. High-tech handicappers should therefore resolve to let computers do the data processing, but complete the information processing and decision-making by themselves.

Figure 6 depicts the data processing function and information processing function as a simple flow.

In the organization the control function is the executive or manager. At the racetrack the control is the handicapper. The control manager, or handicapper, receives the processed data, or informa-

Figure 6

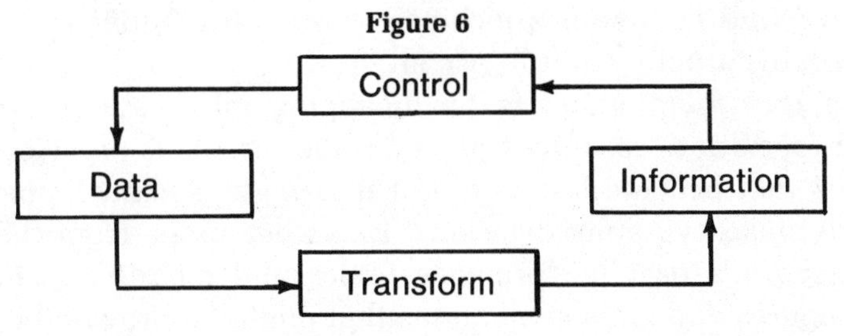

tion, processes it further (information processing), and makes decisions.

THE MIS CONCEPT

The term MIS is an acronym for management information systems. It is a high-tech management concept that refers to the computer strategy for organizing the data processing within an organization, with an emphasis on producing useful information for top managers. This book borrows the concept in the much narrower sense of a highly individualized system that produces information for making handicapping decisions. Some of the information will be computerized; much of it will not be. The handicapper organizes the system and manages the information it produces. Managing information means controlling it in this instance, and evaluating it to make decisions.

A system has three characteristics: (1) a number of component parts, (2) integrated elements, and (3) a common purpose to achieve stated objectives. As the handicapper's objective is to produce all the information pertinent to decision-making in handicapping and wagering, the handicapper's MIS components must be integrated and targeted toward that end.

To develop an MIS, handicappers must acquire the necessary resources. There are five basic types. The easiest to acquire is the computer. Microcomputer storage and processing costs have been dropping dramatically, such that just about anyone can now afford a microcomputer that has relatively large storage capacities. The microcomputer is a crucial component of the handicapper's MIS.

A second is software, the instructions or programs that direct the computer's processing tasks. Computer programs are referred to as applications software, the set of instructions for processing the handicapping data items stored in computer memory. Applications software can be distinquished from systems software, database software, and communications software, and high-tech handicappers will eventually want to use all of it.

The third component is handicapping information resources. The information resources include the products, services, and humans that provide the data and information handicappers will need to make everyday decisions. Examples range from the *Daily Racing Form*'s past performance tables and its results charts to speed figures and class evaluations that might be provided by commercial vendors for a fee.

Some information resources will be computerized; others will not be. Information obtained from the *Form* obviously will not. Data needed to produce speed figures or class ratings or pace ratings or trainer stats or dosage indexes and the sets of instructions that tell the computer how to process the data should be computerized, as the processing tasks are complicated, time-consuming, and error-prone. Some information about past performances that can be processed mentally and quickly, and changes substantially from one race to a succeeding race, such as early speed points or form defects, should not be computerized, although it might later be contained in special databases.

Source information, such as the selection, preference, and elimination guidelines of various systematic methods; reference material, such as the chef-de-race table or my own international stakes catalogue; and bulky data banks, such as trainer statistics and probability values of the numerous handicapping characteristics should indeed be computerized. The computer facilitates instant retrieval and updating of source materials.

Beyond material information resources, the handicapper's MIS can include other handicappers who supply special information. These might be handicappers who produce speed figures or trip notes. They may be body language experts or bloodlines specialists. They might also be computer specialists, and must be where the requisite technical skills are lacking. Handicappers who are not programmers will want computer programmers as resources of the MIS. These can be informal, friendly, collegial consultations among handicapping associates, perhaps as part of an information exchange. But the technical expertise will be vital to managing the computerized components of the MIS. High-tech handicappers need not develop the skills or expertise, but they must know how to organize it.

A fourth component of the information system is the database.

This is a data resource that ultimately includes *all* the data items relevant to the handicapping process. A small, manageable version of the ideal database is appropriate and realistic to begin. The data items are structured and stored in a way that permits both (a) the processing required by applications programs and (b) the spontaneous inquiries made by handicappers during the problem-solving phases of handicapping. The physical construction of the database will be difficult, time-consuming, technically complicated, and not inexpensive, but it is indispensable. The logical construction of the database is both elementary and fun.

The final component of the MIS is an information-oriented handicapper. That kind of handicapper sees information management and fully informed decision-making as the key to success in the information age. Handicappers who become information managers will see themselves as responsible for identifying their information needs, organizing a system that produces all the relevant information, using the information appropriately, and working to refine and improve the information and the system over time.

To sum up, the handicapper's MIS has five integrated components.

1. A microcomputer
2. Software
3. Information resources
4. A database
5. An information-oriented handicapper

The handicapper's MIS therefore involves information, technology, and human resources, the three linked in a system that hardly resembles the lonesome, self-reliant handicapper pouring over the past performances at the track or in the privacy of a den. In practice, the information management approach changes the handicapper's physical environment drastically. That environment now includes a computer work station, outside information services, a network of handicapping and computing specialists, and the racetrack itself.

Participation on a local handicapping information team, where members each provide the others with special types of information or services, is a highly recommended expediency. Information management and decision-making can remain individualized. Maybe only one or two members of the team will own microcomputers. A time-sharing schedule might be developed and each member's ap-

Figure 7
The Handicapper's MIS

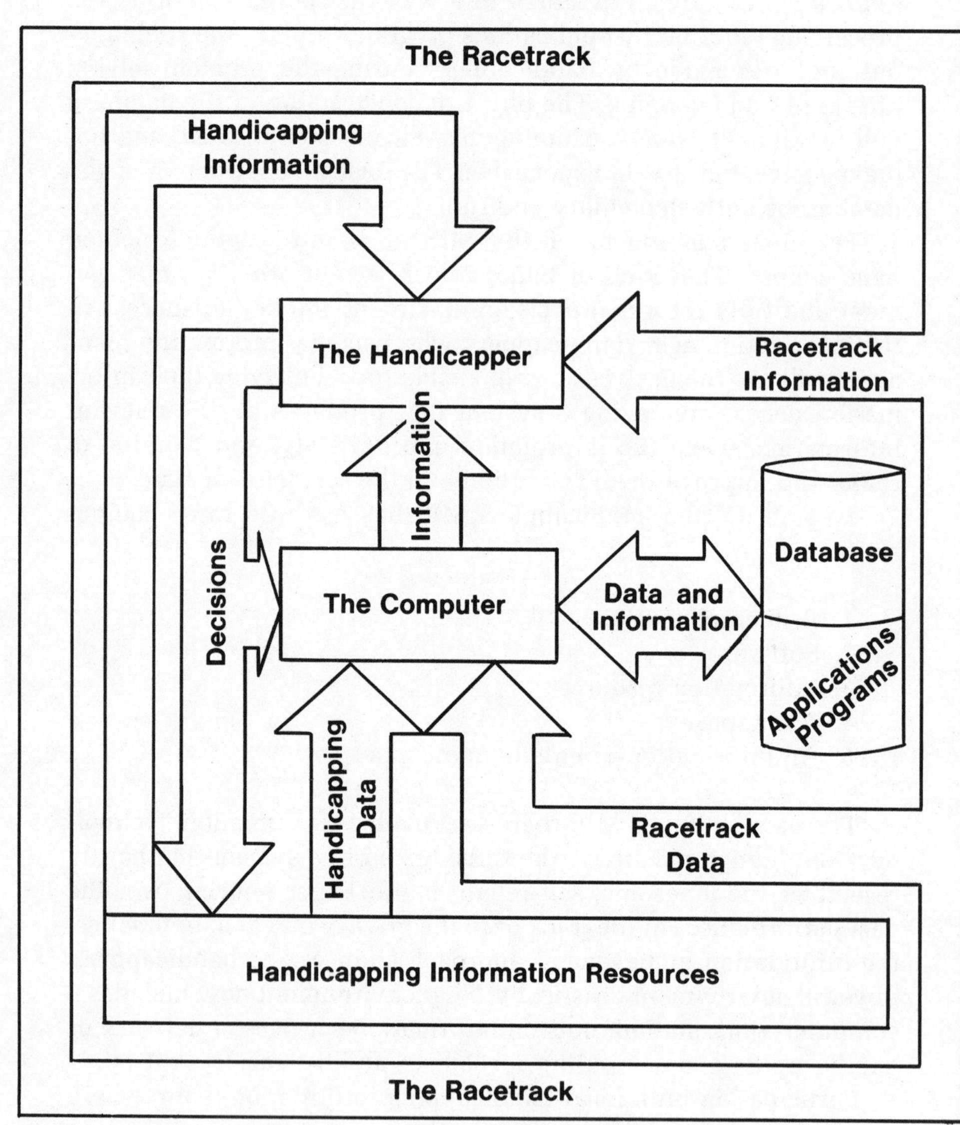

plications programs run separately; some programs may be shared. Databases can be shared and queried individually.

Examine Figure 7, an MIS model for handicappers. The model suggests that data and information for making handicapping decisions will be organized from three sources: the racetrack, the handicapper's information resources, and the computer. Notice that data and information flow both to the computer and to the handicapper. Data processed by the computer eventually flows to the handicapper as well. The handicapper completes the information processing and makes the decisions. As shown, the handicapper as manager controls all the system components.

Racetrack data and information includes track conditions, biases, body language, odds fluctuations, trip notes, and even soft information, the discussions and conversations among handicappers and insiders at the races. At first these data flow directly to the handicapper, who processes it on the spot and converts it into information for evaluating the horses at hand. Most racetrack data and information eventually passes into the computer to be stored in the database. Track conditions, biases, body language, and trip notes usually will become part and parcel of the handicapper's database.

The handicapping information resources, including the humans, likewise pass data to the computer or information to the handicapper, the latter normally bypassing the computer, unless it must be stored or processed. The handicapper uses the information to make decisions that can be fed back into the computer for further processing or retrieval, or to the information resources for recordkeeping.

A reasonable timeline for developing the MIS is a year. The actual physical implementation takes longer, perhaps another year. MIS development is often seen to have a life cycle of four phases. The MIS life cycle is examined in greater detail in the book's concluding chapter.

THE DSS CONCEPT

While MIS is the organizing concept for utilizing computer technology to marshal information within a corporate context of information management, another approach borrowed from the organization concentrates on the specific decisions top managers must make. It is termed DSS, meaning decision support systems, computerized in part; and the idea transports well to an information management approach to handicapping.

Decision support systems are information subsystems intended to produce the information required to make specific types of decisions effectively. The decisions can usually be operationally defined, in terms of their information requirements. If handicappers need to decide which horse is likely to control the early pace, for instance, the DSS concept means they must first produce *all* the data items and information known to determine reliably the probable results of early pace contests. Thus, while the MIS concept supports information management broadly and strategically, DSS supports tactical decision-making.

Proponents of DSS advise top corporate managers to focus on predictable problems and decisions that are commonplace, and build decision support systems that produce types of relevant information. Handicappers should do the same. A decisions model of the handicapping process asserts that certain types of decisions will be confronted again and again. A decision support system can be elaborated to inform each type of routine decision.

To inform routine decisions as to which horses will control the early pace, handicappers should specify first the information requirements associated with those decisions and next the means of obtaining the information.

Decision	**Information** *Requirements*	*DSS*
Which horse will likely control the early pace?	Speed points	*Daily Racing Form* data items
	Fractional times	Pace figures or velocity ratings
	Pace analysis	*Daily Racing Form* past performances
	Class appraisal	Class ratings

In other columns, the DSS might be defined more specifically, in terms of (1) input requirements, (2) processing tasks, and (3) output. To obtain speed points, the input requires the running position and the beaten lengths at the first call of the last three rateable races.

The processing task is a simple mental operation, addition. The output is a numeric from "0" to "8."

Processing can be done mentally or by computer applications. Output can be qualitative or quantitative, subjective or objective. High-tech handicappers can freely limit the information requirements and *DSS* resources to suit personal tastes, beliefs, or interests. To clarify the early speed, for example, I rely on speed points and nothing else.

The illustration reflects its proponents' argument that DSS can best attack problems and clarify decisions that in part demand quantitative analysis and in part subjective evaluation. Many handicapping problems and decisions require both quantitative and qualitative information for their solutions.

Table 3 lists information requirements and DSS resources for ten handicapping decisions that are common to an information management approach to handicapping.

The table indicates that well-defined decision support systems can constitute significant subsystems of the handicapper's information system. The MIS is concerned with supporting the entire handicapping process, whatever its logical elements and system components. Decision support systems focus on typical handicapping problems or decisions. They solve problems and inform decisions that have become everyday concerns. High-tech handicappers benefit by specifying the taxonomy of problems and decisions they grapple with all the time, often without having access to the information of interest.

High-tech handicappers should realize that reference to a database as a decision support system must be interpreted as an ideal. Conceptually, the database denotes *all* the data items relevant to the individual's total handicapping process. This is an abstraction, especially in the beginning. In practice, databases will be small, containing some handicapping data items, but not containing many more. Databases will be incomplete, not fully functioning. Where databases remain incomplete, other MIS components must serve as DSS substitutes. This means more manual data processing and more physical activity will be required to obtain the desired information.

Table 3
The Information Requirements and Decision Support Systems (DSS) for Shaping Handicapping Decisions Common to the Information Management Approach

Decisions	*Information Requirements*	*DSS*
1. Is there a single selection at an acceptable price?	Data items and information relevant to all the standardized systematic methods of handicapping	Speed handicapping model Class handicapping model Comprehensive handicapping model Trip handicapping model Pace handicapping model
2. Which horse is the class of the field?	The race conditions.	*Racing Form* information. Conditions of eligibility.
	The selection and elimination profiles of horses well suited to the typical class demands of eligibility conditions.	Database.
	Past performances. Eligibility conditions previously completed.	*Racing Form* data Results charts or database
	Key races in past performances. Ability times.	*Racing Form* information Applications program
	Demonstrated class levels of horses in today's field.	*Racing Form* information or applications program (class ratings)
	Recent class appraisal ratings, manner of performance.	Database
	Back class.	*Racing Form* data or database
3. Which horse has the highest speed figures?	Par times	Database
	Final times	*Racing Form*
	Daily variants	Personal records or database
	Beaten-lengths chart	Database
	Speed figure charts	Database
	Speed figures	Applications program
	Previous speed figures	Database
4. Which horse will likely control the early pace?	Speed points	*Racing Form* data items
	Fractional times	Pace figures (applications program)
	Pace analysis	*Racing Form* information
	Class appraisal	Class ratings

Table 3 *(continued)*

Decisions	*Information Requirements*	*DSS*
5. Which horses figure best on speed-pace or class-pace analysis?	Speed figures Pace figures Class ratings	Applications programs and the database
6. Is there a successful trainer pattern in the field?	Trainer statistics Trainer patterns information	Database or vendors Database or vendors
7. Which horses have powerful impact values on key handicapping factors?	Probability statistics	Database
8. Which horses will benefit most from improving form? Which might decline?	Form defects Form advantages Probability statistics Trainer patterns	*Racing Form* data *Racing Form* data Database Database or vendors
9. Which horses are obvious overlays and underlays?	Odds fluctuations Probable outcomes scenarios Mathematical estimates of expected advantage	Racetrack data The MIS Applications program
10. Which horses represent keys in exotic wagers?	Probable outcomes scenarios Daily Double, Exacta, and Quinella payoff projections Mathematical estimates of expected payoffs for odds combinations in exotic pools Odds lines	The MIS Racetrack data Exacta-Perfecta gauge Racetrack data

Chapter 7

THE HANDICAPPER'S COMPUTERWORLD

Computer processing is usually indicated if the [handicapper] uses large volumes of data, performs complex or lengthy calculations, and needs rapid response, high accuracy, or management information.

Raymond McLeod, Jr.* Professor, Texas A & M University. In "Management Information Systems," published by Science Research Associates, a subsidiary of IBM, 1983.

A BIT IS NOT A BYTE, and a word is not a word, but a group of characters, which are letters, numbers, or special designations usually consisting of eight bits, or a single byte.

A charming aspect of computer technology is its vocabulary. The terminology is new, unfamiliar, and susceptible to the same rapid change that characterizes the technology. It puts non-technical people off, but should not. The terminology sounds more complicating than it is and is indispensable to intelligent communication. This chapter defines and discusses several common computer concepts and terms. Handicappers should learn the material well and master the definitions of terms, even if they do not intend to purchase personal computers. They probably will experience a change of heart, and besides, the new learning amounts to a fundamental educational charge of the information age.

In addition, novices at computer science, also the uninitiated, the uninvolved, and the unintended, however averse to electronics, data processing, or even machines, should have patience with this and the succeeding two chapters, which will attempt to drag racegoers

* The technical material in this chapter is based on Professor McLeod's excellent text, which the author used during a twelve-week course at UCLA Extension, titled "Introduction to Management Information Systems and Systems Analysis," Fall, 1983.

into the computer age in explicitly recommended ways. The patience is a virtue for three reasons.

First, the author is a novice at computer science too. The advice dispensed here will be straight line and straightforward, the recommendations promoted in a rather hard-sell campaign without due consideration of the alternatives. If handicappers are told to buy a microcomputer having 256K of storage or more and featuring the MS/DOS or UNIX operating systems, the reasons will be well thought out, but there are several sensible alternatives. Discussion of technical terms and concepts will be brief and crisp. Experienced computer devotees might feel shortchanged by the brevity, but newcomers will be spared. Technical discussions can become tedious, messy, and misleading.

Second, keep in mind that computer technology is largely a faster, better substitute for manual and mental versions of the handicapping processes you already engage. Nothing the computer does cannot be done mentally or manually by handicappers without computers, though much of it would be greatly impractical. Computer technology enhances productivity and effectiveness in combination.

Third, handicappers should understand absolutely that computerized information systems of the kind advocated by this book can be designed and operated expertly by persons having no technical skills with computers, save the abilities to type on the keyboard and to follow instructions. The handicapper's role is conceptual and logical but not necessarily technical.

DISTINGUISHING THE CONCEPTUAL AND THE TECHNICAL

That third point deserves elaboration. The pedestrian image of the computerworld as the technician's province is only partially correct. The important work preceding system development, applications programming, and even operating the computer is entirely conceptual and logical. Before databases can be developed technically or constructed physically, for instance, they must be structured logically. Before applications programs can be written technically, they must be developed conceptually. Handicappers who want to buy only prepackaged software will be abandoning the golden opportunity to define the handicapping computerworld for themselves.

Rather than buy computer programs that ask for data input that may not be relevant either to the handicapper's familiar methods or to the kind of race analysis at hand, handicappers should resolve to develop their own applications programs. Beyond the conceptualization of the information to be obtained, all that is needed is to write out the computer's instructions in a logical sequence. Instead of performing as the computer asks, handicappers should instruct the computer as to what they expect from it.

My handicapping friend Vic Stephens has taken the proper approach. Stephens became a devotee of William L. Scott's "ability times," a measure of class and speed in combination. Stephens has collected windfalls on "double-advantage" horses that go postward at overlay odds. Calculating numerous ability times by hand, however, can be time-consuming, repetitive, and tedious, and the technique demands concentrated mental energy. But the technique can be programmed for computer processing simply. Stephens not only owns a microcomputer but also has access to a son-in-law with programming skills in several languages. Stephens therefore wrote out the logical instructions by which his micro could compute "ability" figures. He presented the sequence of instructions to his son-in-law, the technician. The two consulted on the program logic and the sequence of instructions. The son-in-law went away. He returned promptly with a short applications program that requires Stephens to enter four data items (numbers) for each horse he wishes to rate. The computer does the calculating and spits out the ability time figures. Stephens rates a full field within minutes. A horse or two can be computed within seconds. No mistakes.

The logical instructions and corresponding programming language for calculating ability times in BASIC appear in Table 4.

Handicappers everywhere can imitate Stephens on the logical pole of the approach. Determine the output requirements and merely write out the logical sequence of instructions you want the computer to follow. Next, consult a computer programmer. They are everywhere nowadays, and many are racegoers or handicappers. Consult on the program logic and its sequence of instructions. The programmer can write the program in a computer language compatible with the handicapper's personal computer. Test the program.

Table 4
Ability Times*
Logical Instructions and Computer Program in BASIC

Logic *Handicapper: Vic Stephens*	*Computer Program (Partial)* *Programmer: A. L. Wilson*
1. Input	190 INPUT "RACE LENGTH= ":L
• Length of a race	200 IF L=999 THEN 1840
• Half time	210 PRINT
• Final time	220 PRINT "HT,FT,LBH,LBF= "
• Lengths behind at finish	230 REM HT=HALF TIME
• Lengths behind at half	240 REM FT=FINAL OR 6F TIME
	250 REM LBH=LENGTHS BEHIND
	260 REM AT HALF
	270 REM LBF=LENGTHS BEHIND
	280 REM AT FINISH OR 6F
	290 PRINT
	300 INPUT HT,FT,LBH,LBF
2. Calculation Steps	320 GOSUB 1950
• Calculate final quarter time	370 IF L=5 THEN 1010
• Determine beaten-lengths adjustment factor	380 IF L=5.5 THEN 900
	390 IF L=6.0 THEN 880
• Calculate gain-loss factor	400 IF L=6.5 THEN 710
• Calculate energy adjustment factor	410 IF L=7.0 THEN 440
	1290 GL=LBH−LBF
• Determine ability time	1520 REM P=GAIN/LOSS FACTOR
• Convert ability time to points	1600 EA=1.0
	1710 REM END OF ENERGY ADJUST
	1720 AB=AGAT+EA
	1730 PTS=140−(AB*5)
3. Output	1770 PRINT "ABT= ";N
• Ability time in minutes, seconds, and fifth-seconds	1780 PRINT
	1790 PRINT "PTS=";PTS
• Ability points	1800 PRINT

* Handicappers enter five data elements; ability times and points displayed within seconds.
Programmer A. L. Wilson urges handicappers to provide program design logic in this sequence; (a) output, (b) input parameters, (c) processing steps.

Revise it as needed. Implement. It is that elementary. An alternative is to learn how to program in BASIC, a simple language that can be mastered in a one-semester course.

Structuring a handicapping database logically will be more complicated, and the interplay between handicappers and technicians more continuous and problematic, but the same consultative approach applies. Handicappers provide the concepts, logic, and design instructions; technicians supply the technical assistance.

The point surely is that handicappers must retain personal control over the conceptual and logical structure of all the handicapping information that computer programs and databases are expected to produce. The first precept of high-tech handicapping is therefore the same as with noncomputerized handicapping, namely that individuals should do their own handicapping and not depend on the prepackaged applications of others.

THE MICROCOMPUTER BOOM

From the early 1950s until the 1980s computer users have already witnessed five generations of computer families, each distinguished by advances in circuitry, primary storage, secondary storage, and input/output devices. Software innovations have followed suit. The computer evolution is summarized for your information in Table 5.

The microcomputer boom began as recently as the 1970s when IBM replaced the magnetic cores in primary storage with semiconductor chips. The chips are so small that the microcomputer has often been called "a computer on a chip." All the system logic and arithmetic circuitry are housed on a single semiconductor chip, as shown in Figure 8. A second figure, Figure 9, shows the physical form of the chip to dramatize its minute size. The chip symbolizes the computer's evolution toward smaller size, denser storage, more powerful processing capability, and lower costs. Soon after, by 1975, the first personal computers were marketed to individuals. The first was the Altair 8800, sold by Mits, Inc., in kit form for $400. A year later the young electronics entrepreneurs Steve Wozniak and Steve Jobs built the first *Apple* in their garage, a historical phenomenon that today remains the runner-up in microcomputer market shares to IBM. In 1977 came the Commodore Pet and the Radio Shack TRS-80, and in 1979–80 the personal computer market exploded. Today there are dozens of models available, plus a thousand or more separate sources of peripherals and software.

The main issue between hardware and software is their communication compatibility. Software must be developed in a programming language the machine can understand. Applications software for the Apple, for example, does not work on the IBM personal

Table 5
Computer Hardware and Software Evolution

Generation	*Dates*	*Circuitry*	*Primary Storage*	*Systems Software*	*Applications Software*
First	To 1958	Vacuum tubes	Magnetic cores	Assemblers, Compilers, Sort/Merge Routines, Input/Output Controls	Accounting Application, Scientific Computations
Second	1958–1964	Transistors	Magnetic cores	Same	Same
Third	1964–1970	Integrated Circuits, Small-Scale Integration	Magnetic cores	Operating systems, Compilers, and Interpreters	Report writers, Operations research
Fourth	1970s	Large-scale integration	Semiconductor chips	Virtual storage, Database management systems, Compilers, Query languages	Data retrieval routines, Decision support models
Fifth	1980s	Very large-scale integration	Semiconductor chips	Data communications, User-friendly languages	Same as fourth but more sophisticated

SOURCE: Management Information Systems, by Science Research Associates, IBM, 1983.

Figure 8
A Computer on a Chip

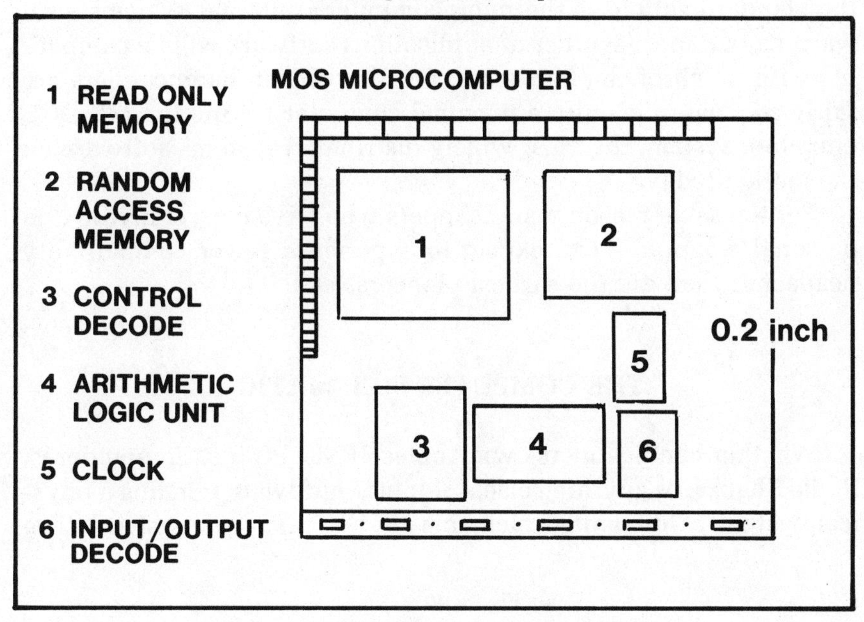

Figure 9
The Semiconductor Chip Is Extremely Small

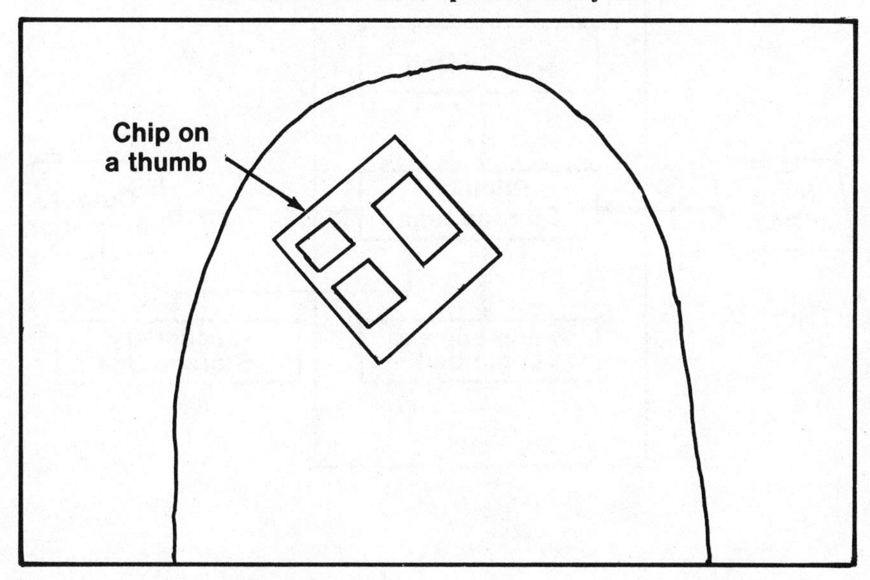

computers. Software developed for the original Radio Shack TRS-80 cannot be used elsewhere. Fortunately, the trend today is toward the standardization of the microcomputer's internal systems software, so that most sources of applications software will be compatible with much of the hardware. For this reason handicappers are fairly advised to acquire a personal computer having the MS-DOS operating system, the most widely distributed systems software on the market today.

For the same reason, handicappers who have not yet invested in personal computers can expect to experience fewer compatibility headaches than did the earliest pioneers.

THE COMPUTER SCHEMATIC

Whether handicappers own Apples, IBMs, PC Jr.s, Commodores, Radio Shacks, or anything else, computer hardware remains a physical system of integrated elements that is easily understood. The

Figure 10
The Computer Schematic

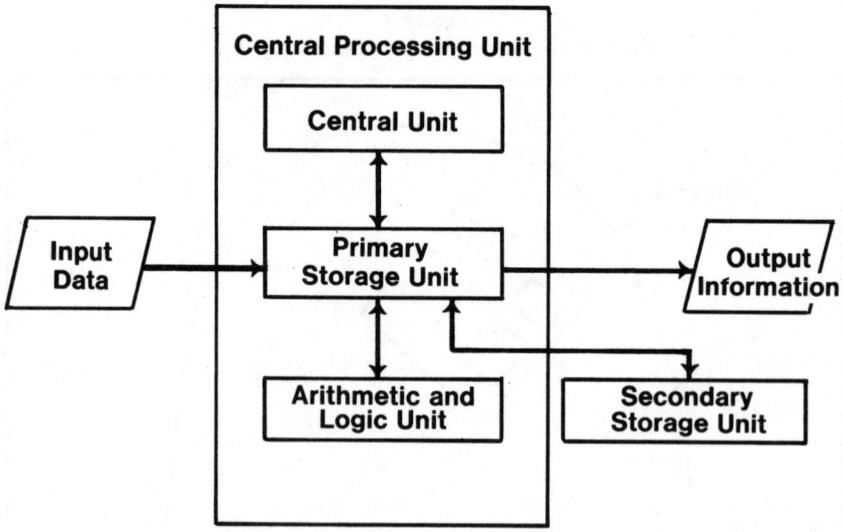

diagram in Figure 10 shows the basic architecture of all computers. It is called a computer schematic.

A computer has one or more input devices that send data or information to the transformation and control part of the system. The central control is called the central processing unit, or CPU. Data and information are stored and manipulated here. The CPU contains three subparts. It has a primary storage area where data and computer programs can be stored. Any calculations or logical decisions, such as making speed figures or ranking horses, are made in the arithmetic and logic unit. A control unit manages the entire system flow. It can move data in and out of primary storage and send other data into the arithmetic and logic unit. When data and information have been processed, it is transmitted from the CPU to the output devices.

Input, processing, output—it's practically that simple. Storage limits, however, complicate the basic system. Notice the link in the computer schematic between the primary storage unit and a secondary storage unit. When primary storage is filled, a secondary storage unit, such as a magnetic disk or magnetic tape, maintains data that can be transmitted to primary storage and there processed. The secondary storage unit may be connected to the computer system, or put "on-line," or it may be disconnected, termed "off-line." Handicappers will need secondary storage units. Those relying only on applications programs will need only floppy disk units, the soft, cheap disks that resemble 45 rpm records. Handicappers who construct databases, as this book recommends, will prefer the advantages of larger magnetic disk units, usually called hard disks.

COMPUTER STORAGE

As shown in Figure 11, the storage portion of the CPU has five uses, preferably referred to as the five *conceptual* areas of storage. These areas are not physically distinct but are logically distinct.

As data or information enter primary storage from an input device, probably a terminal keyboard, it is placed in the *input* area. In the *applications program* area is the list of handicapping instructions that tell the computer what to do with the entered data. The applications program performs the necessary data movements, calculations, and logical decisions, and places the processed data or information in the *output area*. The new data or information are

Figure 11
The Five Conceptual Areas of Primary Storage

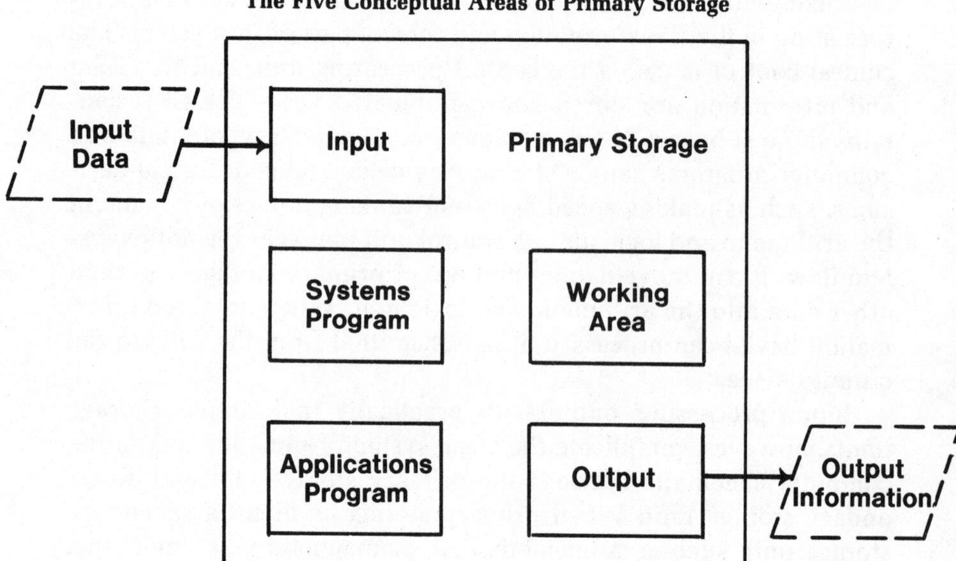

transferred from the output area to an output device, such as a terminal screen or a printer. Handicappers will generally want handicapping information displayed on a terminal screen, but for some purposes will also rely on printers. Someday perhaps the past performance tables contained in a database miles away will be printed out on printers attached to personal computers in the home, not to mention that handicapping information will be shared among subscribers who belong to a computerized handicapping communications network. But that is futureworld.

Most applications programs require a separate *working area* for storing immediate totals, as with speed figures, pace analysis, trainer stats, or any brand of figure handicapping, as well as constants and descriptive characters. Finally, the execution of applications programs is controlled by other software, primarily the operating system, denoted as the systems program in Figure 11. The systems programming language and the applications programming language must be compatible for the computer to produce the desired output.

The sizes of storage areas differ sharply from one computer model to another. As considerable storage space is required by the operating systems' programs and by the instructions of applications programs, not to mention the space requirements of database programs or communications programs, which might be on-line, handicappers are well advised to demand primary storage space of 256K (kilobytes) or more. This would normally facilitate a comprehensive operating system, database management system software, a communications program, and perhaps three applications programs. Handicappers unaware of the power of the database concept are also unaware that they will need all of the above kinds of programming to stay ahead in the information age.

A handicapper like Andy Beyer might store speed handicapping application programs, a speed-trip handicapping database, and portions of a trainer database. Bill Quirin might store a speed and pace applications program, a speed-pace-trip database, and his master turf sires database. Handicappers who employ speed, class, and pace in combination might store applications programs to process the data for each of those rating methods plus a larger database. Handicappers who proceed as this book recommends will want large, expandable storage capacity for all types of handicapping data items, notwithstanding their modest beginnings. As will be seen, high-tech handicappers should construct the database first and develop applications programs later, the reverse of conventional practice.

SECONDARY STORAGE

Secondary storage supplements the limited capacity of primary storage, or main memory. Secondary storage devices are usually hooked to the CPU—primary storage—by cables, and the data contained in secondary storage are channeled into primary storage for processing by systems software. The most popular form of secondary storage for microcomputers is the floppy disk, or diskette. When larger storage capacities and faster data transfer is desired, owners of personal computers can opt for *hard disks* as secondary storage devices. Hard disks can be removable from the main unit and attached by cable, or they can be stored inside the CPU cabinet.

Handicappers who wish to construct sizable expandable databases would be best served by hard disks as secondary storage devices. Floppy disks normally store about 150 to 800 kilobytes of data; hard disks store 5 to 40 megabytes (a megabyte is one million bytes). Small databases can be developed using floppies, but the desired qualities of flexibility, speed, and expandability are sacrificed, and ultimately the integrating function of the database in the information management approach can become problematic. Several floppy disks will always be off-line. Handicappers can consult the book's chapters on databases to determine whether the higher costs of hard disks can be justified for their personal MIS.

Handicappers should also be aware that there are two basic types of secondary storage, sequential and direct. The two are vastly different. With sequential storage, the data records (groups of data items on a particular topic) are arranged and stored in a sequence the computer can search only in that sequence. All the records must be processed sequentially, in order, and the computer cannot move immediately to the record of interest. This presents processing complications, notably the amount of time required to handle simple processing tasks that require the retrieval of several records located in different areas of secondary storage. Sequential storage is associated with the batch processing of earlier computer business applications, such as payroll, sales, and inventory. The medium of sequential storage is tape.

Direct storage was developed to overcome the processing limitations of sequential secondary storage. Direct storage allows an access mechanism to move directly to the data records of interest. The device making this possible is referred to as a *direct access storage device*, or DASD. The medium of direct access secondary storage is a magnetic disk. The time and flexibility of the processing function is enhanced tremendously using disks.

Handicappers will obviously prefer magnetic disks as secondary storage devices, but tapes do have a cost-effective use of importance to many handicappers. Tapes are the preferred form of *historical* storage. When data and information on horses and trainers is no longer needed for daily decision-making, when it is no longer current, the computer can transfer these data from disks to tapes, and the tapes can be stored on a shelf or in the closet.

When the past performance tables finally become part of an electronic database, handicappers can store these on tapes as well. The complete history of past performances for all horses on the circuit for several years will be copied on tapes, economically stored, and readily retrieved. The entire rooms and closetfuls of rotted *Racing Forms* in the private residences of devoted handicappers will become a past glory.

Handicappers who rely on floppy disks for secondary storage need hardware with two disk drives. One drive can hold applications software, the second the handicapping database. The Apple II computer system in Figure 14, page 182, contains two disk drives, shown just to the right of the terminal display screen. Floppy disk storage capacity can vary widely but can go as high as one megabyte and is expanding all the time. Magnetic tape cassettes are often offered as an alternative to floppies, but these offer sequential storage and therefore sequential processing only. Retrieval time will be longer, processing will be slow, and data items cannot be manipulated with imagination and flexibility, as high-tech handicappers would surely prefer.

BITS AND BYTES AND THE CPU

The data, information, and program instructions stored in the computer are represented in coded form. The computer is an electronic device, of course, and thus represents numbers, letters, and special characters with groups of tiny electronic elements. Each element is called a bit—a contraction of *binary digit*. Each bit is represented as either a *zero* or a *one*. Each bit is either *off* or *on*, its binary states.

As indicated in Figure 12, numbers and letters are represented by groups of bits. The most common computer codes are 8-bit codes, which represent the number and letters with groups of eight bits. Each group of bits represents a *character*, and a group of bits representing a character is called a byte. In other words, a *byte* is a group of bits representing a character. The common term kilobyte (K) means one thousand bytes, or a thousand characters. Thus a microcomputer having 128K of primary storage can hold 128,000 characters in main memory.

A *"word"* is not a word but a group of characters that the CPU processes as a unit. The "word" size varies, but microcomputers normally process either eight-bit words (a single character) or 16-bit words (two characters). The latest microcomputers can process 32-bit words, or four characters, at once. This improves processing time terrifically but is of little practical benefit to handicappers. Pro-

Figure 12
Two popular 8-bit computer codes for the numbers and letters

0	11110000	10110000
1	11110001	10110001
2	11110010	10110010
3	11110011	10110011
4	11110100	10110100
5	11110101	10110101
6	11110110	10110110
7	11110111	10110111
8	11111000	10111000
9	11111001	10111001
A	11000001	11000001
B	11000010	11000010
C	11000011	11000011
D	11000100	11000100
E	11000101	11000101
F	11000110	11000110
G	11000111	11000111
H	11001000	11001000
I	11001001	11001001
J	11010001	11001010
K	11010010	11001011
L	11010011	11001100
M	11010100	11001101
N	11010101	11001110
O	11010110	11001111
P	11010111	11010000
Q	11011000	11010001
R	11011001	11010010
S	11100010	11010011
T	11100011	11010100
U	11100100	11010101
V	11100101	11010110
W	11100110	11010111
X	11100111	11011000
Y	11101000	11011001
Z	11101001	11011010

cessing time is not much of a practical issue for owners of microcomputers, which, after all, are users of small systems. Some handicappers might prefer micros that process 16-bit words or 32-bit words, and therefore work two to four times as fast as has been the microcomputer norm, but by and large high-tech handicappers will be perfectly satisfied with an 8-bit processing capability.

THE OPERATING SYSTEM

The best way to conceptualize a computer's operating system is to think first of a human operator, a person who decides which programs to run, sets up the input and output devices, sets the control switches, and actually operates the machine. The operating system is merely an array of software that performs those tasks. The operating system permits the computer to control itself, without human intervention. It is referred to as systems software.

Handicappers should know that operating systems software consists of three types of programs, and each type performs critical functions.

The *control* program, sometimes called the supervisor or executive, manages the computer as it executes programs. It stores programs in *input,* schedules data for processing, and calls data into primary storage when the computer is ready to handle it. It also can retrieve data from secondary disk storage and maneuver it through various channels into main memory.

Language-processors include special functions, such as sorting and merging, the two basic data processing tasks: the librarian and utilities. The *librarian* obtains applications programs from secondary storage when they are needed. Handicappers can store dozens of programs in this way and send them to primary storage whenever they choose. *Utilities* convert data from one medium to another. When the racing season has ended and handicappers wish to preserve the historical record, a disk-to-tape utility allows them to copy data records from the disk files onto a magnetic tape file.

ON-LINE PROCESSING

On-line means simply that a device is connected to a computer. Any device that is not connected to the computer, such as a floppy disk, is off-line. On-line systems have each of their parts linked to the CPU. With on-line systems, as soon as data is entered into primary storage the computer performs the necessary transactions and produces its outputs. The processing occurs *now*. Moreover, all the files can be immediately updated. Another transaction can be handled immediately, and all files will be updated again. Handicappers have access to up-to-date information. This is not a paramount concern, of course. Handicapping facts and stats can be updated weekly, as horses do not run every other day. Class indexes, dosage indexes, trip notes, body language, and bias notes can be updated periodically, as handicappers prefer. Trainer stats can be updated monthly.

Figure 13 shows how a horse's latest speed figure is calculated and the speed files simultaneously updated. By entering a horse's name or number, its latest race's final time, beaten lengths, the class-distance par, and the daily variant, handicappers can use a speed analysis application to produce a new figure and update the speed figure files immediately. The horse's new figure would be

Figure 13
On-line Processing of a Horse's Speed Figures

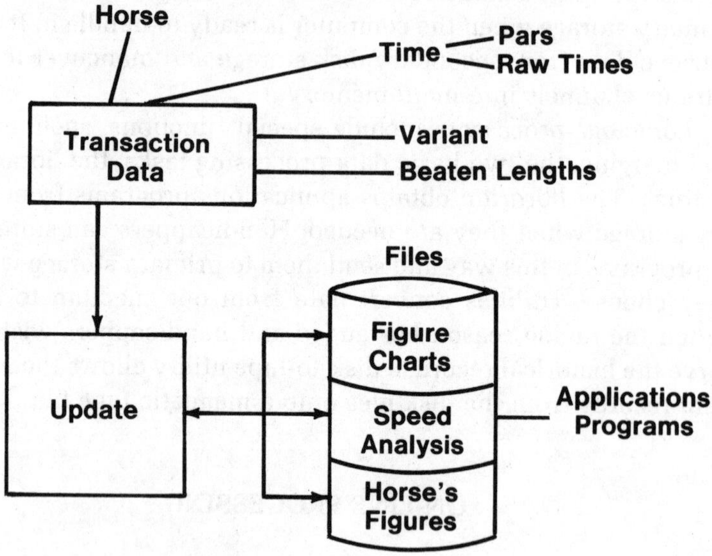

added to the updated record. Figures more than a year old might be eliminated from the files. If speed figures were ranked from high to low for each class-distance category during a season, that file would be updated instantly as well, and handicappers would know at the push of a button the relative significance of each horse's figures.

With on-line processing, handicappers can be secure that their data records and files are as current as the last transaction completed. To the extent that data has been recorded on floppy disks that are stored off-line, updating cannot occur simultaneously with processing.

THE COMPUTER CONFIGURATION

Microcomputer systems consist of a CPU and various peripherals, which include input/output devices and secondary storage units. The arrangement of the system is called a computer configuration. The computer configuration selected depends on the information needs of the users. Handicappers should not mistake the point. The distinguishing feature of a computer configuration is not the devices attached to the CPU but the performance capabilities of the system. The system need only satisfy identified needs. The information needs of systematic method players and information managers will differ greatly. Information managers will need database and communications devices and a different kind of computer configuration.

Handicappers generally have both explicitly defined information needs, the input and output of systematic methods, and situational information requirements that are not entirely predictable. Handicappers who become information managers will have large storage needs, as well as heavy data manipulation needs. Handicappers who become members of handicapping information teams, and many who do not, will want printed output regularly. Eventually, many handicappers will prefer data communications devices, peripherals that link them to other computers, databases, and information sources. All of this has diverse implications for the basic computer configuration.

As a first preference, the high-tech handicapper's computer configuration would use terminal input (keyboard), at least 256K of primary storage, a standardized operating system, a hard disk unit for large expandable secondary storage, and printed output from a slow-speed character printer. The performance capability of the system facilitates the on-line storage of all possible handicapping data items, the on-line processing of multiple applications programs, database interrogation, and the production of information on both terminal display screens and printed materials. The printed materials can be carried to the racetrack or to sessions with other members of the handicapping information system.

Where the hard disk unit is omitted, a CPU unit with dual disk drives for two floppy disks is much preferred to the single disk drive. This permits the simultaneous on-line processing of applications programs and spontaneous calls to the database. Secondary storage can be achieved altogether on floppy disks, of course, a lower cost alternative to the hard disk unit. The trade-off is a reduced capability of simultaneous on-line processing and updating. Floppy disks cost a few dollars, the hard disk unit several hundred.

The Apple II computer system shown in Figure 14 is perfectly acceptable as a computer configuration for high-tech handicappers. The CPU and the input keyboard are contained in the base unit on the left. Above the base unit is a terminal display screen, called a

Figure 14

Courtesy of Apple Computer, Inc.

cathode ray tube (CRT), the output device. At the right are two floppy disk drives for secondary storage. A second output device, a nonimpact printer, is not shown but is readily available. Another version of the Apple II system is cheaper but has a single disk drive only. This is inconveniencing for various storage, processing, and operating purposes.

Handicappers will differ individually in their preferences for terminal screen sizes and colors, although larger screens and green color are often preferred. The Apple II, for example, displays 24 lines of 80 characters each, and the TRS-80 displays 16 lines of 32 characters each.

The computer configuration recommended to handicappers here will cost a ballpark $2,500, perhaps less, as personal computers continually become more price competitive. The payoffs at the racetrack will be well worth the money spent.

SOFTWARE

The computer program is a set of instructions that tell the computer what to do. A small program may contain only eight to ten instructions. A large one may contain several thousand instructions. A single instruction is required before the computer will perform any specific action, such as read a page, multiply a number by another, write a line on a printer, or get information and display it.

The computer program is written in a programming language. Several languages can be used to write computer instructions, the most popular among nontechnical users being BASIC, which means *Beginner's All-Purpose Symbolic Instruction Code.* Handicappers who are not programmers can become programmers by learning BASIC. The education takes weeks, lasts a lifetime. A programming language consists of an array of different types of instructions, or commands, perhaps a hundred. BASIC has roughly fifty commands. Basic instruction in BASIC, in combination with consultation engaging trained computer programmers, will facilitate the programming of whatever applications handicappers now prefer, or will, anytime in the future.

Early experience has indicated that most racegoers who own microcomputers will be content to purchase prepackaged software and merely insert the programs into the disk drives. This is reasonable procedure only when the programs offer data management capability or produce handicappng *information* that is helpful in solving problems or making decisions. It constitutes inexplicable procedure when the programs imitate systematic handicapping methods that make selections or rank horses, unless the set of program instructions corresponds to personal handicapping procedures, without any basic variations. This is almost never the reality. Handicapping is a highly individualized pastime, its indispensable attraction to most regular practitioners.

By recommending that handicappers learn BASIC, this book does not pretend they should become ardent computer programmers, refusing to settle for anything but personal applications programs. That would be naive. Handicappers should expect to purchase several software packages through the years, and at great personal advantage. Some will promote handicapping applications; others will not, but will support the handicapping and wagering purposes. A nonhandicapping applications program that all handicappers should appreciate is the spreadsheet. This is the large table of rows and columns generally associated with accounting and mathematical applications, but of wide general use besides. The popular VisiCalc program contains 63 rows and 254 columns, such that large spreads of information can be displayed in a tabular format. The spreadsheet format, commands, and data manipulations are easily understood by the nonbusiness person and the average racegoer.

All handicappers can identify more or less with manual spreadsheets. Paper and pencil worksheets are used for several important handicapping applications. Compiling trainer stats is an example. Another is keeping financial records. Speed handicappers use worksheets to calculate par times, variants, and adjusted times. Completing spreadsheets by hand, of course, is slow and problematic. Changes to values within the columns and rows usually affect the numbers in other columns and rows, as well as all totals. This necessitates lots of erasing and recalculating, and contributes to occasional errors. The whole manual process is time-consuming and frustrating.

The microcomputer ends that misery. The computer can erase, rewrite, and recalculate values, rows, columns, and entire spreadsheets within seconds. No errors. Row and column intersection cells can be related to one another, as when the totals put in one column are the result of adding several other column values. The row and column cells can also contain labels, text, or even formulae. Handicappers can use the spreadsheet to compile and update trainer statistics in numerous categories. Totals and percentages are updated at all times.

Spreadsheets are particularly useful for keeping money management records. A prime reason regular knowledgeable handicappers, even talented ones, tend to lose more money than they should is that they do not keep adequate financial records. A prime reason they do not is because the daily record-keeping of multiple types of pari-mutuel investments is annoyingly ineffícent and time-consuming. The microcomputer wipes away those excuses. All types of bets and financial records are recorded, processed, and totaled at a touch for each race. Totals are constantly updated. Moreover, important financial data manipulations become handy. Handicappers can readily determine their success rate at odds of 2 to 1 and lower—a mark of expertness is the ability to make a profit of 25 percent or more on short-priced selections—or 5 to 1 and higher. Exacta profits or losses are easily monitored. Handicappers can establish win percentages and dollar nets on maiden races, nonwinners allowance races, feature races, and claiming races. They can establish how well they are doing on dirt versus turf, in sprints versus routes, with fillies and mares versus the males.

If portfolio wagering is conducted, with several types of bets in the portfolio, the portfolio can be managed carefully, and unsuccessful types of wagers can be eliminated. New bets can be added and studied. Where systematic method wagering is conducted in the straight pools, perhaps using a fixed percentage formula, once a baseline performance has been established, a number of "what if" simulations can be carried out. Necessary formulae can be deposited in the row-column cells. Spreadsheet commands allow the formulae cells to manipulate the data in other cells. The simulations will guide thoughtful handicappers into greater profit-making zones. For example, what if the base bet had been increased by one percent? What if the original capital investment were increased by $5,000? What if the win percentage were boosted by three percentage points? What if the average odds on winners were improved by a half point? In additional simulations, the amounts wagered, win percentage, average odds on winners, and numbers of bets can be held constant, and the money management methods varied. Which performs the best? What are the advantages and risks of each? To be sure, the spreadsheet application is a powerful computer program for high-tech handicappers.

Handicappers who become information managers will want three additional types of software. One is futuristic, a communications package. This will permit handicappers with microcomputers to communicate regularly with other computers and search other databases. Additional hardware peripherals will be required, notably a low-cost item called a modem, or acoustic coupler, which attaches to telephones as well as to computers and converts sound waves to binary digits and vice versa. These bonanzas may be less than a decade away.

High-tech handicappers of today can be expected to purchase two systems software packages, a database management system (DBMS) and query languages. The two are regularly combined in a single package. These programs permit handicappers and the CPU to control and retrieve the records in a database and to interrogate the database as they please. The two are recent developments in microcomputer software that promise to become the most powerful functions in high-tech handicapping, much like the tail wagging the dog. The next chapters discuss the database concept and the handicapper's view of the handicapping database, the centralizing and most exciting component of a high-tech handicapping information system.

HANDICAPPING APPLICATIONS PROGRAMS

Within the grim view of computerized selection models of handicapping stated in Chapter 4, this book favors two powerful computer models of high-tech handicapping, the Sartin Methodology and The Sports Judge. Extensively tested, both get impressive win percentages at profitable averaged odds on the winners. They have demonstrated the ability to beat the races.

Applications programs that provide handicapping information, not selections, can also be useful, notably to handicappers having scarce interest in developing high-tech information systems of their own, but dogged tired nonetheless of performing the routine manual and mental gymnastics of full-dress handicapping. If well-documented, reasonably priced programs can supply speed figures, class ratings, dosage indexes, or pace ratings that handicappers can live with, having the information at less sweat facilitates greater productivity in handicapping.

A particularly powerful, properly priced pari-mutuel wagering application program is the Z-System, which provides place and show overlays at strong winning probabilities several races a day as a result of little input effort on hand-held computers. The wagers can be sizable, but so can the long-term expected profits.

As intimated in Chapter 4, whether handicapping applications programs deliver selections or information, the best way for high-tech handicappers to get an honest line on the stuff is by consultation with leading high-tech consumer protection magazines and not advertisements in publications such as *Daily Racing Form*.

Chapter 8

THE HANDICAPPER'S DATABASE

The next generation of high-tech millionaires will be the inventors and purveyors of specialized databases.

JOSEPH KIRKMAN, CEO
Peopleware Systems International
Laguna Beach, CA

WITHOUT HESITATION or qualification, this book proclaims the computerized database approach to information management and problem-solving in handicapping to be the most potentially valuable concept in the evolution of the game. Without hesitation or qualification as well, it asserts that the biggest winners of the information age at all racetracks will be talented handicappers with access to handicapping databases.

These modern players will have unabridged access to otherwise unavailable data, information, and handicapping relationships that clear the way to numerous true overlays, the only horses that enable handicappers to beat the races. They will be equipped to solve intricate, obscure handicapping problems and make final decisions that not only will elude most of the crowd, but also will baffle their fellows who have stayed tied too tightly to systematic methods and conventional wisdom. They will make regular killings in the exotic pools especially, and have real chances to earn a year's corporate salary within weeks, when knowledge, skill, and luck roll along in harmony for the short term. Those kinds of assertions were not possible only a few years ago, but are today at the tip of the high-tech, information age. To be sure, getting there will not be easy. Handicappers will need to develop an array of new skills, both technical skills and new ways of thinking. But the efforts demanded will not be seriously resisted by handicappers who can understand the issues and appreciate the possibilities.

4th Golden Gate

5½ FURLONGS. (1.02⅘) MAIDEN CLAIMING. Purse $6,000. 2-year-olds. Weight, 118 lbs. Claiming price $16,000.

Bam's Rascal
Gr. c. 2, by Shady Fellow—Misty Banner, by Green Banner
Br.—Bam Stable (Cal) 1982 0 M 0 0
Own.—Bam Stable 118 Tr.—Hess R B $16,000
Lifetime 0 0 0 0
Jun 11 GG 5f ft 1:023 hg Jun 6 GG 3f ft :364 hg Jun 1 GG 5f ft 1:071 h May 26 GG 4f ft :493 hg

A Faster Piaster
Dk. b. or br. f. 2, by Piaster—Lady Go Round, by In Zeal
Br.—Keough Mr-Mrs S R (Cal) 1982 2 M 0 0
Own.—Hi Card Ranch 115 Tr.—Mortensen Dee $16,000
Lifetime 2 0 0 0
27May82-4GG 5f :221 :463 :594ft 57 118 73½ 85¾ 76½ 79¼ McGurn C1 ©M20000 75 Proper'sPepper,T.J.'sDy,TripleAct 12
20May82-6GG 5f :222 :464 :594ft 30 117 66¾ 78½ 813 814 ChapmnTM5 ©[S]Mdn 71 BitOfCri,MidnightGrce,GrciousRbl 10
Jun 13 GG 4f ft :481 h Jun 8 GG 3f ft :38 h May 13 GG 5f ft 1:022 h May 8 GG 5f ft 1:021 h

De Novo
Ch. g. 2, by Pat K—Sunny Sailing, by Sunny Outcome
Br.—Blain Rebecca (Cal) 1982 1 M 0 0
Own.—Blain Rebecca 118 Tr.—Larsen Lavar $16,000
Lifetime 1 0 0 0
11Jun82-4GG 5½f :222 :463 1:064ft 22 118 11 2hd 44 68 Winland W M8 [S]Mdn 63 Homers Brother,It'sABear,L'Ctricle 8
Jun 3 GG 5f ft 1:033 hg May 27 GG 4f ft :501 hg May 20 GG 3f ft :37 h Apr 24 GG 2f ft :253 hg

Pet Me
B. c. 2, by Petrone—Milady Grace, by Ky Pioneer
Br.—Marshall J (Cal) 1982 2 M 0 0
Own.—Marshall J 118 Tr.—Hess R B $16,000
Lifetime 2 0 0 0
16May82-4AC 2f :223ft 16 118 4 86¼ 85½ Fuentes A P9 Mdn — ArticultdProof,EstrnBttor,FbldImg 9
25Apr82-4AC 2f :222ft 12 118 8 74¾ 76½ Fuentes A P4 Mdn — BigDddyDon,ArticultdProof,FrHous 8
Jun 12 GG 5f ft 1:05 h Jun 6 GG 3f ft :364 hg May 29 GG 4f ft :502 h ●May 13 AC 3f ft :35 hg

Hazit
B. g. 2, by Flush—Chizzy Wink, by Johns Jest
Br.—Cassatt Judy (SC) 1982 1 M 0 0
Own.—Caravelli & Tribulato 118 Tr.—Conner Jack $16,000
Lifetime 1 0 0 0
5May82-6GG 5f :221 :461 :584ft 73 118 118½1116 1014 920 Campbell B C11 Mdn 70 TwlghtCrr,CnsltngSrgn,HmrsBrthr 11
Jun 12 GG 4f ft :491 hg Jun 5 GG 5f ft 1:04 hg May 31 GG 5f ft 1:023 h May 23 GG 4f ft :494 h

Kevin The Great
Dk. b. or br. g. 2, by Blazestone—Hoist Belle, by Hoist Away
Br.—Shindel M (Cal) 1982 4 M 0 1 $438
Own.—Shindel M 118 Tr.—Caton Dent $16,000
Lifetime 4 0 0 1 $438
26May82-3TuP 5f :223 :471 1:003ft *7-5 119 42¼ 41½ 52¼ 42 Sanchez J M2 M10000 72 Thor Son, Gar, Glory Spot 6
12May82-3TuP 4½f :232 :471 :533ft *9-5 119 6 52¼ 34 36½ Sanchez J M10 M10000 •77 InattentiveFox,Jlim,KevinTheGret 10
28Apr82-3TuP 4½f :224 :454 :521ft 8½ 119 5 77 610 69¼ Stallings W E2 Fut Trl 81 BeuTwister,SydneyVonGlick,Wlthy 9
14Apr82-3TuP 4½f :221 :461 :53 ft 11 119 5 914 712 64½ Stallings W E5 Mdn 82 TwoBreds,TypiclPro,Tringle'sFirst 10
Jun 13 GG 4f ft :50 hg Jun 6 GG 4f ft :511 h Apr 26 TuP 3f ft :364 h

Sporty's Star
B. c. 2, by El Diplomatico—Star Cos, by Mr Thong
Br.—Sport o King Ranch Inc (Cal) 1982 2 M 0 0
Own.—Valov J 118 Tr.—Wheeler R C $16,000
Lifetime 2 0 0 0
28May82-1Fno 4½f :231 :541ft *2¼ 1135 9 98 88¾ 76 Pedroza M A8 M15000 87 SusitnaSdie,Checoth,D.B.B.'sDrem 10
12May82-5Fno 4½f :231 :531ft *9-5 118 8 819 818 820 Stubblefield DP1 Mdn 78 BountifulArt,ChinofStrs,‡TnklTrsur 8
May 4 Fno 4f ft :471 hg Apr 27 Fno 3f ft :394 h

King's Fleet
Dk. b. or br. g. 2, by Native Fleet—Brown Glaze, by Donut King
Br.—Arons Joyce L (Cal) 1982 1 M 0 0 $65
Own.—Lotspeich M E 118 Tr.—Wahl Wayne $16,000
Lifetime 1 0 0 0 $65
28May82-1Fno 4½f :231 :541ft 9½ 118 6 53¾ 67 52¾ Memis G S10 M15000 90 SusitnaSdie,Checoth,D.B.B.'sDrem 10
May 18 Fno 3f ft :36 h May 15 Fno 3f ft :36 hg May 11 Fno 4f ft :514 h May 5 Fno 3f ft :39 h

Rudolph B. c. 2, by Olympiad King—Fragrant Flower, by Windy Sands
Br.—Boone Judge Mrs W B (Cal) 1982 2 M 0 0
Own.—Rudolph E 118 Tr.—Nolan William $16,000
Lifetime 2 0 0 0
21May82-6GG 5f :22 :46 :58^{3}ft 35 118 8$^{4\frac{1}{2}}$ 8^9 9^{14} 8^{17} Stallings W E^1 Mdn 74 CnsltngSrgn,HmrsBrthr,McKnlPrk 11
5May82-6GG 5f :22^1 :46^1 :58^{4}ft 24 118 10$^{8\frac{1}{4}}$ 9^{13}11^{15} 8^{20} Tohill K S^9 Mdn 70 TwlghtCrr,CnsltngSrgn,HmrsBrthr 11
Jun 12 Pln 4f ft :49^4 h May 17 Pln 4f ft :49^3 h May 3 Pln 3f ft :37 hg Apr 26 Pln 5f ft 1:02^1 h

Pardon Me Mac B. g. 2, by MacArthur Park—Pardon Me Miss, by Tulyar
Br.—Johnson & Rancho Paraiso (Cal) 1982 0 M 0 0
Own.—Johnson R C 118 Tr.—Murphy Chuck $16,000
Lifetime 0 0 0 0
Jun 14 Pln 4f ft :49^3 h Jun 3 GG 5f ft 1:04^3 b May 29 GG Tr. 5f ft 1:01^4 hg May 22 Pln 4f ft :48^1 b

An especially instructive illustration is in order. It comes by way of professional handicapper Ron Cox, of San Francisco, who surrendered a teaching career to racetrack pursuits, and while making an income at the windows, has for years provided comprehensive information services to handicappers on the Bay Meadows-Golden Gate Fields circuit.

The race was the fourth at Golden Gate on June 17, 1982, an awful maiden-claiming dash for two-year-olds. The past performances of the horses are listed here, but handicappers of genuine skill could not be expected to spot the generous winner. Cox himself would have been lost without the computer-generated information supplied by handicapping colleague Bob Healy.

By happy coincidence, the race was programmed the afternoon that handicapping author William L. Scott decided to apply his newly discovered form analysis methods to the nine-race card at Golden Gate. That the fourth race that day was unpredictable by conventional methods is reflected by Scott's unrestrained opposition to taking it seriously. Handicappers should pause now to read Scott's summary of the race in Figure 15. Speed and class handicappers can be assured none of these two-year-olds were possessed in the least of either speed or class.

As benefits practiced professionals, Ron Cox pays considerable attention to certain kinds of races that are often pitiably unpredictable by traditional practices. Cox is looking for prices, or more appropriately, for values. A specialty of his is the maiden claiming sprint, for which he has developed a speed handicapping method he calls "gap" handicapping, a technique that relies on identifying larger-than-standard differences in the class-distance par times of maiden claiming price ranges. The method works well at all tracks outside of New York and southern California. Cox's method concentrates on maiden claiming horses dropping down. Cox's windfall here, however, appeared from a vastly different information source. It came from Bob Healy's microcomputer.

Handicappers should examine Table 6, a portion of a computer printout of trainer statistics for trainer R. B. Hess, who started two horses in this cheap sprint. The complete printout reveals how well Hess has performed at Golden Gate during 1982 overall, in Exacta races, with maidens, with claimers, with starters, and with allowance and handicap horses, as well as how well the jockeys Hess uses have performed on his horses. Only two weeks were left at Golden Gate on June 17, enhancing the data's reliability.

Figure 15
Author William L. Scott's analysis of a race at Golden Gate, unplayable by classical handicapping or systematic methods but not by information management.

4th Golden Gate

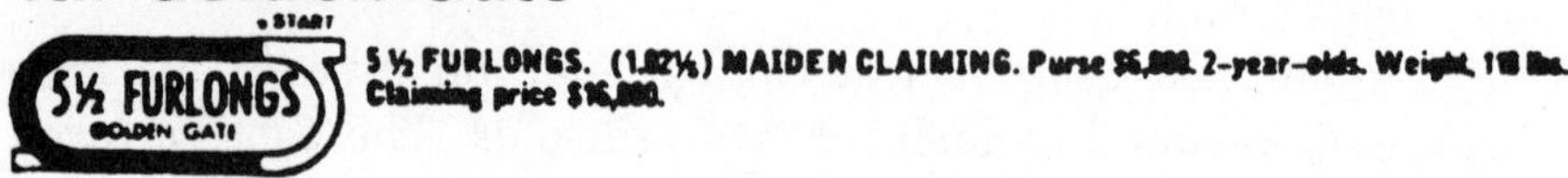

5½ FURLONGS. (1.02⅘) MAIDEN CLAIMING. Purse $6,000. 2-year-olds. Weight, 118 lbs. Claiming price $16,000.

The fourth race was a maiden claimer for two-year-olds at five and a half furlongs. This early in the two-year-old season, if young horses are already running for a claiming price as moderate as $16,000 you should realize that playing almost anything would be a wild gamble. If something did look exceptionally good on our form rules, we would not be compelled to pass. But here, neither the favorite, Bam's Rascal, nor the second betting choice, Pardon Me Mac, had ever raced before. Would you want to risk your money on either of these two?

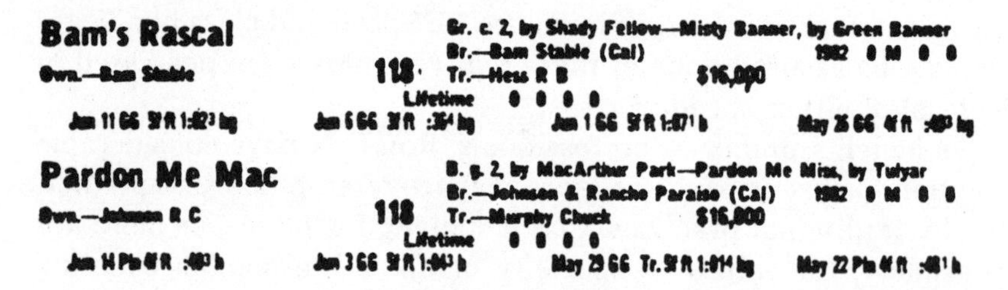

Bam's Rascal — Gr. c. 2, by Shady Fellow—Misty Banner, by Green Banner
Own.—Bam Stable — 118 — Br.—Bam Stable (Cal) — 1982 0 M 0 0
Tr.—Hess R B — $16,000
Lifetime 0 0 0 0
Jun 11 GG 5f ft 1:02³ hg — Jun 6 GG 3f ft :36⁴ hg — Jun 1 GG 5f ft 1:01¹ b — May 26 GG 4f ft :49³ hg

Pardon Me Mac — B. g. 2, by MacArthur Park—Pardon Me Miss, by Tulyar
Own.—Johnson R C — 118 — Br.—Johnson & Rancho Paraiso (Cal) — 1982 0 M 0 0
Tr.—Murphy Chuck — $16,000
Lifetime 0 0 0 0
Jun 14 Pln 4f ft :49³ b — Jun 3 GG 5f ft 1:04³ b — May 29 GG Tr. 5f ft 1:01⁴ hg — May 22 Pln 4f ft :48¹ h

While both of these colts show five furlong workouts, which is to their credit, the fact that their owners have placed a very modest price tag on them in their initial start tells you all you need to know. In *Investing at the Racetrack,* the critical nature of the unknown factor was stressed, and advice there was not to play any horse whose form was unknown. While that harsh rule has been softened because of the shining value of the five furlong workout, a first-time maiden claimer starter is so unknown, even with the good workout, that you might as well try to pin the tail on the donkey as to fathom the winner. If a young, unraced horse showed a really impressive five furlong workout, he would not be running for a price tag. It is as simple as that.

While no one likes to pass three races in a row, this is still a key discipline factor in making money at the racetrack. You have to keep away from the dubious gambling propositions. While you may be able to make an argument for playing the second or third races on this card, which would be foolish enough, there is no way to soundly play this event. You might even expect one of the entries to romp in with the kind of big price that is shown.

FOURTH RACE **5 ½ FURLONGS. (1.02⅘) MAIDEN CLAIMING. Purse $6,000. 2-year-olds. Weight, 118 lbs. Claiming price $16,000.**

Golden Gate

JUNE 17, 1982

Value of race $6,000, value to winner $3,300, second $1,170, third $840, fourth $540, fifth $150. Mutuel pool $96,448.

Last Raced	Horse	Eqt.	A.	Wt	PP	St	¼	⅜	Str	Fin	Jockey	Cl'g Pr	Odds $1
16May82 4AC8	Pet Me		2	118	4	7	$6^{1\frac{1}{2}}$	4^{hd}	2^2	1^2	Munoz E	16000	21.70
11Jun82 4GG6	De Novo		2	118	3	1	$1^{\frac{1}{2}}$	1^3	1^5	2^3	Winland W M	16000	3.30
	Bam's Rascal		2	118	1	5	$4^{\frac{1}{2}}$	5^2	5^1	3^{nk}	Caballero R	16000	2.60
5May82 6GG9	Hazit		2	118	5	8	$7^{\frac{1}{2}}$	$7^{1\frac{1}{2}}$	$6^{1\frac{1}{2}}$	4^2	Campbell B C	16000	37.50
	Pardon Me Mac		2	118	10	3	$3^{1\frac{1}{2}}$	2^{hd}	3^1	5^{hd}	Lamance C	16000	2.90
26May82 3TuP4	Kevin The Great		2	118	6	6	8^3	8^3	8^2	6^{hd}	Stallings W E	16000	9.00
27May82 4GG7	A Faster Piaster		2	115	2	2	5^2	6^{hd}	7^{hd}	7^{nk}	McGurn C	16000	22.60
28May82 1Fno5	King's Fleet		2	118	8	9	9^1	9^2	9^6	8^2	Troestch R	16000	8.60
21May82 6GG8	Rudolph	b	2	118	9	4	2^{hd}	3^1	4^{hd}	9^{13}	Tohill K S	16000	15.70
28May82 1Fno7	Sporty's Star	b	2	118	7	10	10	10	10	10	StubblefieldDP	16000	23.40

OFF AT 2:36. Start good. Won driving. Time, :22⅘, :47, 1:00⅘, 1:07⅘ Track fast.

$2 Mutuel Prices:			
5-PET ME	45.40	14.40	7.40
3-DE NOVO		5.40	3.80
1-BAM'S RASCAL			3.60

B. c, by Petrone—Milady Grace, by Ky Pioneer. Trainer Hess R B. Bred by Marshall J (Cal).

PET ME, reserved early, rallied slightly wide into the stretch, lugged inward through the drive despite left-handed urging and was up in time. DE NOVO dueled for the early lead, drew clear on the turn, opened a long lead into the stretch but could not last. BAM'S RASCAL, never far back, finished evenly in the stretch run. HAZIT, outrun early, came out for the drive but lacked the needed response. PARDON ME MAC attended the pace from the outside to the stretch and weakened. KEVIN THE GREAT was wide and showed little. KING'S FLEET was outrun. RUDOLPH had speed to the stretch and faltered.

Owners— 1, Marshall J; 2, Blain Rebecca; 3, Bam Stable; 4, Caravelli & Tribulato; 5, Johnson R C; 6, Shindel M; 7, Hi Card Ranch; 8, Lotspeich M E; 9, Rudolph E; 10, Valov J.

Scratched—Brandon's Folly (11Jun82 4GG5); Lucky People (10Jun82 6Hol); Time To Star.

Handicappers should scan the trainer-jockey stats for maiden races and return to the past performances tables for the June 17 race. Now what do they see?

Cox was able to use the trainer-jockey data to find a $45 winner far beyond the scope of conventional handicapping practice. He zeroed in on these data items:

1. Hess's regular rider is Raul Caballero, who until June 17 had ridden 83 of the trainer's 118 starters and had won with 6 of

Table 6
Trainer Statistics for R. B. Hess at Golden Gate

Trainer		*Total*				*Maiden*			
Jockeys	*NDH**	T	*W*	*P*	*%W*	T	*W*	*P*	*%W*
Hess, R. B.	(12)	118	18	12	15.3	41	6	2	14.6
Baze, R.	(0)	1	0	0	0.0	1	0	0	0.0
Caballero, R.	(6)	83	11	10	13.3	23	3	2	13.0
Dillenbeck, B.	(0)	1	0	0	0.0	1	0	0	0.0
Munoz, E.	(4)	13	5	2	38.5	3	2	0	66.7
Nicolo, P.	(1)	1	1	0	100.0	0	0	0	0.0
Noguez, A.	(0)	1	0	0	0.0	0	0	0	0.0
Simpson, D.	(0)	6	0	0	0.0	6	0	0	0.0
Sorenson, O.	(0)	4	0	0	0.0	0	0	0	0.0
Winick, D.	(1)	8	1	0	12.5	3	1	0	33.3

* Number of different horses that were winners

his 12 different winning horses. Caballero's win percent for Hess was 13.3, approximating the trainer's average. Caballero was named on a Hess two-year-old in the 4th June 17, Bam's Rascal, the race favorite.

2. No other jockey had ridden more than a single winner for Hess, excepting Eric Munoz, who mounted four different winners for the trainer. Munoz' win percentage on 13 Hess starters was 38.5, about two-and-one-half times the trainer's average.
3. With maidens, Caballero was 3 for 23 on Hess horses, duplicating his average. Munoz was used only three times with maidens, but he had won twice.

On the last point, Cox also recalled that Munoz' two maiden winners not only paid big mutuels but had shipped to Golden Gate from other race tracks. Munoz' horse for Hess in the 4th at Golden Gate on June 17 had shipped there from Caliente, the Mexican track just south of San Diego, where it had lost twice at two furlongs against straight maidens.

From these diverse data items, Cox intuited that Hess was loading his longshots into the competition from races outside of the regular circuit and using Munoz to pull the trigger. With regular rider Caballero on a Hess favorite today and Munoz again poised on a long-priced Hess shipper, Cox noticed a repeating pattern, and he played it.

Does the data support the inference and the bet? On both counts, absolutely. If the price were low, the data would support the inference but not the bet. But the odds were 21.7 to 1, and that practically demands the bet.

This is a gorgeous illustration of information management and a database approach to problem-solving in handicapping, pulled off by a practiced handicapper up-to-date on the information fronts of his pastime and assisted by a colleague with a computer. The case is hardly an isolated example. The rewards are obvious. By any methodological standard or conventional logic, the Golden Gate race is unplayable, as Scott insisted. But Cox is practicing high-tech handicapping in the information age. He searches the trainer database, looking for intriguing patterns, trying to shed light on a horrid field. It works.

In deciding to back *Pet Me*, Cox was hardly relying on deductive reasoning skills. His decision instead was highly intuitive, a type of reasoning a substantial body of research has shown to be the thinking skills much preferred by many experts in various fields: expertise and intuitive thinking do go hand in hand. Their association often counts big in the information management approach to handicapping.

Numerous opportunities of this kind await handicappers on all circuits who have access to databases and the savvy to use the information smartly. Both the knowledge and experience must be present. Neither is sufficient without the other.

The first priority is a high-tech imperative, designing and constructing the personal handicapping database. Handicappers should not be misled. It is an absorbing chore that begins small, grows over time, and takes years to complete. The update and revision functions never cease. Constructing the actual physical database is a long-term project, far more technically complicated than developing applications programs. Yet high-tech handicappers cannot afford to be without the database capability in the end. The commitment to the database does make the difference. It is the future in this specialized field.

WHAT IS A DATABASE?

No high-tech term has been mangled more severely than the database. Many owners of microcomputers boast frequently about their databases, but very few actually have one. A database is not merely a large body of data about a subject, such as handicapping, horses, trainers, or pedigrees. In fact, data filed in a computer according to the instructions of an applications program are *not* a database or part of one. Those data constitute the files of an applications program, or several applications programs, and are stored in a file structure. No one can retrieve or use the data, except by following the program's instructions. That is practically the contradiction of the database and its purposes. Thus, the fasionable instinct to refer to any large repository of computer-based information as a database is wrong-headed and misleading.

The guru of databases in the computer industry is James Martin, a research scientist with IBM for nineteen years and currently an independent consultant to corporations that are struggling to convert a generation of applications programs locked in "file structures" to database storage and technology. Martin has defined the database as follows:

> A collection of data designed to be used by different programmers is called a database. We will define it as a *collection of interrelated data stored together with controlled redundancy* to serve one or more applications in an optimal fashion. The data are stored so that they are *independent* of programs which use the data. A common and controlled approach is used in adding new data and modifying and retrieving existing data within the database.

Consider a second definition, supplied in a text by Texas A & M professor Raymond McLeod, Jr., on management information systems:

> Essentially the *database approach* involves the use of logically integrated files to meet the information needs of [handicappers]. The idea is not to build one giant file containing everything—that would be impossible. Rather, the contents of files are interrelated in order both to reduce redundancy and to facilitate the retrieval of data from the various files to meet information requirements.

The term redundancy, which appears in both definitions, refers to the use of the same data items in several applications programs, such as speed handicapping, class handicapping, pace analysis, or ability times. Each would contain, for example, the data item: *adjusted final times.* The database controls the redundancy and eliminates it entirely for most data items. Adjusted final times would be stored once but retrieved repeatedly for many different applications. This enhances the storage capacity and flexibility of the system. Data items are no longer repeated in the physical system for program after program.

Yet, the crucial characteristic of the database is *data independence.* Data are stored independent of a computer program's instructions. That is, the data exists outside of a computer program's file structures. Once data independence is established, a number of diverse applications programs can use the same data items again and again, and in varying combinations. Handicappers building a database can develop future applications programs as they please. The programs might focus variably on class, speed, pace, form, pedigree, trainers, or combinations of handicapping factors. Regardless, handicappers simply dip into the existing database to find the data items required to run the program.

Another characteristic of the database indispensable to handicappers is the *logical integration* of the data files. This means simply that the database is structured so that all the files can be interrelated in logical ways. This permits handicappers to query the database or ask spontaneous questions, such that the computer can search the interrelated data files, combine the appropriate data items, and answer the question.

Imagine the forlorn handicapper who just doesn't know whether Horse A, an intriguing longshot, has ever won at today's route following a layoff of six months or longer. The past performance tables do not show a clue. High-tech handicappers query the database: Get all the horses with first initials A to L that have won at eight furlongs or beyond following a six-month layoff. Within seconds handicappers realize whether Horse A has won "fresh" at the route before.

This capability is impossible by utilizing the file structures of applications programs. Those data can be controlled and manipulated only according to the sequence of a computer program's instructions. Handicappers need more flexibility than that. The problem-solving phase of the information management approach is troubled consistently by unpredictable questions. The query function of database approach solves those troublesome problems.

In sum, a true database is often characterized by three coexisting attributes:

- Data independence
- Logical integration of the data files
- Controlled redundancy of data items

THE DATA HIERARCHY

The high-tech handicapper's main task is the *logical* design of a personal database. A few organizing principles can guide the design work. One is to begin with a manageable amount of data. Many handicappers will want to collect data not trapped in the past performance tables, such as speed figures, trip notes, and trainer stats. Another organizing guideline is that data exists in a hierarchy. Handicapping data should be organized according to that hierarchy.

The smallest unit of data that has meaning to users is called a data item. It is a single piece of data, or a datum. The past performance tables contain dozens of data items, which handicappers can define in terms of their actual or logical *meaning*. Final time is a data item that has obvious meaning perhaps, but *adjusted final time* begs for finer definition. *Key race*, an abstraction from the running lines, also represents a data item from the past performances. It might be defined as a horse's best recent race against the most advanced competition. It might also be defined as a race containing two or more next-out winners. Handicappers who design databases will be providing the *logical* meanings of numerous data items.

Notice that data items are supposed to have meaning to users (handicappers). Thus the first call position of the second most recent race is not a data item, as it has little meaning for handicapping purposes. But the data item called speed points, which includes the first call position of the second race back, would be a greatly meaningful data item. Handicappers should understand that they are dealing here with meaningful data that have logical definitions and not meaningless items of data.

Data items related to particular topics are assembled into *records*. The array of data items in the past performance tables comprise the performance record for a specific horse. Similarly, the data items describing a jockey's performance would comprise the record for a jockey. Records are meaningful data items that have been grouped together. Data items can be grouped together in various ways to form different records. The data items associated with the points of call in the past performance tables can be grouped together to form a pace record for a horse. They can also be combined to reflect form defects and form advantages.

When the records for a topic are grouped together, they constitute a *file*. Thus the past performance records of all the horses competing at a racetrack constitute the past performance *file* for the track. Files can be partitioned logically, as handicappers prefer. The past performance file might be partitioned into the file for four-year-olds and up, the file for three-year-olds, and the file for two-year-olds. Handicappers might also construct the filly and mare file, the colts and geldings file, the turf horse file, or any other logical arrangement of records. Note that the same data items and records reappear *logically* in more than one of the above files. In the actual physical arrangement and storage of the database, however, data items would appear just once.

The hassle of defining data items so that they have common interpretation and use among database users has been overwhelming in the large corporations where for years squadrons of computer programmers have filed data to serve the purposes of their applications programs. The same data items, filed variously, have obtained various definitions throughout the firm. The conversions to database systems are now costing millions and lasting years, due in considerable measure to these logical inconsistencies. Few corporations have been unscathed.

Handicappers who develop personal databases do not face severe problems of data item definition or file conversion, although data stored in current applications files should be converted to database files. Participants on handicapping teams face stickier conversion problems. The problem of defining data items in commonly acceptable ways is sure to rear its ugly head. Needless to say, not all handicappers define data items such as speed points, pace ratings, adjusted final times, daily variants, or average earnings in the same way, not to mention higher-order abstractions in handicapping such as class, sharp form, or related distances.

THE DATABASE ORGANIZATION

The contents of handicapping databases should evolve incrementally from a master plan of development. For many handicappers early stages of development might concentrate on personal databases that reflect well-defined analytical purposes. Class handicappers naturally will look to data items that reflect thoroughbred class as well as related factors that influence demonstrated class, notably form and pace. Speed handicappers will build their first databases on speed, pace, and early speed relationships, and probably trip data, as trips can seriously affect speed figures. A few handicappers might choose to develop giant handicapping databases during a relatively short, intense timeframe, say a year, but most will attach separate smaller databases as time goes by.

My personal biases take me along an opposite direction. I intend to begin with specialized databases that emphasize data relationships not contained in the past performance tables or results charts. By specialized databases I mean trainer statistics, pedigree data, trip notes, and a special kind of performance record for horses.

Whenever the starting points, the ultimate objective is to build giant databases bulging with all the data items of handicapping. Separate databases can be organized by subjects or by applications (handicapping methodology). Subject databases will be larger than applications databases and will contain far more data items. The development timeline will therefore be longer. The update function will be more consuming. Applications databases, alternately, contain only the data items relevant to the handicapping schemes of interest. They will be narrower in scope and complexity, and easier to build. The update function will not be nearly so demanding. The large organization builds subject databases almost exclusively, as it has too many applications to handle. Each application draws from several subject databases. Handicappers do not suffer the problem of too many handicapping applications. The fundamental handicapping applications can be well defined and delimited. Moreover, all the applications databases use data items associated with horses. Thus one developmental approach builds two subject databases at first, for horses and trainers, and adds data items to each whenever applications programs are masked against the subject databases, but several data items necessary to run the applications are missing.

Below are five subject databases and six applications databases by which handicapping data items, records, and files can be organized.

Subject databases
- Horses
- Trainers
- Jockeys
- Owners
- Breeders

Applications databases
- Class appraisal
- Form analysis
- Speed handicapping
- Pace analysis
- Trainer evaluations
- Pedigree evaluation

Several applications databases are often combined to form a single larger database, as when class, form, and pace data items are combined to form a class appraisal database. If speed, pace, and trip applications data items were then combined to form a speed handicapping database, several of the pace data items would be the same as those contained in the class handicapping database. Some pace data items might be defined differently in the two databases. The duplications would be *logical* but not *physical*. The same data items would be stored only once in the physical system. Theoretically, all the handicapping data items might be contained in two giant-sized subject databases, for horses and trainers. This would prove unwieldy, impractical, and unmanageable for development purposes. As points of departure, high-tech handicappers should probably determine to build first versions of the largest two subject databases, horses and trainers, and one of the two fundamental applications databases, either the class handicapping database, embracing form and pace data items besides, or the speed handicapping database, embracing pace and trip data items besides. Other separate databases can be added later.

PHYSICAL AND LOGICAL DATA ORGANIZATION

The data items, records, and files to be included in each database will be defined logically by individual handicappers. Each handicapper includes whatever data items he prefers to include. The organization of a database is therefore a *logical* expression of handicapping relationships, as seen by the individual. Moreover, the logical structure of the database does not resemble the physical representation or physical storage of the data items in the computer.

Handicappers who will develop databases need to impress upon themselves at the outset this critical distinction between logical and physical data organization.

Logical data organization is the handicapper's view of the handicapping data. Physical data organization is the arrangement of the handicapping data in computer storage. The two views are separate and greatly distinct. Handicappers can forget about physical data organization and computer storage during the logical phases of data organization. Experienced handicappers in that way will be capable of structuring a comprehensive handicapping database, as the task is entirely conceptual and logical. It is not technical. Later, when the logical databases have been structured and the data items operationally defined, handicappers can consult with computer technicians as to the physical representation and storage. The point is that technical considerations come later, not earlier, in database design.

REDUNDANCY*

Figure 16 shows a logical view of a computerized information system without a database. It consists of several applications programs. Several data items are repeated in several programs. Figure 17 shows the same system with a database. The applications programs are masked to the database, drawn from a single pool of items. Handicappers should study the two systems carefully. The first is characterized by data redundancy and inflexibility. Data items are trapped in their programs. They cannot be moved about.

Figure 16
A system without a database: redundancy and inflexibility. Imagine 1000 such files. James Martin, *Principles of Data-Base Management*, © 1976, pp. 37, 38.

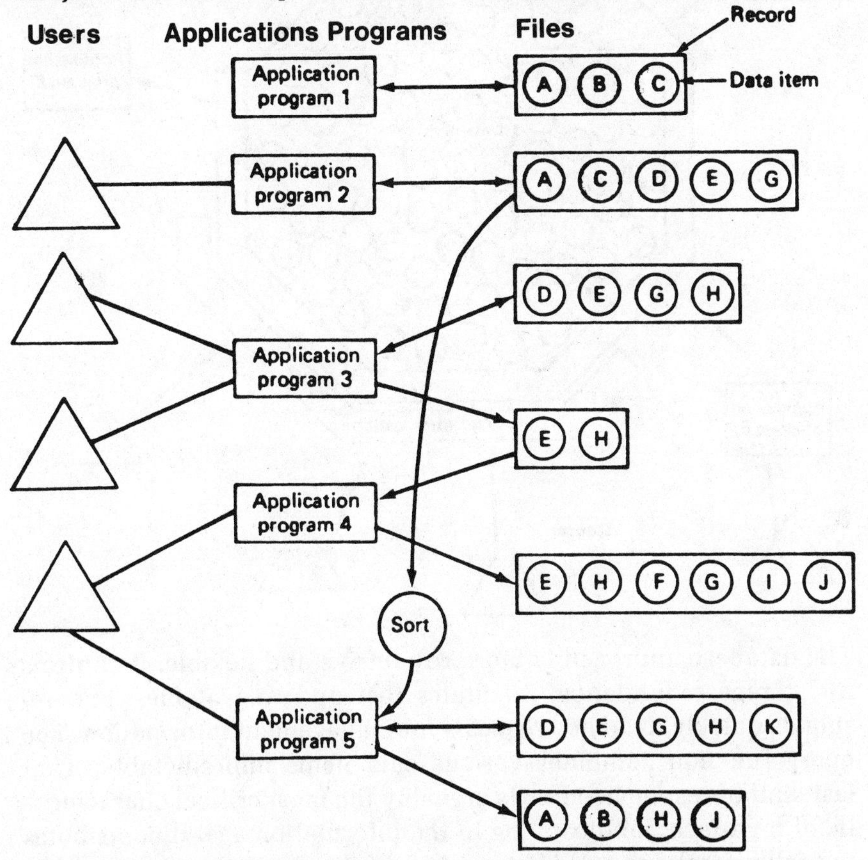

*Much of the technical material of this section is based on James Martin's text, *Principles of Data-Base Management*, Prentice-Hall, Englewood, New Jersey, 1976.

Figure 17
The system of Fig. 16 with a database: more responsive, lower costs.

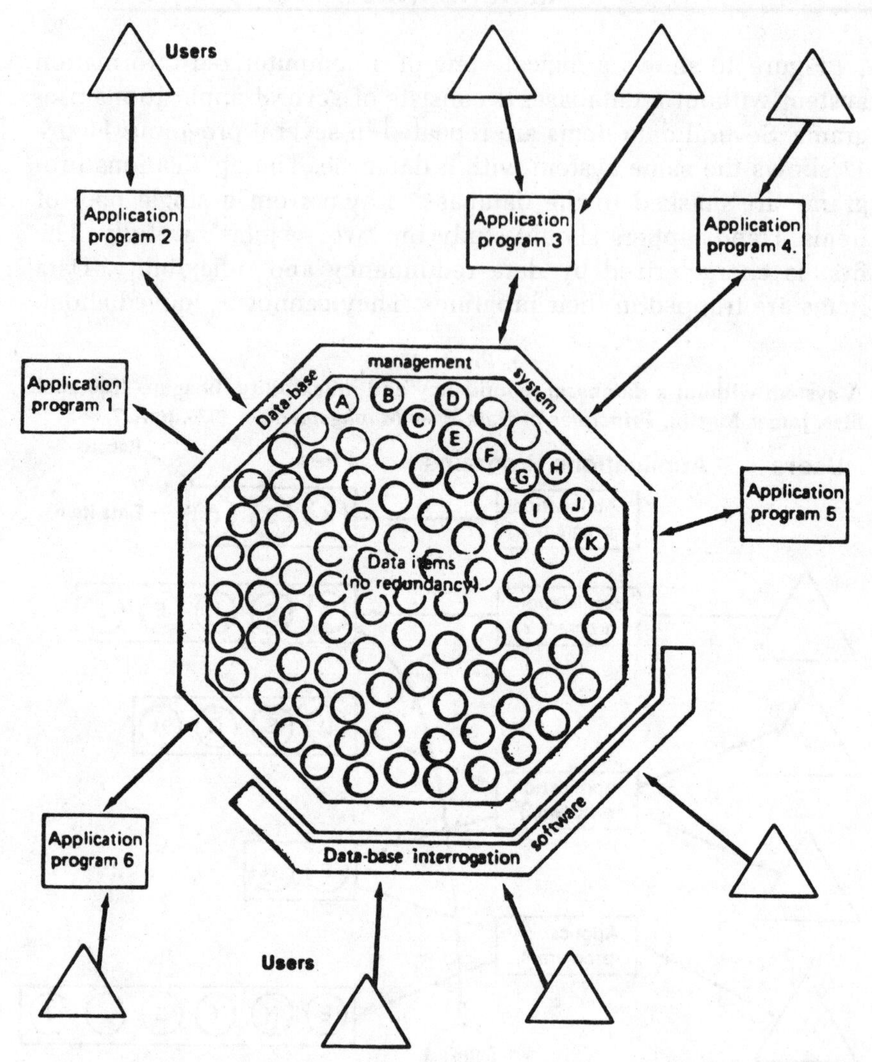

The database approach is more responsive and flexible. It controls for storage redundancy, facilitates fast updating of files, and responds quickly to *ad hoc requests* for management information. The query function combines various data items unpredictably. That last kind of responsiveness is arguably the most critical characteristic of high-tech handicapping in the information age. It contributes to problem-solving and informs decision-making.

In high-tech handicapping systems many of the handicapping data items will be duplicated in multiple applications programs. Without a database approach the items will therefore be duplicated numerous times in storage. Handicappers need consider only a horse's name or a race's date. This duplication of data items in multiple applications programs is called *redundancy*.

As Figure 17 reveals, the database approach controls the redundancy. Uncontrolled redundancy has several disadvantages:

1. Critical storage capacity is reduced, sometimes severely
2. Multiple update functions are continually needed to make the files current
3. To the extent the system as a whole defines the same data items differently or updates the files at different times, the system can give inconsistent or inaccurate information

Redundancy is particularly harmful in a system that requires relatively large amounts of updating. Computerized handicapping information systems require repeated updating, even daily updating. Imagine the high-tech handicapper's chagrin at updating the numbers, times, figures, and statistics contained in multiple files.

In the database approach the computer assembles various records from the same pool of data items. The same records or data groupings can be used by multiple applications programs. With a database approach several members of a handicapping team could fish various data items from the same pool to write various applications programs. The data items remain independent of the computer programs that are using them.

AD HOC REQUESTS

Ad hoc requests are spontaneous unpredictable queries to the database for information needed to solve specific handicapping problems or make decisions. The query is sometimes referred to as a *call* to the database. In the large organization, information that is generated by query is considered management information, or information that managers (handicappers) can use to make *ad hoc* decisions. Handicappers should think of the query capability in the same important way. Handicappers can make as many "calls" to the database as they please. Obviously, the larger the database, or the more data items contained in it, the greater the handicapper's flexibility in making queries, or asking questions.

Has Horse A ever won at today's new distance? Has it won only when helped by a favorable track bias? The database knows.

Specialized software can extract from a properly structured database any combination of data items needed to respond to a handicapper's query or call. Handicappers who value the query function of the database approach will prefer larger databases containing as many data items as feasible. Theoretically, if all the possible handicapping data items were stored in the database, all plausible queries or calls would be answered immediately.

A query or call to the database can be embedded in an applications program, and many are (get Horse B's highest speed figure) but need not be. Other specialized software permits handicappers to interact directly with the database, without the interference of an applications program. The ability to manage the database and interrogate it without writing computer programs has become the great leap forward in management information systems in the large organization, because the information can be retrieved spontaneously. It is no less a leap for high-tech handicappers.

SCHEMATA AND SUBSCHEMATA

The handicapper's logical view of the database organization is called the *schema*. To say it differently, the schema is a formal representation of the database and its structural relationships. The schema describes the logical organization of the database. It includes a chart of the types of data that are used. It gives the names of the data types and it specifies the structural relationship among them. The schema does not provide the values for any data items, but is a broader framework into which the data values fit. The database framework remains relatively constant, while the data item values change repeatedly, perhaps daily. Figure 18 shows a schema for a class handicapping database. Five class relations are specified: horses, conditions of eligibility, speed and pace figures, ability times, and stakes credentials. For each relation several data types are shown. Other relations and data types are possible, such as earnings and consistency, but are not included in this logical view of

Figure 18
A Schema or Logical View of a Class Appraisal Database. The Schema Includes the Types of Data To Be Included and the Attributes of Interest.

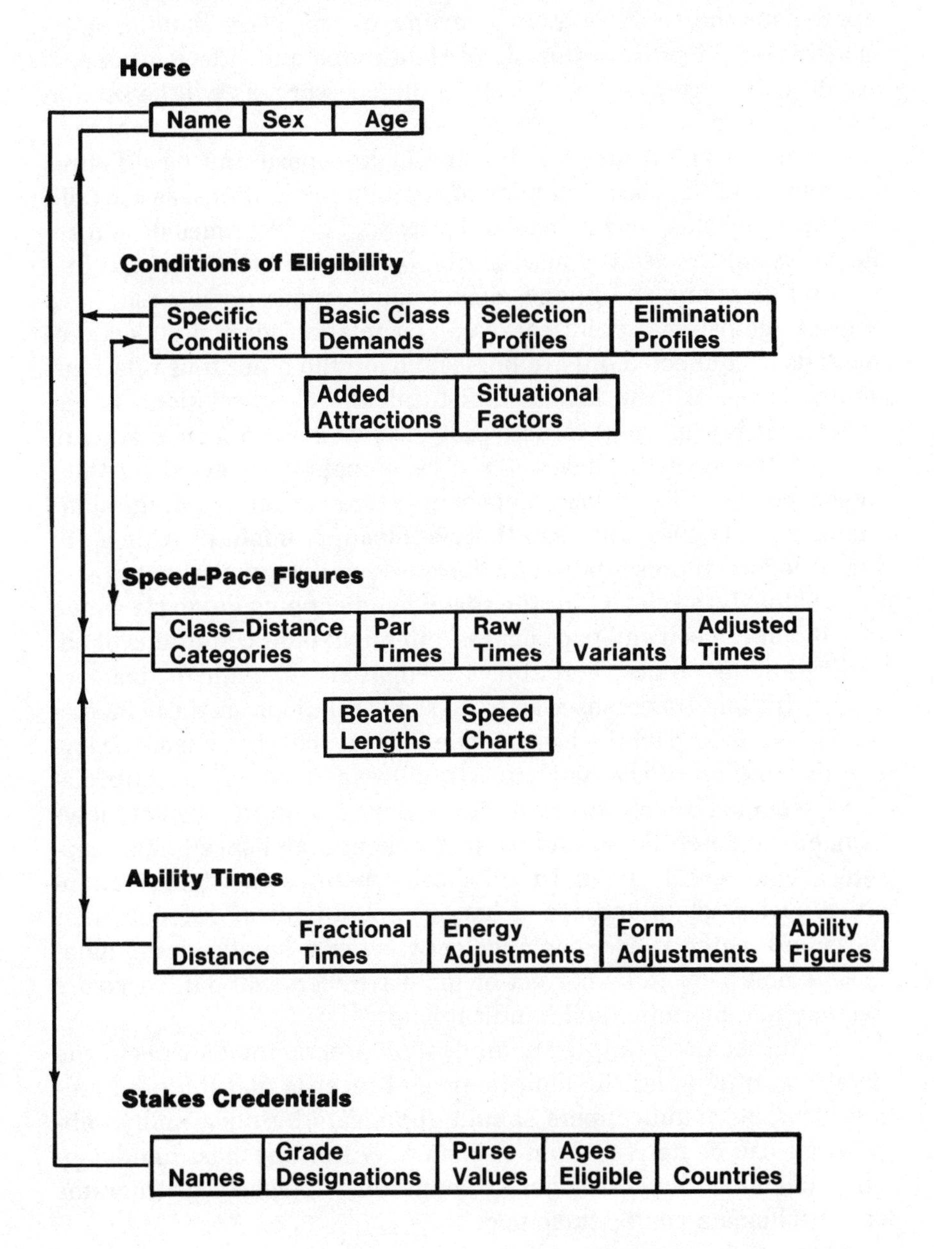

class. The data types can be broken down into numerous more specific data items.

The solid lines show how the five class relations in the class handicapping database should be interrelated. Handicappers can appreciate that a full-blown charting of the class handicapping schema would portray dozens of data items and interrelations. It would look messy and is ultimately unnecessary, as will be seen in the next chapter.

By the structure indicated in the figure, speed and pace figures are related to the class demands of eligibility conditions, as are ability times. By this schema speed figures and ability times do not reflect class of themselves necessarily, but should be interpreted in a context of racing conditions. Speed handicappers, of course, have argued against that point forever. The relation termed stakes credentials is connected only to horses. So are the other four relations. Thus, *horses* are the logical link to all the class relations in the schema. It is important to note that the schema is a *logical* view or representation of the indexes and relationships reflected by thoroughbred class. The "class" data is not stored or related in the same manner in the computer. Another handicapper might provide a different logical representation of the class handicapping database.

Obviously, a schema for the speed handicapping database would also include speed and pace figures, but might omit conditions of eligibility, ability times, and stakes credentials. It might instead include trips and track biases as basic speed relations, as these factors are known to impact the figures. The point is that the schema or logical description of the database structure, as well as the interrelations specified among the data items, depends on the logical views of handicappers. Class handicapping schemas and speed handicapping schemas will vary in their logical structures and data relationships, just as class and speed handicappers tend to differ in their ideas and methods. No single logical view of a handicapping database is held here to be correct or inclusive. Personal databases are best defined by individual handicappers.

A subschema is simply an application programmer's view of the database. It includes the data items and records that the programmer uses. As handicappers should appreciate by now, many subschemas can be derived from a schema. Within the class handicapping schema outlined in Figure 18 numerous applications programs or subschemata can be imagined.

One application might determine whether horses satisfied the class demands of a particular variation of race conditions. A second might produce a horse's top three pace ratings. A third might attempt to relate speed figures to the class demands of eligibility conditions, identifying the horses that have earned the highest figures under the most competitive conditions. Each applications program constitutes a subschema of the database. Each would use a different subset of the data items representing our logical view of thoroughbred class.

In designing the logical structure of a database, theoretically high-tech handicappers can ignore the concept of a subschema. The preferred procedure is to design the database first and develop applications programs later. Once the database has been constructed, if certain data items required by a basic applications program are missing, the database design has been faulty, but the database remains flexible and extensive. New data items can be added at any time.

In practice, of course, databases will be constructed gradually. If handicappers will be relying on basic applications programs to find winners and the processing requirements of the method are severe, as with speed handicapping, they should surely include the data items required by the application program in the earliest stages of database design. Thus, the first phase of database design for a speed handicapper would represent the data items and data relationships required to produce the kinds of adjusted times or speed figures the handicapper fancies. On the other hand, if a handicapping method does not require the processing speed and capability of a computer, handicappers might continue to apply their methods manually and concentrate on building databases that can supply specialized information that will supplement the information provided by basic methods.

The next sections of this chapter outline schemata or logical views of the database organization for four separate handicapping databases and conclude with brief discussions of the specialized software that permits the amazing linkages and manipulations of practically all the data items in the separate databases. Happily, there is no need to chart the schemata and the interrelations of the data types, as was done in Figure 8.5. The most advanced of the database software now allows database design specialists to structure all data relations in the same way. The logical view of all data relationships can be that of a table consisting of rows and columns. The data relationships are held to be two-dimensional and *flat*, as if belonging to *flat files*. The table is called a *relation* and the table columns are referred to variously as domains, data types, or data items. The class relations and data types of Figure 8.5 can therefore be represented *flatly*, without connecting lines and arrows to show the interrelations as follows:

Table 7
Class Appraisal Database Schemata in *Flat, Two-Dimensional* Form

Horse (Name, Sex, Age)

Conditions of Eligibility (specific race conditions, basic class demands, selection profiles, eliminations profiles, extra added attractions, situational factors)

Speed-Pace Figures (class-distance categories, par times, raw times, daily variants, adjusted times, speed beaten lengths, speed charts)

Ability Times (distance, fractional times, energy adjustments, form adjustments, beaten lengths, ability figures)

Stakes Credentials (names, grade designations, purse values, age eligibility, countries)

In the next chapter handicappers will learn how current database technology allows users to link data items to one another in multiple ways. To be sure, the numerous data items for the above class relations could be related to each other *logically* in the database so that numerous diverse data items could be combined to respond to a particular query. The specialized database software controls and manages the physical linkages among the data items.

Suppose handicappers wanted to identify the horses that had run the fastest times under the most advanced nonwinners' allowance conditions. A call to the database might say: List the horses that have earned speed figures of 95 or higher under NW3xMC allowance conditions (for nonwinners three times other than maiden or claiming) or NW4xMC allowance conditions. A short list of horses would appear on the terminal screen. The list might include the potential winner in today's restricted stakes feature. Or, handicappers might want to query the database: Get the *listed* stakes of France, by purse values. Any logical handicapping question can be answered from the corresponding logical design structure of the database. This capability distinguishes high-tech handicappers from manual handicappers as the reliance on handicapping applications programs does not, and represents the most powerful information edge at the modern racetrack.

THREE SEPARATE SCHEMATA OF HANDICAPPING

The trend among progressive handicappers toward collecting information that is not trapped in the past performance tables sets a tone for database development. High-tech handicappers want fast access to data, information, and relationships not apparent in the past performances or results charts. As always, they seek hidden overlays.

The database is filled with them. Handicappers will have to decide which relations interest them most, what types of data to collect, and how to organize the data items so that the most intriguing relationships can be clarified quickly. That much is subjective.

Database design technology permits the data items in separate databases to be rated logically (see Chapter 9) and the logical data relationships can be retrieved in combination. As noted, the first step in logical database design is to draw the schemata or logical descriptions of the database. Charting is no longer necessary. All that is needed is to identify the relations of most interest and the types of data for each. Below are schemata for three separate handicapping databases. The three are called Racehorse, Trainer Performance, and Racetrack, and are arranged in *flat* form. The next chapter presents the tabular view of the schemata, where the specific data items are identified, data values can be inserted, and all the data item relationships established.

Handicappers will note the Racehorse database is not a past performance record for a particular horse, but rather an array of key performance indicators for every horse on the grounds. A tabular view of a Racehorse database will be shown in the next chapter. Below are the three schemata.

Racehorse

Horse (name, age, sex)

Class (competitive level, class rating, key race, back class, top ability figure, stakes designation)

Speed (speed figure, pace figure, pace rating, velocity rating, final quarter)

Form (wins when fresh, longest layoff/won, workout pattern, form cycle)

Distance (best distance, sprint winds, route wins, distance range)

Pedigree (dosage index, turf I.V., mud rating)

Handicappers would define the data types in sensible ways. The abstract concepts certainly must be carefully defined. *Competitive level* might be defined as the class level of the best race within ninety days for claiming horses, within the past season for nonclaiming horses. Best race could mean won, finished in the money, or finished within two lengths (sprint) or three lengths (route). *Key race* might refer to the eligibility conditions of the horse's best effort in the past two seasons, or a recent race from which two or more winners have emerged. *Distance range* might mean the low and high number of furlongs at which the horse has won or placed. Stakes designation might mean Gr. 1, Gr. 2, Gr. 3, L (listed), O (Open), or R (Restricted). Data types cry out for precise definition. The exercise gives the database a much sharper logical definition.

Trainer Performance

Trainer (name, starters, sprint %, route %, basic rating, specialty, weakness, regular jockey, longshot jockey, major owner)

Jockey (jockey name, sprint stats, sprint %, sprint ROI, route stats, route %, route ROI)

Statistics (win %, claim %, alw %, stakes %, sprint %, route %, turf %, 3-yr. %, 2-yr. %, 4-up %, favorites %, odds 8–17%, 1st starters %, ROI)

Patterns (class drop T.W. %, class rise T.W. %, repeaters T.W. %, short layoff T.W. %, long layoff T.W. %, stretch out T.W. %, jockey shift T.W. %)

Collecting trainer statistics and identifying profitable trainer patterns are perhaps the most fashionable exercises today among regular handicappers, who must hunt for overlays in the face of increasingly small and noncompetitive fields. Trainer pattern data can be detectable throughout a single long season, as the Cox anecdote at the beginning of this chapter indicates. Conspicuous overlays that fit the patterns can be supported, notwithstanding the unreliability of a small sample.

Performance data is another matter. A three-year baseline is mandatory, a five-year baseline preferable. Does that mean the trainer database will take three to five years to construct? In most instances, yes, but handicappers on several circuits enjoy a lengthy head start.

Some professionals have marketed trainer statistics to regulars for years now. Greg Lawlor, of San Diego, who supplies these data for the southern and northern California circuits, is a leading example. Lawlor markets a five-year baseline.

A survey of Lawlor's trainer statistics for Santa Anita shows why the five-year baseline should prevail. So many categories and subcategories of performance can be important that the raw numbers in many cells remain too low to permit reliable statistical inference.

Table 8 shows Lawlor's 1978–1983 Santa Anita baseline for southern California trainer Richard Mandella. He provides the same profiles for the forty trainers who win 75 percent of the races. Lawlor uses so many data categories (84) that even when all odds levels and all types of races are grouped he finds fewer than thirty starters in more than half his data categories (47). Statisticians usually would consider thirty race outcomes a number that reflects a normal distribution of events. That guideline figure might be higher for races that are influenced by so many variables and interactions. When Lawlor uses subcategory data, as when he studies maidens at odds of 5 to 1 or lower, he rarely approaches a normal distribution of events. And this is a five-year baseline. The implication is plain. Subcategory data are best restricted to identifying interesting trainer patterns, which are not normally distributed but should not be used to evaluate trainer performance, which is normally distributed.

Handicappers benefit by grouping trainer data into broader categories for performance evaluation and not by subdividing categorical data into odds levels and types of races. At the subcategory levels, they can remain alert for high rates of return on the invested dollar, especially when inspecting smaller subcategories of data. The subcategories that are paying big are potentially a part of successful trainer patterns.

Table 8
Trainer Statistics for Southern California Trainer Richard Mandella, 1979–1983

Mandella Richard
SYSTEM: TOP TRAINERS

SANTA ANITA

	ALL ODDS	*M(5 DOWN)*	*M(5–12)*	*C(5 DOWN)*	*C(5–12)*	*A(5 DOWN)*	*A(5–12)*	*12*
	STS W% ROI	*STS W% ROI*	*STS W% ROI*	*STS W% ROI*	*STS W% ROI*	*STS W% ROI*	*STS W% ROI*	*STS W% ROI*
1ST TIME STARTER	46/ 28/270	19/ 47/150*	11/ 18/150	---/---/---	---/---/---	---/---/---	---/---/---	16/000/000
1ST TIME STARTER (MCLM)	15/ 0/ 0	5/ 0/ 0	4/ 0/ 0	---/---/---	---/---/---	---/---/---	---/---/---	6/000/000
LAST RACE 61–150 DAYS	16/ 12/ 78	---/---/---	1/100/999	---/---/---	2/ 0/ 0	2/ 50/125	2/ 0/ 0	8/000/000
LAST RACE 151 DAYS	16/ 12/ 43	1/100/360	1/ 0/ 0	---/---/---	---/---/---	2/ 50/165	4/ 0/ 0	8/000/000
1ST RACE AFTER CLAIM	2/ 50/180	---/---/---	---/---/---	2/ 50/180	---/---/---	---/---/---	---/---/---	---/---/---
SPRINTS (OVERALL)	159/ 15/ 84	14/ 14/ 56	6/ 16/143	21/ 0/ 0	12/ 16/161	43/ 37/124	23/ 8/ 61	34/000/000
1ST LAST RACE	43/ 18/ 76	---/---/---	---/---/---	4/ 0/ 0	---/---/---	16/ 37/126	7/ 14/100	12/000/000
1ST; LAST RACE WAS MAIDN	22/ 22/ 73	---/---/---	---/---/---	---/---/---	---/---/---	13/ 38/124	3/ 0/ 0	6/000/000
2ND LAST RACE	26/ 11/ 58	6/ 16/ 58	2/ 50/430	4/ 0/ 0	2/ 0/ 0	7/ 14/ 42	3/ 0/ 0	2/000/000
3RD LAST RACE	30/ 30/142*	2/ 50/220	---/---/---	5/ 0/ 0	3/ 33/316	12/ 50/179**	3/ 33/240	3/000/000
4TH OR WORSE LAST RACE	60/ 8/ 71	6/ 0/ 0	4/ 0/ 0	8/ 0/ 0	7/ 14/141	8/ 37/112	10/ 0/ 0	17/000/000
LAST RACE 1–8 DAYS	10/ 20/165	---/---/---	2/ 0/ 0	2/ 0/ 0	2/ 50/475	1/ 0/ 0	2/ 50/350	1/000/000
LAST RACE 9–15 DAYS	56/ 19/ 80	5/ 20/ 70	3/ 33/286	10/ 0/ 0	3/ 0/ 0	15/ 46/134*	5/ 20/144	13/000/000
LAST RACE 16–30 DAYS	63/ 12/ 56	8/ 12/ 55	1/ 0/ 0	6/ 0/ 0	7/ 14/141	15/ 40/141*	12/ 0/ 0	12/000/000
LAST RACE 31–60 DAYS	30/ 13/122	1/ 0/ 0	---/---/---	3/ 0/ 0	---/---/---	12/ 25/103	4/ 0/ 0	8/000/000
2ND RACE 61–150 DAYS	14/ 7/ 67	---/---/---	1/ 0/ 0	2/ 0/ 0	1/100/950	3/ 0/ 0	1/ 0/ 0	5/000/000
2ND RACE 151+ DAYS	9/ 22/ 82	1/100/440	---/---/---	1/ 0/ 0	---/---/---	2/ 50/150	4/ 0/ 0	1/000/000
UP IN CLASS	60/ 15/ 71	3/ 0/ 0	2/ 50/430	7/ 0/ 0	2/ 50/475	18/ 33/106	9/ 0/ 0	15/000/000
SAME CLASS	70/ 18/ 84	9/ 22/ 87	3/ 0/ 0	4/ 0/ 0	4/ 25/247	22/ 36/123	13/ 15/109	13/000/000
DOWN IN CLASS (ALL)	30/ 6/ 24	2/ 0/ 0	1/ 0/ 0	11/ 0/ 0	6/ 0/ 0	2/100/370	1/ 0/ 0	7/000/000
DOWN 1 LEVEL ONLY	14/ 14/ 52	1/ 0/ 0	---/---/---	6/ 0/ 0	2/ 0/ 0	2/100/370	1/ 0/ 0	2/000/000
DOWN 2 LEVELS ONLY	7/ 0/ 0	---/---/---	---/---/---	3/ 0/ 0	3/ 0/ 0	---/---/---	---/---/---	1/000/000

DOWN 3 OR MORE	9/ 0/ 0	1/ 0/ 0	1/ 0/ 0	2/ 0/ 0	1/ 0/ 0	---/---/---	---/---/---	4/000/000
MAIDEN TO MCLM	2/ 0/ 0	1/ 0/ 0	1/ 0/ 0	---/---/---	---/---/---	---/---/---	---/---/---	---/---/---
SAME DISTANCE	145/ 15/ 87	14/ 14/ 56	6/ 16/143	19/ 0/ 0	12/ 16/161	40/ 35/115	19/ 10/ 74	30/000/000
DIFFERENT DISTANCE	14/ 14/ 52	---/---/---	---/---/---	2/ 0/ 0	---/---/---	3/ 66/246	4/ 0/ 0	4/000/000
WINNING FAVORITE	18/ 16/ 90	---/---/---	---/---/---	---/---/---	---/---/---	6/ 16/ 60	5/ 20/140	3/000/000
BEATEN FAVORITE	22/ 4/ 15	9/ 11/ 38	---/---/---	4/ 0/ 0	2/ 0/ 0	4/ 0/ 0	1/ 0/ 0	1/000/000
2ND RACE AFTER CLAIM	---/---/---	---/---/---	---/---/---	---/---/---	---/---/---	---/---/---	---/---/---	---/---/---
3RD RACE AFTER CLAIM	2/ 0/ 0	---/---/---	---/---/---	1/ 0/ 0	---/---/---	---/---/---	---/---/---	1/000/000
SHIPPER NOT OF SO CAL	18/ 16/175	---/---/---	---/---/---	2/ 0/ 0	---/---/---	4/ 50/185	2/ 0/ 0	9/100/000
2 YEAR-OLD FILLIES	1/ 0/ 0	---/---/---	---/---/---	---/---/---	---/---/---	1/ 0/ 0	---/---/---	---/---/---
2 YEAR-OLD COLTS	2/ 50/165	---/---/---	---/---/---	1/ 0/ 0	---/---/---	1/100/330	---/---/---	---/---/---
3 YEAR-OLDS	54/ 14/ 79	8/ 0/ 0	5/ 20/172	5/ 0/ 0	2/ 50/495	13/ 38/145*	7/ 0/ 0	11/000/000
4 YEAR-OLDS & UP	102/ 15/ 85	6/ 33/131*	1/ 0/ 0	15/ 0/ 0	10/ 10/ 95	28/ 35/112	16/ 12/ 88	23/000/000
FEMALES	92/ 18/104	9/ 11/ 38	1/ 0/ 0	5/ 0/ 0	4/ 25/247	31/ 35/123	18/ 11/ 78	21/000/000
MALES	67/ 11/ 56	5/ 20/ 88	5/ 20/172	16/ 0/ 0	8/ 12/118	12/ 41/127	5/ 0/ 0	13/000/000
2ND TIME STARTERS	16/ 0/ 0	8/ 0/ 0	4/ 0/ 0	1/ 0/ 0	---/---/---	---/---/---	---/---/---	3/000/000
SPR TO RTE TO TODAY	7/ 14 /37	---/---/---	---/---/---	1/ 0/ 0	---/---/---	2/ 50/130	---/---/---	4/000/000
TURF TO DIRT	4/ 25/120	---/---/---	---/---/---	---/---/---	---/---/---	2/ 50/240	1/ 0/ 0	1/000/000
DIRT TO TURF	8/ 25/103	---/---/---	---/---/---	1/ 0/ 0	---/---/---	3/ 33/ 86	1/ 0/ 0	2/000/000
"OFF" TRACK TODAY	52/ 13/ 52	5/ 0/ 0	---/---/---	6/ 0/ 0	6/ 0/ 0	14/ 42/144*	9/ 11/ 77	11/000/000

		STRENGTHS — SPRINTS	*WEAKNESSES — SPRINTS*
TOP SPRINT JOCKEY: ToF	70/ 17/122		
TOP SPRINT JOCKEY: McC	33/ 24/ 67	1st Time Starters 5-1 Odds Down — 47% Win Overall — $190	4th or Worse Last — 8% Win Overall
TOP SPRINT JOCKEY: DlE	29/ 24/120	3rd Last Race — Overall 30% Win — $252	2nd Back 61-150 — 7% Win Overall
TOP SPRINT JOCKEY: HwS	25/ 16/ 99	Allowances 5-1 Odds Down — Overall Win 37% — $206	Class Drops — 7% Win Overall
TOP SPRINT JOCKEY: VzP	24/ 16/102		Beaten Favorites — 4% Win Overall
TOP SPRINT JOCKEY: PcL	18/ 22/ 58		Odds 12-1 Up — 2% Win Overall

Table 8 *(continued)*

Mandella Richard
SYSTEM: TOP TRAINERS

SANTA ANITA

	ALL ODDS	M(5 DOWN)	M(5–12)	C(5 DOWN)	C(5–12)	A(5 DOWN)	A(5–12)	12
	STS W% ROI	STS W% ROI	STS W% ROI	STS W% ROI	STS W% ROI	STS W% ROI	STS W% ROI	STS W% ROI
1ST TIME STARTER	2/ 50/ 70	---/---/---	1/ 0/ 0	---/---/---	---/---/---	---/---/---	---/---/---	1/000/000
1ST TIME STARTER (MCLM)	---/---/---	---/---/---	---/---/---	---/---/---	---/---/---	---/---/---	---/---/---	---/---/---
LAST RACE 61–150 DAYS	2/ 50/125	1/100/250	---/---/---	---/---/---	---/---/---	---/---/---	1/ 0/ 0	---/---/---
LAST RACE 151 DAYS	2/ 0/ 0	---/---/---	---/---/---	---/---/---	---/---/---	2/ 0/ 0	---/---/---	---/---/---
1ST RACE AFTER CLAIM	3/ 33/120	---/---/---	---/---/---	3/ 33/120	---/---/---	---/---/---	---/---/---	---/---/---
ROUTES (OVERALL)	118/ 11/ 70	8/ 12/ 50	5/ 0/ 0	10/ 10/ 33	5/ 20/142	27/ 22/ 76	29/ 6/ 61	28/000/000
1ST LAST RACE	20/ 20/ 63	---/---/---	---/---/---	3/ 33/110	1/ 0/ 0	3/ 66/203	7/ 0/ 0	2/000/000
1ST; LAST RACE WAS MAIDN	5/ 0/ 0	---/---/---	---/---/---	---/---/---	1/ 0/ 0	1/ 0/ 0	3/ 0/ 0	---/---/---
2ND LAST RACE	24/ 12/ 65	3/ 0/ 0	---/---/---	---/---/---	1/ 0/ 0	12/ 16/ 57	5/ 20/176	2/000/000
3RD LAST RACE	18/ 11/ 42	1/ 0/ 0	1/ 0/ 0	1/ 0/ 0	1/ 0/ 0	8/ 25/ 95	4/ 0/ 0	2/000/000
4TH OR WORSE LAST RACE	56/ 7/ 84	4/ 25/100	4/ 0/ 0	6/ 0/ 0	2/ 50/355	4/ 0/ 0	13/ 7/ 69	22/000/000
LAST RACE 1–8 DAYS	16/ 18/ 62	2/ 0/ 0	---/---/---	2/ 0/ 0	---/---/---	4/ 50/167	2/ 0/ 0	4/000/000
LAST RACE 9–15 DAYS	35/ 8/ 96	3/ 0/ 0	3/ 0/ 0	3/ 33/110	2/ 0/ 0	8/ 12/ 41	7/ 0/ 0	7/000/000
LAST RACE 16–30 DAYS	45/ 13/ 68	3/ 33/133	1/ 0/ 0	2/ 0/ 0	2/ 50/355	11/ 27/ 96	13/ 7/ 69	11/000/000
LAST RACE 31–60 DAYS	22/ 4/ 40	---/---/---	1/ 0/ 0	3/ 0/ 0	1/ 0/ 0	4/ 0/ 0	7/ 14/125	6/000/000
2ND RACE 61–150 DAYS	5/ 0/ 0	---/---/---	---/---/---	1/ 0/ 0	---/---/---	---/---/---	1/ 0/ 0	1/000/000
2ND RACE 151+ DAYS	7/ 0/ 0	1/ 0/ 0	---/---/---	---/---/---	1/ 0/ 0	---/---/---	2/ 0/ 0	3/000/000
UP IN CLASS	35/ 14/ 62	1/ 0/ 0	1/ 0/ 0	4/ 25/ 82	2/ 0/ 0	6/ 33/101	8/ 12/112	8/000/000
SAME CLASS	64/ 7/ 70	5/ 0/ 0	3/ 0/ 0	3/ 0/ 0	---/---/---	19/ 15/ 49	19/ 5/ 46	14/000/000
DOWN IN CLASS (ALL)	20/ 15/ 81	1/100/400	1/ 0/ 0	4/ 0/ 0	4/ 25/177	2/ 50/255	2/ 0/ 0	6/000/000
DOWN 1 LEVEL ONLY	12/ 25/135	1/100/400	1/ 0/ 0	2/ 0/ 0	2/ 50/355	2/ 50/255	2/ 0/ 0	2/000/000
DOWN 2 LEVELS ONLY	5/ 0/ 0	---/---/---	---/---/---	2/ 0/ 0	---/---/---	---/---/---	---/---/---	3/000/000
DOWN 3 OR MORE	3/ 0/ 0	---/---/---	---/---/---	---/---/---	2/ 0/ 0	---/---/---	---/---/---	1/000/000
MAIDEN TO MCLM	---/---/---	---/---/---	---/---/---	---/---/---	---/---/---	---/---/---	---/---/---	---/---/---

SAME DISTANCE	87/ 8/ 39	6/ 16/ 66	3/ 0/ 0	7/ 0/ 0	3/ 0/ 0	21/ 9/ 58	24/ 8/ 74	18/000/000
DIFFERENT DISTANCE	31/ 19/158	2/ 0/ 0	2/ 0/ 0	3/ 33/110	2/ 50/355	6/ 33/140*	5/ 0/ 0	10/ 1 /
WINNING FAVORITE	6/ 16/ 31	---/---/---	---/---/---	1/ 0/ 0	---/---/---	1/100/190	1/ 0/ 0	---/---/---
BEATEN FAVORITE	11/ 27/208	2/ 0/ 0	2/ 0/ 0	---/---/---	---/---/---	2/ 50/255	4/ 50/445	1/000/000
2ND RACE AFTER CLAIM	---/---/---	---/---/---	---/---/---	---/---/---	---/---/---	---/---/---	---/---/---	---/---/---
3RD RACE AFTER CLAIM	1/ 0/ 0	---/---/---	---/---/---	---/---/---	---/---/---	---/---/---	---/---/---	1/000/000
SHIPPER NOT OF SO CAL	11/ 9/ 80	---/---/---	---/---/---	2/ 0/ 0	1/ 0/ 0	2/ 0/ 0	3/ 33/293	2/000/000
2 YEAR-OLD FILLIES	---/---/---	---/---/---	---/---/---	---/---/---	---/---/---	---/---/---	---/---/---	---/---/---
2 YEAR-OLD COLTS	---/---/---	---/---/---	---/---/---	---/---/---	---/---/---	---/---/---	---/---/---	---/---/---
3 YEAR-OLDS	27/ 14/145	5/ 20/ 80	4/ 0/ 0	3/ 33/110	1/ 0/ 0	4/ 25/127	4/ 0/ 0	6/000/000
4 YEAR-OLDS & UP	91/ 9/ 48	3/ 0/ 0	1/ 0/ 0	7/ 0/ 0	4/ 25/177	23/ 21/ 67	25/ 8/ 71	22/000/000
FEMALES	49/ 14/ 78	---/---/---	3/ 0/ 0	3/ 33/110	1/100/710	12/ 25/ 85	17/ 11/104	11/000/000
MALES	69/ 8/ 64	8/ 12/ 50	2/ 0/ 0	7/ 0/ 0	4/ 0/ 0	15/ 20/ 68	12/ 0/ 0	17/000/000
2ND TIME STARTERS	3/ 0/ 0	---/---/---	1/ 0/ 0	---/---/---	---/---/---	---/---/---	---/---/---	1/000/000
SPR TO RTE TO TODAY	19/ 10/ 29	1/ 0/ 0	1/ 0/ 0	---/---/---	---/---/---	4/ 50/137	5/ 0/ 0	6/000/000
TURF TO DIRT	11/ 27/194	1/ 0/ 0	---/---/---	1/ 0/ 0	1/ 0/ 0	4/ 25/ 90	4/ 50/445	---/---/---
DIRT TO TURF	9/ 0/ 0	1/ 0/ 0	---/---/---	---/---/---	---/---/---	1/ 0/ 0	1/ 0/ 0	4/000/000
"OFF" TRACK TODAY	29/ 17/162	1/ 0/ 0	2/ 0/ 0	5/ 0/ 0	1/ 0/ 0	6/ 33/130*	5/ 20/180	8/000/000

TOP ROUTE JOCKEY: ToF	30/ 13/132
TOP ROUTE JOCKEY: HwS	16/ 18/ 72
TOP ROUTE JOCKEY: PcL	15/ 6/ 47
TOP ROUTE JOCKEY: DlE	11/ 27/ 77
TOP ROUTE JOCKEY: VzP	10/ 10/ 33
TOP ROUTE JOCKEY: SbR	18/ 22/ 58

STRENGTHS — ROUTES

Won Last Race — 20% Win Overall — No Profit
1 Class Level Drop — 25% Win Overall — Modest Profit
Beaten Favorite — 27% Overall — $237
Allowance 5-1 Odds Down — 22% Win Overall — No Profit

WEAKNESSES — ROUTES

Maiden Winners Last — 0 for 5
Last 4th or Worse — 7%
Last Race 31-60 Days — 4%
Same Class — 7%
Dirt to Turf — 0 for 9

Source: JGL Enterprises, San Diego, California

Trainer data, of course, can sometimes be explosively valuable. Notice trainer Mandella's win percentage (28) and rate of return (270 percent) with first starters in sprints at all odds, which appear at the upper left of the profile. Mandella is a genuine expert with first starters, and handicappers who have supported his debuting maidens religiously have earned approximately two-and-one-half times their investment.

Most of the data types in the trainer schema appear self-explanatory. The initials of the data type, as in *class drop T.W. %*, refer to the total number of class drops recorded, the number of wins, and the *percentage* of wins.

Racetrack

Track (name, circumference, stretch, par variant)
Surface (date, official footing, variant, bias)
Class (overnight purse average, stakes money, Gr. 1 stakes, Gr. 2 stakes)

The data values of *bias* might rely on the covenient notation provided by Andrew Beyer in *The Winning Horseplayer*. Track biases were noted by seven letter designations and symbols:

GR	Good rail	S	Speed favoring track
GR+	Very strong good rail	S+	Very strong speed favoring track
BR	Bad rail		
BR+	Very strong bad rail	C	Track that favors closers

Unlike the class appraisal application schema presented earlier, the racehorse, trainer performance, and racetrack schemata are subject schemata. Data are organized for a subject, not a method of handicapping. The next schema is also a subject matter schema, similar to the past performance records of horses but actually quite different. I call it "performance histories."

PERFORMANCE HISTORIES

A telltale database would be keyed to the races of specific horses, similar to the past performance tables of *Daily Racing Form*. A complete past performance history could be summarized, at least for horses aged two, three, and four. Older claiming horses could be represented by their previous twenty races. The data items, however, would include significant racing values not trapped in the customary past performances. Handicappers could call these data records into view whenever the horses raced, seeing a complete supplemental past performance history for every horse. The schema is presented below.

Performance History

Horses (names, dates, races, distance)

Class (eligibility conditions, purse value, class rating, ability time rating, manner of performance, par time)

Speed (adjusted final time, speed figure, pace figure)

Trips (position, trouble, bias, body language, track condition)

An updated database of this kind for every horse on the circuit can be transformed into windfalls at the windows. The complete histories of younger nonclaiming horses is vital to understanding overall patterns of development, as well as cycles of improvement and decline. The histories are not difficult to compile.

Handicappers need not include all horses in the database. All winners would be included, as might second and third finishers and other horses of special interest. Handicappers, notably those working alone, must decide what the input and update demands of the databases will be. Members of handicapping information teams can divide the workload, possibly sharing two or more microcomputers, but this puts tremendous emphasis on teamwork and communication.

THE DATA DICTIONARY

A data dictionary is a database concept and tool that is analagous to the purpose of English-language dictionaries. It is a catalogue of all the data items in the database, giving their names, logical definitions, and physical characteristics, including storage locations. Thus, handicappers and technicians will again need to collaborate. The data dictionary also tells how the data items are being used, as in what applications programs or in what databases. It's a tracking system for data items, which itself is computerized and easily updated.

Data dictionaries become imperative for two reasons. First, high-tech handicappers, computer programmers, and database technicians need periodically to consult the dictionary to establish whether certain data items are already in the physical storage as well as whether the data items have been defined consistently. The data items are defined logically and in computer language. Has *early speed* data been defined in the existing database as it has been in the exciting new applications program about to be implemented? If not, new data items must be added to the database before the new program can be run, or the application program must be modified. Otherwise, faulty output will emerge, as when handicappers expecting adjusted fractional times receive velocity ratings instead, or the new program simply won't run. Standardization of the data items should be seen as a beneficial nuisance where multiple applications programs may eventually be used.

Second, the computerized data dictionary is typically integrated into the structure of the specialized software that controls and operates the database. In this way the data dictionary facilitates the computer's retrieval of data items in physical storage and plays a helpful role in responding to queries. This purpose is less important in personal computer systems, but high-tech handicappers will benefit nonetheless. At a glance, before operating the computer, handicappers who become familiar with the data dictionary will know whether the data items associated with their queries are in fact contained in the database. Clever high-tech handicappers will learn to use the data dictionary to guide their interrogations of the database.

THE DATABASE MANAGEMENT SYSTEM (DBMS)

What is a database management system? It is a package of software that manages a database. It consists of several interrelated languages, programs, and capabilities that are the most effective means to store, control, and retrieve the contents of the database for various purposes.

Users can think of this specialized software system as a *map* among the schema or logical description of the database, applications programs that use the database, and the physical storage arrangement in the computer. The DBMS also typically contains a command file and language for nontechnical users that allows them to query the database as often as they please.

Database management systems should be viewed as indispensable to high-tech handicapping and are relatively expensive. The DBMSs for the mainframe systems of corporations can cost more than $100,000. Micro systems use DBMSs that cost between $200 and $500. Regardless of cost, this software item merits the highest priority.

Database management software is preprogrammed. The programs are available commercially either from computer manufacturers or from software firms. No area of research and development is more hectic in the microcomputer industry today, reflecting the importance and growing popularity of databases. Until very recently database management systems for microcomputers were not even available. The data resource capacity of small computers was judged too small to warrant the development costs, and widespread use of a DBMS among personal computer owners was hardly imagined. The microcomputer boom and the tremendous increase in small computers' storage densities changed all that forever. Today the software frontier of small computer systems is the database management system. Two outstanding packages currently available are described below.

Handicappers should realize that using a database management system appropriately and expertly requires training and work on the part of the users. This book strongly recommends enrollment in a hands-on university-level course that teaches users to understand and implement database management systems for microcomputers.

Whether handicappers use the DBMS software discussed in this book or new more powerful products, which without doubt will appear shortly, the system should satisfy certain criteria that allow handicappers to meet their information needs. A database management system must support:

1. *Data independence*, or the storage and retrieval of data items independently of the files of applications programs
2. The speedy handling of *ad hoc requests* to the database
3. *Controlled redundancy*, such that the DBMS knows where

any repeated data items are located as well as how to search the files for the fastest retrieval of the desired items
4. *Data items relationship versatility,* or the capability of combining data items from separate databases and files in multiple ways
5. *Data security,* or the control of individual access to databases, including passwords for organizing and changing the files

The classic database management system for microcomputers is called dBase II ($450; Ashton-Tate, Culver City, CA). It was the first great success in the field, and for a time all other DBMSs for micros will be automatically compared to dBase II.

dBase II has two levels of performance capability, but for high-tech handicappers Level I performance is more than sufficient. Handicappers can create files, even from existing applications programs, and once the files have been created they can add new records, edit the data items, eliminate data items and records, search the files, and print reports. All of the records, or specific data items within records, can be displayed on the terminal screen or reported to paper. dBase II is characterized by efficient record management and the manipulation of large quantities of data and information quickly and powerfully.

For high-tech handicappers who are well endowed with computer technical skill, dBase II also contains its own programming language, ADL, for Applications Development Language. This facilitates the speedier development of applications programs from the database. High-tech handicappers can write their own programs by using the DBMS.

Knowledge Man ($500; Micro Data Base Systems, Lafayette, IN) is currently the most powerful database management system for microcomputers. It integrates both database and spreadsheet capabilities, so consumers in effect receive spreadsheet software and a DBMS for their money. Knowledge Man consists of several interrelated applications having a single user interface, which is a command language of more than two hundred commands. Handicappers will have to learn the commands, but with practice the language development becomes facile. The command language is more like English than computerese. More important, Knowledge Man sets up the kinds of relational databases handicappers want, and *ad hoc inquiries* to the database are a salient feature of the system. Knowledge Man will be described in greater detail in the next chapter, which discusses relational databases.

High-tech handicappers should not underestimate the importance and power of the database management system. They should resign themselves absolutely to mastering the system they select. DBMSs are not for computer technicians. They are for ordinary computer users. Neither dBase II nor Knowledge Man require technical expertise for their mastery. In a real sense, DBMS users are learning how to master a challenging, fairly complicated game having its own language and symbols, such as handicapping itself does, but nothing more than that. And as with the development of handicapping knowledge and skill, the long-term benefits and advantages will be unusually gratifying.

QUERY LANGUAGES

If handicappers wish to ask questions of databases from computer terminals and receive immediate responses, they must resort to *query languages*. Query languages are special database software programs and are generally so simple that practically anyone can employ them effectively.

Query languages exist and have been growing in importance because it is highly desirable that computer users who cannot program should be able to query databases and extract from them the information they need. The market for query languages is bustling with the top corporate executives in the nation who are generally unsophisticated with computers and who do not ask for the kind of prepackaged operational and administrative information transmitted by applications programs. These executives tend to want the unanticipated information they often do need in a hurry—tomorrow the race will be over. Thus, the image of handicappers as information managers takes concrete form through the spontaneous use of query languages and databases.

Query languages are many and often are associated or integrated with a database management system. The language usually consists of a command file and structure that permits a variety of questions and searches. Knowledge Man, for example, has its own query language, consisting of more than two hundred commands.

An IBM query language, called IQF, for Interactive Query Facility, which goes hand in hand with an IBM database management system, is not much more than a small version of the English language. Using IQF or its facsimile, handicappers would simply enter English-language statements from a computer terminal. IQF maintains its own database of words and phrases that might be used and also has a database containing all the data items that can be combined. If IQF were a handicapping database—it is not—users could enter commands such as the following.

> From the trainer files, please list the trainers who have won Gr. 1 stakes with three-year-olds during the past five seasons; list each trainer's number of starters in these races, and list the names of the winning horses, the stakes they won, and the purse values of the races won.

Handicappers who rely on query languages not integrated with a database management system may prefer languages that use similar English-language statements. The languages can normally be tailored to any application, even handicapping, and even can begin with a small vocabulary that grows with its repeated use.

As is true of evaluating handicapping applications programs, the best way to keep in touch with the rapid development of DBMSs and query languages for microcomputers is to read *PC Magazine* or other consumer protection periodicals.

THE DATABASE FUNCTIONS IN THE HANDICAPPER'S MIS

Without question, the database function contributes the most progressive, most powerful advances of high-tech handicapping. It drives what I have called the information management approach to handicapping. The database as a handicapping tool applies throughout the handicapping process. It serves these several functions:

1. Has the capability of organizing and storing all of the handicapping data in a single repository that is independent of computer applications programs

2. Supplies information during the early analytical phase of handicapping that can clarify issues and relationships and that can identify problems
3. Represents the data item resource for multiple handicapping applications programs, including all future applications programs that might not be adequately understood today
4. Answers spontaneous questions of all kinds and complexity during the later problem-solving phase of handicapping
5. Provides the *comprehensive information* upon which more confident, informed decisions about handicapping and wagering will ultimately depend
6. Serves as the central repository of team handicapping activities

In the not-too-distant future national and regional handicapping databases will form the hubs of widely dispersed communications networks among handicappers. Regular high-tech handicappers will subscribe to those databases, which they will repeatedly contact by telephone. They will therefore enjoy access to the most handicapping information available anywhere. Portions of the larger databases can be downloaded into personal databases in the home, right across telephone lines. Eventually *Daily Racing Form* will market its past performance tables and results charts electronically as well, and high-tech handicappers will subscribe to those important databases. Instead of buying a daily newspaper on a single circuit, which can be surprisingly difficult to obtain in many locations, high-tech handicappers may prefer to have the day's past performances at selected tracks around the country printed out by personal computers in their homes. The possibilities are no less than fantastic.

Handicappers must also understand that when national and regional databases are constructed, the data items and data relationships of greatest interest will be specified by the handicappers involved in the systems design work. Handicappers who subscribe will receive the performance records, speed figures, class ratings, and trainer stats of other handicappers. As welcome and helpful as remote handicapping databases will be to most of us, those who use them will suffer some loss of their individuality as handicappers, which after all is the single most magnificent attraction of the pastime, the ingredient that makes handicapping the personally satisfying and rewarding experience and challenge that it is.

But so what? The information will have overcompensating value and will not be otherwise available. Individual handicappers will still make the final decisions, and these will be better informed.

CHAPTER 9

THE RELATIONAL VIEW OF DATA

Tables, tables, tables, tables.... It's simplicity itself.

HAROLD EINSTEIN, Instructor
Management Information Systems
UCLA Extension

HIGH-TECH HANDICAPPERS will be relieved to discover that the multifarious, complex relationships of handicapping can be expressed in a simple and elegant way. The key is the relational nature of data on a subject. Handicappers who have experienced how abruptly short form can nullify superior class or how track bias can accentuate cheap speed will hardly be startled by assertions about the relational aspects of handicapping information. All the data items of handicapping relate to all others, some strongly, many moderately, and some marginally. The book is predicated on that reality and the multitudinous data relationships that are plausible.

These data relations can be formulated by the use of fascinating databases called *relational databases.* No matter how complicated the data relationships espoused by handicapping theories and methods are, they can *all* be reduced to two-dimensional relationships and *expressed logically in tabular form.* All the data items and significant relationships are expressed in carefully laid tables. To put it differently, high-tech handicappers who have developed a schema or logical view of a handicapping database can depict all the logical relationships within the schema merely by erecting tables, tables, tables, and more tables. This chapter constructs the tables for the schema of handicapping presented in Chapter 8. The procedure is not only simplicity itself, it's fun.

THE RELATIONAL DATABASE: SIMPLICITY ITSELF

Table 9 is a tabular expression of the *Racehorse* relation presented in the previous chapter. Several data items (columns) have been added. The table is understood as a rectangular array of rows and columns containing a number of two-dimensional data relationships. The rows contain data records. In Table 9 the data records represent the horses named in the far left column. The columns contain the attributes or values of the data items that identify the columns. In setting up database tables, handicappers need adhere to a single guiding principle only. The tables must be set up in a way that no information about the relationships between the data items is lost. In practice, the principle is relatively easy to follow, as the tables have certain mathematical properties:

1. *Each entry in a table represents one data item.* There are no repeating data items, such as two speed figures or two class ratings
2. *The tables are column-homogeneous.* All the data items in a column are of the same kind
3. *Each column is assigned a distinct name*
4. *All rows are distinct.* Duplicate rows are not allowed
5. *Both the rows and columns can be viewed in any sequence at any time without affecting the information provided*

These few rules protect the logical structure and integration of the tables, and therefore of the resulting database. They allow handicappers to relate the data items in tables to one another in multiple ways. The guidelines will become familiar with practice, which returns our attention to Table 9.

The table is referred to as a *relation,* or an array of two-dimensional logical data relationships. Handicappers must remember the tables are only *logical representations* of the database and are kept entirely separate from the physical representations of data in the computer. Each relation has a name. The relation of Table 9 is called *Racehorse.* It consists of an inordinate number of data items and data relationships useful for evaluating the past performances of horses but which cannot be unearthed in the conventional past performance tables.

Handicappers can fairly regard the table as buried treasure.

Table 9
RACEHORSE

	NAME	*AGE*	*SEX*	*Competitive level*	*Class rating*	*Back class*	*Key race*	*Ability figures*	*Stakes designations*	*Speed figure*	*Pace figure*	*Pace rating*	*Race shape*	*Wins when fresh*	*Longest layoff/win*	*Form cycle*	*Best distance*	*Distance range*	*Dosage index*	*Turf I.V.*	*Mud rating*
	MY COUNTESS	4	F	NWI	21	C32				93	85	228	AA	0	23	N	8.5	8–9	10.0	NA	B
	MAX'S LADY	4	F	NWI	23	NA			0	96	93	251	SA	0	9	N	8.5	8–8.5	7.36	1.80	A
RECORD	SOCIAL WHIRL	4	F	C45	20	C20	C50	26	0	93	92	256	SA	0	27	R	8.5	7–8.5	11.44	0.61	NA
	PETITE FLEUR	3	F	MONU	15	NA			0	93	88	250	SS	0	27	N	8.5	8–8.5	3.55	1.15	B
	SARATOGA ROXIE	5	M	C50	22	NWI			0	94	90	254	AA	1	64	F	8	8	15.15	0.44	B
	GOLDEN SCREEN	4	F	MON	30	NA			0	91	91	243	AS	0	NA	NA	8.5	8–8.5	4.66	0.75	P
	BUENA FE II	4	F	C100	9	NA			0	97	93	272	AF	0	42	N	6.5	6–8	NA	NA	A
	PRINCESS POLEAX	4	F	C12		M12			0	97	94	145	SS	0	16	N	8.5	8–8.5	6.79	2.22	A

PRIMARY KEY DOMAINS

A *relation* of handicapping data items called *Racehorse*.
A database constructed of relations or tables is called a *relational database*.
Racehorse has eight rows or data records. Each record consists of twenty *domains* or columns. Each domain represents a data item type.
The table values are data items.
The relation has a *primary key*, the *name* of the horses. The primary key identifies a record uniquely.

The table contains records of eight horses. The complete relation or table would contain similar records for *all* the horses active on the circuit or at least all the winners and close runners-up or as many horses as handicappers chose to limit the size of the relation. The table can be expanded as additional horses win or finish close, or contracted as horses become inactive. None of the data relationships will be affected.

Examine the record for Social Whirl, in row three. You will find twenty data items in the record. Each column of data items for Social Whirl is called a *domain*. As handicappers decide, the data items in a domain or column are defined operationally, and the values in the cells of the domain depend on those definitions. A domain is homogeneous; it consists of the same data items. The values in the cells can change, however, as often as necessary. Of the twenty domains in *Racehorse,* at least nine will change periodically. Furthermore, horses will be added to and eliminated from the table constantly.

Many of the data item headings across the top of the table have been taken from the *Racehorse* schema of the previous chapter. They represent five handicapping relations, of class, speed, form, distance, and pedigree. As handicappers prefer, the class relation might be portrayed separately. It would consist of six domains, including competitive level, class rating, back class, key race, ability figures, and stakes designations for each horse. Similarly, the speed, form, distance, and pedigree relations might be depicted separately. Other data items and domains might be added. That would modify the schema of the handicapping database. That's perfectly acceptable. The database development activity is at this point strictly a conceptual and logical exercise, but conceptual clarity is critical prior to the actual physical representation and storage of data. The task is to represent the data relationships of interest as completely and accurately as possible.

Each relation has a primary key. The primary key uniquely identifies the record of interest, or table row. In *Racehorse,* the primary key is the *name* of each horse. No other domain or column value uniquely identifies these records. Handicappers should pause to understand that. All the data items in the record are keyed to the *name* of the horse. When searching for data records, the computer needs to identify the primary keys. For each table they construct therefore, handicappers will identify a primary key, or a domain of data items that uniquely identifies the data records.

Assume the computer has put a call to the database requesting the record of Social Whirl. What do handicappers know about the horse at a glance?

Regarding class, handicappers know that this four-year-old filly has had a basic competitive level just below the $50,000 claiming category, an upscale level from her three-year-old season, when she competed well at $20,000 (back class). The filly's key race, defined here as the best race against the most advanced competition, has been against $50,000 horses, and its best ability time figure is 26. The class rating of 20 refers to the size of purse in thousands that Social Whirl has been able to compete for impressively.

Of Social Whirl's speed, her top figure has been 93, but her preferred race shape (SA) means a slow early time, and average final time is her cup of tea. This filly needs an ordinary, favorable pace and average final time to show her best stuff.

She has not yet won fresh, her longest layoff before winning being twenty-seven days. Social Whirl's form cycle has been rated *R*, meaning here that she has shown the capacity to *repeat* good efforts once she demonstrates acceptable form. Her best distance has been 1 1/16M, which is also the far end of her distance range, and Social Whirl hasn't looked impressive at six furlongs.

As would be expected of cheaper claiming types, Social Whirl's pedigree index for dosage and turf racing looks uninspiring. A *mud rating* is not available, because the filly has no history on the "off" going.

As can be seen, the *Racehorse* table is designed to present handicappers with benchmark information about the capacities, preferences, and potential of the most interesting horses on the circuit. Not all horses active need to be included in the table. It's a judgment call. If handicappers felt the necessary information about Social Whirl was essentially available in the past performances or from other manual sources or that her record was just insignificant, *Social Whirl* could be eliminated from the Racehorse relation. The *DBMS* would eliminate her record from physical storage, leaving additional room for more intriguing horses.

By means of the database management system (DBMS), handicappers can retain separate logical views of the physical database and set up any number of records or data item relationships that the computer can find, combine, and display. The relational view of data becomes extremely flexible and extremely powerful.

For example, should the hardware or physical representation of the data in the system be changed, the logical view of the data remains unchanged. When logical views are modified, the physical structures can persist. This is data independence. Within the computer industry the relational view of data and the relational database concept have evolved from a single overriding consideration, namely the convenience of the majority of applications programmers and users (handicappers).

This point is worth pursuing, as handicappers are fortuitously among the major beneficiaries.

As databases became increasingly popular in the large organizations, especially among top managers who were technically inept yet sufficiently imaginative to perceive the potential for obtaining complicated, spontaneous, unpredictable management information quickly and easily, simply by retrieving it with a query, perhaps in multiple forms, from a single large repository, the database technologies for describing the data were increasingly supposed to satisfy three purposes:

1. Can be understood and manipulated easily by users with no training in programming
2. Makes it possible to add to the existing logical structures of data or the applications programs that have been using the existing data relationships
3. Permits the maximum flexibility in making unanticipated or spontaneous inquiries from terminals

As stated, the handicapper's task is the logical representation of the handicapping database. That begins with drawing the schema, which begins with an understanding of handicapping and a practiced experiential sense of the data relations for which meaningful applications programs (subschemata) might be written, and concludes with the construction of tables that represent to the computer all the logical data item relationships throughout the database. Different handicappers, to be sure, will construct very different schemata, subschemata, and tables, which, to repeat, are nothing more than various logical expressions of handicapping relations. In the next sections this book presents a number of tables for the handicapping database. There could be dozens of others. In a sense, anything goes, as all handicapping data is relational, yet it pays to remain logical, at least in the beginning. Imaginations can be set free later.

To begin, we construct a table of data items similar to what handicappers confront every day in the past performance tables. The difference is crucial. Several of the data items used here are *not available* in the conventional past performances. Neither, therefore, are the significant data relationships the data items can be combined to display. I label the relation generically, as Performance History, because it can display the entire past performance history of a thoroughbred, and not just its previous ten races. Each relation or table is named after the horse it represents. It is a dynamite portion of any high-tech handicapper's database.

PERFORMANCE HISTORIES AND SECONDARY KEYS

The columns of Table 10 reflect considerable information about horses not found in *Daily Racing Form*'s past performance tables. Similar performance histories are potentially collectible for every horse on the circuit. Handicappers may choose to add domains, or more likely, drop a few, thereby modifying the schema or logical definition of the database. Note that in Table 10 the primary key now consists of two data items, Fan Club's number and its race dates. The number alone would identify any race record in Fan Club's history, but the number and race date specify a unique data record. When two data items are needed to represent a record uniquely, the primary key is called a concatenated key.

Fan Club, a three-year-old, has a history of nine races, or nine data records. Each record is displayed in a row. For each record the table contains twenty-two types of data items, of which twelve are performance data items, clustered around three handicapping factors: class, speed, and trips.

Excepting the latest race and four performance domains, the data values have been omitted. The domains with data items specified are eligibility conditions, ability figures, speed figures, and track bias. Notice the four are each identified as secondary keys. This is meaningful. Secondary keys have two functions: (1) identify data items that are pivotal for formulating data relationships and (2) guide the computer to the locations of records in physical storage. Handicappers are concerned with the first purpose. As tables are being constructed logically, the question should be asked repeatedly: Which of these data items form the basis for identifying a number of significant relationships in handicapping? Theoretically,

Table 10
PERFORMANCE HISTORY: *FAN CLUB, F, 3*

Horse #	*Date of Race*	*Race #*	*Race-track*	*Race Distance*	*Eligibility Conditions*	*Purse Value*	*Ability Time Figure*	*Manner of Performance*	*Race Variant*	*Par Time*	*Final Time*	*Variant*	*Adjusted Final Time*	*Speed Figure*	*Pace Figure*	*Race Shape*	*Trip Position*	*Trouble*	*Track Bias*	*Body Language*	*Official Track Condition*
0001	20 AUG	1ST	DM	6	C16	8	32	A	F5	110^{4}	109^{4}	F4	110^{3}	87	92	AF	IT	0	S	D	F
0001	30 JUL	2ND	DM	6	C25		30							81					S		
0001	18 JUL	5TH	HOL	6	C25		28							79					O		
0001	21 JUN	1ST	HOL	6	C25.5		27							78					O		
0001	23 MAR	1ST	SA	6.5	C25		28							79					GR+		
0001	8 MAR	1ST	SA	6	C255		28							79					GR		
0001	29 FEB	1ST	SA	6.5	C32		20							61					C		
0001	15 FEB	4TH	SA	6	M32.5		27							79					GR		
0001	2 FEB	4TH	SA	6.5	32.5		22							66					BR		
PRIMARY KEY	PRIMARY KEY				SECONDARY KEY		SECONDARY KEY							SECONDARY KEY					SECONDARY KEY		

The Performance History for *Fan Club*. The relation contains 9 records and 22 domains, of which 17 are "Performance" domains. The *Fan Club* relation has a primary key and four secondary keys.

all data items in a relational database can be identified as secondary keys. Nonetheless, the logical exercise pays intellectual dividends. It clarifies the relational nature of handicapping information.

Handicappers will recognize that the four domains selected as secondary keys in Fan Club would be essential data items for providing informational linkages about class, speed, and trips in combination.

Conditions of eligibility signify the class demands and class limits of today's race. Ability figures reflect a horse's demonstrated class, specifically an index of its late speed, as modified by considerations of form and early pace. These data items can be usefully related to each other and to similar data about other horses. They might be related in turn to other types of data items. Secondary keys can be thought to establish "chains" of data items throughout the database. The chains clarify complicated handicapping data and relationships, and facilitate the query function.

The secondary key for the speed data contained in Table 10 is the speed figure. Speed figures can be linked to eligibility conditions and to trip data. The secondary key for the trip data contained in Table 10 is track bias, precisely because biases can impact so strongly on speed figures. They can also inflate ability figures.

These four secondary keys have been judged here to reflect the most significant information handicappers covet about the relationships among class, speed, and trips. Identifying the significant secondary keys in relational database tables is another important logical task for high-tech handicappers. We will return to the discussion of secondary keys a little later in the chapter.

PROJECTION

Projection is an enigmatic database term but an elementary concept used to describe the simple operation of splitting a relation into two or more relations. In other words, of splitting a table into a number of smaller tables. Relational databases are characterized by extreme flexibility. The tables can be manipulated in many ways. Theoretically, for instance, handicappers can think of a thoroughbred's performance history as represented by a single monumental table containing the universe of performance data items possible. Similar tables could be imagined for all horses. In practice, the logi-

Table 11
FAN CLUB: CLASS

RACES		Conditions of Eligibility	Purse Values	Ability Times Ratings	Manner of Performance	Race Variant	Finish
20 AUG	1DM6	C16	8	32	A	F5	1^5
30 JUL	2DM	C25		30			
18 JUL	5HOL6	C25		28			
21 JUN	1HOL6	C25		27			
23 MAR	1SA6.5	C25		28			
8 MAR	1SA6	C25		28			

Table 12
FAN CLUB: SPEED

RACES		*Par Times*	*Variant*	*Adjusted Final Time*	*Speed Figure*	*Pace Figure*
20 AUG	1DM6	110^4	F4	111^3	87	265
30 JUL	2DM6				81	
18 JUL	5HOL6				79	
21 JUN	1HOL6				78	
23 MAR	1SA6.5				79	
8 MAR	1SA6				79	

Table 13
FAN CLUB: TRIPS

RACES		*Trip Notation*	*Track Bias*	*Body Language*	*Track Condition*
20 AUG	1DM6	IBITIS	S	DULL	FT.
30 JUL	2DM6		S		
18 JUL	5HOL6				
21 JUN	1HOL6				
23 MAR	1SA6.5		GR+		
8 MAR	1SA6		GR		

cal expression of such tables would be impossibly unmanageable. This theoretical table can be cut and pasted in many ways logically, forming many smaller tables of logical handicapping relations. As long as the design of the table has been proper, projection is simply a cut-and-paste operation by which a large logical table is split into smaller logical tables. The rows and columns can be rearranged as handicappers decide.

Handicappers can consider how the data records (rows) of Table 10 might remain intact, but data item domains (columns) could be rearranged logically to suit the special interests of class handicappers, speed handicappers, and trip handicappers.

Class handicappers might collect the class and trip data. Speed handicappers might collect the speed and trip data. Trip handicappers might collect only trip data.

On page 234 Table 10 has been *projected* into three smaller tables. Notice that a new data item (domain) has been added to Table 11, the class table. *Finish* expresses Fan Club's position at the wire, of course, with lengths won or lengths beaten specified. When a relation is changed, the logical representation of the database (schema) has been modified as well, but neither the physical storage of the data nor any application programs (class applications) that have been using the database need to be changed. Table 12 is a table of Fan Club's speed history. Table 13 is a table of its trip data.

In the three smaller tables the keys remain the same. The secondary keys are still eligibility conditions, ability time ratings, speed figures, and track biases. These keys represent the logical links among the data items in the tables, such that this chain of relationships is almost instantly available to high-tech handicappers who submit queries to the database from their terminals.

TABLES, TABLES, TABLES

On the next pages several tables have been constructed to represent portions of the handicapping database schema presented in Chapter 8. Some of the relations have been split into smaller tables. Several new data items have been added to the schema. Primary keys and several secondary keys have been identified, but some secondary keys have not. Handicappers should examine the tables closely. Attempt to recognize the unidentified secondary keys of most significance or utility to handicappers. These selections are subjective to a degree but secondary keys should provide strong links in chains of data relationships that will have wide general interest to handicappers.

To begin, examine the tables that represent the trainer relation. There are seventeen, no less. Almost any handicapper with a microcomputer can be expected to combine trainer statistics and trainer pattern data. The data in Table 14 and Table 15 were taken from Greg Lawlor's 1979–1983 trainer statistics for Santa Anita Park. Several data values in the fifteen smaller tables of *trainer statistics* and *trainer patterns* have been conveniently contrived, for illustrative purposes.

The trainer relation in Table 14 contains six records for a handful of prominent southern California trainers. The actual table would include practically all the trainers on that circuit. The table contains ten domains, or columns of data items, with *trainer name* the primary key. Although the domains termed *basic rating, specialty,* and *weakness* reflect a subjective element, they are primarily data-based. It may be surprising to learn that trainer Bobby Frankel has a weakness. Across the five seasons Frankel won with 9 percent of his first starters, his poorest showing. John Gosden, an A trainer, is unproductive with sprinters (8 percent). None of the six trainers listed have been rated F, but the rating is of genuine concern to handicappers. It refers to trainers that have never won a race or perhaps to nonwinners of two. Handicappers unfamiliar with trainer data would be amazed at the dozens of trainers at many tracks who belong in category F. Their horses can be summarily dismissed, especially when favored or short-priced.

Which are the secondary keys in the trainer table of most interest to handicappers? My personal preferences are three: *basic rating, specialty,* and *weakness.* I could query the database: List all the A trainers who are weakest with first starters. Get all trainers who are strongest when switching from dirt to turf. List the A, B, C trainers who are weakest on "off" tracks. For the C and D trainers, list their greatest strengths.

Table 15 shows the trainer-jockey sprint and route data items for journeyman trainer Warren Stute and eight jockeys he alternately rides. Despite the low numbers of starters in each case, handicappers would look twice at Stute sprinters under Fernando Toro, the routers carrying Ray Sibille.

Table 14
TRAINER

NAME	Starters	Sprint %	Route %	Basic Rating	Specialty	Weakness	Leading Rider	Longshot Rider	Major Owner
R. FRANKEL	570	23	16	A	DOUBLE CLASS DROP	FIRST STARTERS	PINCAY	CASTANEDA	FIRESTONE
D. FULTON	201	12	8	C	FINISHED 2ND LAST	FIRST STARTERS	MCCARRON	OLIVARES	STEINBRENNER
D. GOSDEN	139	8	20	A	ROUTES	CLASS RISE-SPRINTS	MCCARRON	CASTANEDA	SANGSTER
E. GREGSON	210	16	10	B	BACK IN 2 WEEKS	SURFACE SWITCHES	MCCARRON	TORO	NA
B. HEADLEY	250	14	17	B	BACK IN 8 DAYS	DIRT TO TURF	PEDROZA	TEVAOA	MABEE
D. HOFMANS	190	17	10	B	WON LAST	OFF TRACKS	MCCARRON	NA	NA

PRIMARY KEY

Table 15
TRAINER–JOCKEY

NAME — JOCKEY	Sprint Starts	Sprint %	Sprint ROI	Route Starts	Route %	Route ROI
W. STUTE — P. VALENZUELA	47	23	134	21	09	40
W. STUTE — L. PINCAY	12	33	133	7	00	(100)
W. STUTE — D. PIERCE	8	37	401	12	16	47
W. STUTE — L. ORTEGA	11	0	(100)	6	0	(100)
W. STUTE — F. MENA	9	0	(100)	7	0	(100)
W. STUTE — R. SIBILLE	NA	NA	NA	7	57	335
W. STUTE — F. TORO	6	33	456	5	20	174
W. STUTE — E. DELAHOUSSAVE	8	0	(100)	4	50	202

PRIMARY KEY

Notice that the primary key that uniquely identifies the trainer-jockey records includes both trainer name and jockey name. Either data item standing alone does not identify the record uniquely. Examine the six tables within Table 16 that describe a normal range of trainer statistics. The data items marked as secondary keys reflect the author's biases. I particularly like to distinguish A, B, and C trainers on their demonstrated abilities to win stakes, especially Gr. 1 stakes, and nonclaiming races for three-year-olds. If experienced trainers have not won a Gr. 1 stakes, I like to know that. If a trainer's win percentage with three-year-olds falls significantly below an overall win percentage, I want to know that too. I might query the database: Get the trainers who are 15 percent overall, but show less than 10 percent wins with three-year-olds.

Trainers can also be distinguished interestingly on their per-

Table 16
Six Trainer Statistics Relations

STATISTICS

OVERALL

Name	*Starters*	*Wins*	*Win%*	*ROI*
R. McAnally	433	66	15	99
M. Mitchell			20	
H. Moreno			11	
R. Mulhall			6	
H. Palma			14	

primary key (Name) · *secondary key* (Win%)

CLAIMING (clm)

Name	*Type of race*	*Starters*	*Wins*	*Win%*	*ROI*
R. McAnally	clm	36	8	23	93
M. Mitchell	clm			20	
H. Moreno	clm			11	
R. Mulhall	clm			8	
H. Palma	clm			17	

primary key (Name, Type of race) · secondary key (Win%)

Table 16 *(continued)*

ALLOWANCE (alw)

Name	*Type of race*	*Starters*	*Wins*	*Win%*	*ROI*
R. McAnally	alw	137	19	14	46
M. Mitchell	alw			10	
H. Moreno	alw			10	
R. Mulhall	alw			7	
H. Palma	alw			11	

primary key (Name, Type of race); secondary key (Win%)

STAKES (stk)

Name	*Type of race*	*Starters*	*Wins*	*Win%*	*Gr. Wins*	*Gr. 1*	*Wins ROI*
R. McAnally	stk	130	26	20	15	10	90
M. Mitchell	stk			2		0	
H. Moreno	stk			15		5	
R. Mulhall	stk			4		0	
H. Palma	stk			10		0	

primary key (Name, Type of race); secondary keys (Win%, Gr. 1)

THREE-YEAR-OLDS (3s)

Name	*Age*	*Starters*	*Wins*	*Win%*	*Stks Won*	*ROI*
R. McAnally	3s	47	12	24	3	32
M. Mitchell	3s			7	0	
H. Moreno	3s			8	1	
R. Mulhall	3s			7	0	
H. Palma	3s			12	3	

primary key (Name, Age); secondary keys (Win%, Stks Won)

FIRST STARTERS (1st)

Name	*Type of horse*	*Starters*	*Wins*	*Win%*	*ROI*
R. McAnally	1st	30	1	3	9
M. Mitchell	1st	9	0	0	0
H. Moreno	1st	14	0	0	0
R. Mulhall	1st	21	1	4	15
H. Palma	1st	33	3	10	206

primary key (Name, Type of horse); secondary keys (Starters, Wins, Win%, ROI) formances with first starters. The first starters table indicates that each of four domains have been selected as secondary keys. Many trainers who do well overall are washouts with first starters, à la Frankel. That's nice to know.

The database might be questioned: List the trainers who are 12 percent or better overall, have handled at least twenty-five first starters, but have won with less than 5 percent of the first starters. Table 17 reflects nine classic trainer maneuvers that are known to unearth winners at fancy prices. Most trainers specialize in one, two, or a few of the maneuvers, not so much due to tactical or manipulative actions, although this surely happens, but simply because their abilities and methods contribute to the patterns. As their income depends on getting winners, trainers become highly sensitive to their personal ways of winning, and they tend to repeat the same patterns. Handicappers can prosper alongside the winning trainers, provided they can recognize who does what. The primary keys of the nine tables would be concatenated keys connecting the trainer names to alphabetical values for the domains headed *maneuvers.* In each case the letter designations serve as a shorthand for the name of the relation or table. Individual high-tech handicappers can be expected to construct trainer pattern tables as they see fit.

Tables 18 and 19 describe the data relationships judged important for four relations: Racetrack, Track Surface, Pedigree, and Dosage. Handicappers should examine the tables to identify the data

Table 17
Nine Trainer Patterns Relations

TRAINER MANEUVERS

CLASS DROPS (cd)

Name	*Maneuver*	*Starters*	*Wins*	*Win%*
L. Rettele	cd			
R. Winick	cd			

CLASS RISES (cr)

Name	*Maneuver*	*Starters*	*Wins*	*Win%*
L. Rettele	cr			
R. Winick	cr			

Table 17 (continued)

REPEATERS (re)

Name	*Maneuver*	*Starters*	*Wins*	*Win%*
L. Rettele	re			
R. Winick	re			

SHORT LAYOFF (sl)

Name	*Maneuver*	*Starters*	*Wins*	*Win%*
L. Rettele	sl			
R. Winick	sl			

LONG LAYOFF (ll)

Name	*Maneuver*	*Starters*	*Wins*	*Win%*
L. Rettele	ll			
R. Winick	ll			

STRETCH OUT (so)

Name	*Maneuver*	*Starters*	*Wins*	*Win%*
L. Rettele	so			
R. Winick	so			

SHIPPERS (shp)

Name	*Maneuver*	*Starters*	*Wins*	*Win%*
L. Rettele	shp			
R. Winick	shp			

CLASS DROP-JOCKEY CHANGE (cj)

Name	*Maneuver*	*Starters*	*Wins*	*Win%*
L. Rettele	cj			
R. Winick	cj			

BEATEN FAVORITE (bf)

Name	*Maneuver*	*Starters*	*Wins*	*Win%*
L. Rettele	bf			
R. Winick	bf			

Table 18
Racetrack and Track Surface Relations

RACETRACK

Name	*Circum-ference*	*Length of Stretch*	*Par Variant*	*Overnight Purse*	*Stakes Money*	*Gr. 1 Stakes*
Arlington						
Atlantic City						
Belmont Park						
Bowie						
Churchill Downs						
Del Mar						
Hollywood Park						
Keystone						
The Meadowlands						
Saratoga						

↖ primary key — secondary keys

TRACK SURFACE

Name	*Date*	*Official Condition*	*Variant*	*Bias*
Arlington	Aug 1	ft	F1	0
Arlington	Aug 2	sly	F2	S
Arlington	Aug 3	ft	F2	S
Arlington	Aug 4	ft	F3	+S
Arlington	Aug 5	mdy	S2	BR
Arlington	Aug 6	mdy	S5	+BR
Arlington	Aug 7	sl	S4	C
Arlington	Aug 8	gd	S3	C
Arlington	Aug 9	ft	S2	C
Arlington	Aug 10	ft	F1	0

↖ ↗ primary key — ↖ ↗ secondary keys

relationships that the records represent. In the Racetrack and Pedigree tables, which data items might serve well as secondary keys to other handicapping relationships?

SECONDARY KEYS REVISITED

Two important points about secondary keys deserve some extra attention. First, handicappers are urged to be selective about the

Table 19
Pedigree and Dosage Relations

PEDIGREE

Sires	*Crops*	*Performance Index*	*Average Earnings Index*	*Dosage Index*	*Chief Status*	*Starters*	*Winners*	*Win %*	*Stakes Winners*	*Stakes %*	*Gr. 1 Winners*	*Gr. 2 Winners*	*Turf I.V.*	*Juvenile Rank*	*Mud Rating*

primary key — secondary keys

DOSAGE

Horse	*Age*	*Sex*	*Dosage Profile*	*Dosage Index*	*Center of Distribution*	*Distance Potential (furlongs)*
Althea	3	f		6.60		9
Devil's Bag	3	c		1.00		10
Swale	3	c		1.63		10

primary key — secondary key

data items that become identified as secondary keys. In a relational database, theoretically *all* the data items can be connected to all other data items. That's the promise of a relational database: total flexibility and absolute data item versatility. Yet there is a practical price to be paid. The data items that comprise secondary keys will themselves be stored in an index file in the computer. There may be hundreds of them. The computer uses the index files to facilitate its search for the appropriate data records in physical storage, just as shoppers search the indexes of large department store catalogues for the locations of items they may wish to buy. The larger the number of secondary keys, the greater the amount of storage space required for the secondary indexes. That leaves less storage space for other handicapping data.

More important, an overreliance on secondary keys interferes with the kind of rigorous thinking that constructing a relational database should encourage. Handicappers need to ponder the data relationships of most significance or utility in the tables they construct. If too many data items become secondary keys, almost reflexively, on the premise that all handicapping data is relational, the discriminating purpose of database development gets lost. Overlooked data relationships can be identified later.

Second, secondary keys play a determinant role in formulating the query. Data items that are keyed will be the crucial elements in most calls to the database.

In fact, a reliable technique for identifying useful secondary keys is to rely on the kinds of questions handicappers typically find themselves asking in vain when analyzing the past performances manually. The questions most frequently asked by handicappers, but not answered by the information at hand, reflect the problems they customarily encounter as well as the types of information they judge important for making handicapping decisions.

Do handicappers frequently find themselves wanting to know more than they do about trainers? About the real abilities of horses (class)? About the distance potential of three-year-olds? About the speed and pace capacities of claiming horses? About the trips of beaten favorites? About the recent biases of racetracks? About the body language or trip notes of "good" horses on days they performed so poorly or the track biases on those mysterious days? About the longer histories of older claiming horses? About the conditions and restrictions of allowance and stakes races? The questions never cease, or so it seems, and some are asked frequently enough by handicappers without adequate information resources as to be considered "keys."

The point surely is that handicappers who become alert to the kinds of questions they frequently pose about handicapping data will be better equipped to select the secondary keys of relational database tables. Secondary keys and the query function go hand in hand. The kinds of puzzles handicappers typically leave unraveled provide the clues to the secondary keys they should use, even as the secondary keys they do use determine the kinds of queries handicappers will be prepared to submit to the database.

SEARCHES

The computer uses indexes of primary keys and secondary keys to locate the data records in physical storage. A database has a table of contents, so to speak. It consists of the database keys. The primary key uniquely determines the physical location of specific records, but computers use secondary keys to search for numerous data records that are not unique to those attributes. When relying on secondary keys, computers must search at several areas of storage to find the appropriate records. A top speed figure of ninety-seven will represent several horses whose records are located in various parts of storage. Many times the user's information requirements are sufficiently complicated that secondary keys will be needed to link the data items into chains.

This is usually the case when ad hoc requests are made. For example, the computer receives the command, get me the three-year-olds with speed figures of ninety or higher under NW2 (nonwinners twice) allowance conditions whose trainers have a 12 percent or better win percentage at the route. The primary keys would be the names of horses and trainers, but four secondary keys are needed to link those horses and trainers to the other data items of interest: age, speed figures, allowance conditions, and win percent of routes. Without the capacity to link data items in multiple ways, the computer cannot deliver the desired handicapping information. The power of the information management approach to handicapping will have been sabotaged.

DBMSs FOR EXPERTS

As described in Chapter 8, the collection of software that manages the database is called the database management system (DBMS). A *relational* DBMS supports the development of innumerable tables and facilitates a marvelously flexible manipulation of the data items, records, and files.

Few owners of personal computers on the racing scene today have invested in a DBMS, but instead have poured hundreds of dollars into ill-tested applications programs of one sort or another. In time, the files of those applications programs will have to be converted to database structures. In fact, handicappers without computer experience will benefit if they purchase or develop no handicapping applications programs before first coming to grips with the logical schema of a personal handicapping database and before considering the DBMS they will buy to manage the physical system.

A high-level relational DBMS, which has both data management and programming capabilities will cost between $450 and $750. High-tech handicappers should resign themselves at the outset to paying the price while smiling broadly. The DBMS supplies the ultimate productivity and flexibility demanded by the information management approach to handicapping. Regardless of its cleverness, utility, or effectiveness, no handicapping applications program even approaches the power of the DBMS for resourcefulness in the age of information. The DBMS is the software centerpiece of the handicapper's information system.

The typical DBMS package consists of several interrelated languages and facilities, including a data description language (DDL), a data manipulation language (DML), a query language, and a command file for users to operate the system. The software system operates in ways that can be readily understood conceptually . First, the definition of the data items is communicated to the DBMS by means of the data description language. The DDL is thus used to build the data dictionary. Data items can be added, subtracted, or modified without upsetting the physical system.

Each DBMS has a data manipulation language used to (1) read and write records; (2) sort, merge, add, delete, or change the keys of logical linkages among the records; (3) find the records in physical storage; and (4) retrieve data items, records, and files that satisfy given search criteria. It is important to note that user commands to the DML can be inserted in applications programs whenever those applications need data items, records, or files from the database. The DBMS in that way acts as a buffer betwen the applications programs and the database.

In fact, the DBMS serves a much larger coordinating function in the physical system. Using a DBMS to store or retrieve information in a database involves coordinating the applications programs or query, the DBMS, the operating system (systems software), the channels that store data momentarily and usher it into main memory on signal, and the physical storage devices, mainly disks. The DBMS manages all those operations.

When an applications program interfaces with the DBMS, first the applications program identifies the database file (table), records, or data items involved. The DBMS checks the database and selected files to verify the presence of the data and turns control over to the operating system. Control eventually returns to the DBMS and the applications program. High-tech handicappers will mask multiple applications programs against the database across the seasons, a fascinating interaction within a computerized information system. Even when the database is under construction, many of the applications programs that will employ it are not yet known. They will result from future research. In 1984 alone handicappers were introduced to William Scott's form defects, Bill Quirin's race shapes, and William Ziemba's Z-System of place and show wagering. These ideas can be computerized; Ziemba's practically insist on high technology for their successful application.

Data items in the database (the schema) can be logically interrelated so that selected portions (the subschema) can be retrieved for processing. By creating a data dictionary, we identify the data items composing the database. The instructions of applications programs can be communicated to the DBMS to provide handicapping information. Technically, it works like this:

1. An applications program such as speed handicapping program is being executed. The program requires some data from the database and contains a command that will cause the needed data items to be retrieved. The control unit of the CPU causes each instruction of the applications program to be executed in sequence. When the DML command is

reached, the control passes from the applications program to the DBMS.

2. The DBMS verifies that the data requested has been previously defined in the schema and subschema. In addition, the DBMS uses various indexes to determine the physical storage address of the first item to be retrieved.
3. The DBMS requests the operating system to execute an input operation.
4. The operating system signals the channel to initiate the input operation. The channel causes the data to be accessed, read, and transmitted to a buffer storage area in primary storage. This is a special buffer *storage* used by the DBMS. Control passes from the operating system back to the DBMS.
5. The DBMS transfers the data from the buffer storage to the input area used by the applications program.
6. The applications program processes the data.

A FIELD OF THREE

As this is written, the most competitively intense activity in the evolution of microcomputer software has become the development of true relational database management systems. New, more powerful systems have probably already emerged and will continue to splash onto the market with the customary raves. A review in 1984 of a seven-part series entitled "Data Base Managers" in *PC Magazine* has persuaded the author that there are three high-level DBMSs most advantageous for handicappers. Each is worth buying but will appeal variously to handicappers exhibiting degrees of high-tech aspirations and skills.

The three are *Condor 3, dBase II,* and *Knowledge Man.* Practiced high-tech handicappers will recognize dBase II as the standard bearer in its field, the classic DBMS against which all rivals will be measured. The newest version of this prototypical system, an extension called dBase III, reportedly improves the system's much-remarked power dynamically. Yet handicappers should understand that the power of a DBMS must be balanced against its utility. Power refers to performance capabilities. Utility means ease and range of implementation and use. In general, power and utility connote an inverse relationship to nontechnical users. The more powerful a DBMS, the more complex it likely will be to users, certainly so to computer novices. dBase II is a very powerful system indeed, with its peculiar syntax of commands and programming capability, and for that reason will be preferred mainly by high-tech handicappers with relatively advanced experience and programming skills themselves.

Assuming most high-tech handicappers belong in the newcomer category, either *Condor 3* or *Knowledge Man* will suit fundamental needs better and comprehensively. Both systems boast excellent query capabilities, for example, with *Knowledge Man* offering handicappers more than two hundred commands to the database.

With practice, the DBMS analysts insist, the commands become surprisingly familiar quickly. *Knowledge Man* also comes with an integrated spreadsheet application, so handicappers receive two important functions for the price of one. *Knowledge Man* has no serious operational weaknesses, but its documentation—the instructions and illustrations provided to users who must implement the system—has been judged only "fair." The main complaints cite poorly developed illustrations and unsatisfactory explanations of several higher-order facilities.

Condor 3 stands out for handicappers who are brand new to computer technology. The system has the same essential features of dBase II, but it makes those features much easier to use. First of all, Condor's command language is unadulterated English, not a computer syntax and not an abridged English, as is true of *Knowledge Man*. Condor 3 is, quote, user-friendly. First-time users should experience no problems in employing its query language to sort and merge files, make *ad hoc* requests, or generate reports.

Below are the main features of Condor 3.

Terminology creeps in again during this brief discussion. Handicappers should remember that a database *table* is the logical expression of a file, the term most frequently used when describing events in the physical system. A *field* is the physical location of a data item. That is, the *logical* representation of data is referred to as a table, record, or data item, and the corresponding *physical* representation of the data is called a file, record, or field. In computer discourse the terms *file* and *field* are customarily preferred and enjoy the popular usage. Handicappers should not be confused about high-tech terms that are used interchangeably.

Condor 3 offers the following advantages and benefits.

1. The user interface is strictly a *command set,* and the commands are entered in plain English. Certain commands also invoke small helpful menus of operations that appear near the bottom of the terminal screen.
2. Condor 3 not only will guide handicappers in setting up files (tables) and naming fields (data items) but will let them change the file format or structures at any time, by using a simple FORMAT command.
3. Data entry is a cinch and includes a REPEAT command for the automatic re-entry of previously defined data items.
4. The query language is outstandingly flexible and easy to use. For example, to retrieve the dosage indexes of three-year-olds in the *Racehorse* file (table), handicappers would simply enter: List Horses by Age. 3, Dosage. Index.
5. Output can be *printed* simply by entering PRINT instead of LIST or other query commands, as above.
6. Updating is facilitated—not unimportant in handicapping databases—by a batch-execute command file that offers seven commands in addition to the query commands.
7. A HISTORY file can be created with each update. In effect, the file becomes a historical archive of horses' changing records across a season or several seasons. Useful.
8. While the REPORT generator of Condor 3 is considered problematic and time consuming, it can be circumvented by using two query commands, PRINT BY and TITLE. This allows handicappers to select the fields they want to see on the screen without producing an actual "report" in the report form.
9. Condor 3 comes with excellent documentation. It's relatively easy to learn.

In Table 20 *Condor 3, dBase II,* and *Knowledge Man* have been compared on twenty-eight database features. All three systems leave little to be desired.

Table 20
Three DBMSs for High-Tech Handicappers

Three relational database management systems compared on twenty-eight database features

Features	*Condor 3*	*dBase II**	*Knowledge Man*
Length of time on market	1977	1981	1981
Price	$650	$495	$500
Company	Condor Computer Corp.	Ashton-Tate	Micro Data Base Systems, Inc.
Number of files open at once	Unlimited	2*	Unlimited
Has required fields	Yes	Programmable	Yes
Files built from data dictionary	Yes	User creates	Yes
Index data	Yes	User defines	Yes
Able to modify fields	Yes	Yes	Yes
Error/prompting messages	Yes	Programmable	Yes
Multiple views of database	Yes	Programmable	Yes
Maximum fields per record	127	32	255
Maximum records per file	32,767	65,535	65,535
Demo version available	Yes ($45)	Yes	Yes ($50)
Can add or change indexes	Yes	Up to 7	Yes
Multiple updates on indexes/files	Yes	Yes	Yes
Sorting	32 fields	Multiple fields	Excellent
Flexibility of output	Good	Extremely flexible	Good
Multifile access	Yes	Programmable	Yes
Query output	Excellent	Excellent for experienced users	Excellent
Command languages	English	dBase syntax	Englishlike
Time to make ad hoc inquiries	Seconds	Seconds	Seconds
Time to retrieve random records	1 minute	1 second, with index	1 second, with index

Table 20 *(continued)*

Three relational database management systems compared on twenty-eight database features

Features	*Condor 3*	*dBase II**	*Knowledge Man*
Time to perform sort	2 minutes	2 minutes	2 minutes
Time to make standard report to screen	2 minutes	4 minutes	NA
Minimum configuration hardware	64K, DOS 2 drives	128K, DOS 2 drives	128K, DOS 2 drives
Documentation	Excellent	Good, but incomplete	Fair
Flexibility/ease of use	Excellent	Experience required	Good
Should handicappers buy	Yes	Yes, if experienced	Yes

* As this is written, a more powerful version of dBase II has just been developed by Ashton-Tate. It is called dBase III. For one of several reported updates, dBase III is capable of displaying ten files (tables) on the screen simultaneously. When utilizing a *horse history* database for the great majority of races, ten files covers a full field of horses.

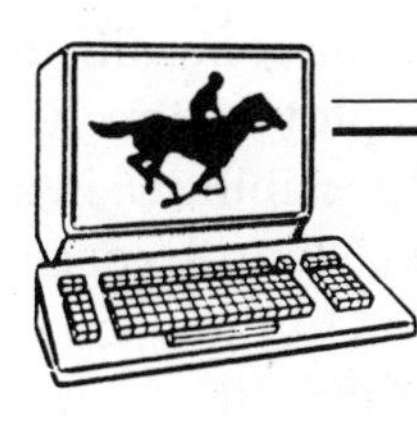

CHAPTER 10

INTUITIVE REASONING SKILLS IN THE INFORMATION AGE

Why do some people perform so much better than others of apparently equal ability? Or, to put it more broadly, what are the ingredients of success and failure?

CHRIS WELLES
"Teaching the Brain New Tricks," in *Esquire*, March 1983

IN A SERIES OF baffling experiments at Carnegie-Mellon University in Pittsburgh, computers with practically unlimited computational capacity were pitted against human chess masters. Cognitive psychologists who were trying to learn how experts think had programmed the computers to search all the possible moves and countermoves on the chessboard for a few moves ahead—an average of thirty-five moves is possible from a given position—and make the best possible move at all times. The human chess masters consistently won. How was this possible? What could account for the chess masters' edge?

The truth is usually elementary, and so it was in this case. The researchers eventually concluded that the experts had stored more knowledge than the computers and processed it much more efficiently. Whereas the computers considered all the possible moves before choosing a best one, the experts did not reason in that logical deductive way. The experts considered only the most probable moves. They screened out irrelevant information immediately. Because considering all possible moves involved a geometric multiplication of options, literally trillions of potential moves, the computers could think only a few moves ahead. But because they did not consider all the possible moves, just those most likely to result in success, the human experts were able to think several steps ahead of the computers. The experiments contributed much insight regarding expertise and mental performance that can be generalized to the best of handicappers. Experts typically employ extensive knowledge and experience to reduce the number of possibilities in a complex problem-solving situation to a highly select few. Experts deal in probabilities, not possibilities.

Moreover, their vast knowledge and experience permit experts to size up new situations quickly and sense or infer how best to proceed. Experts do not depend upon formal logical-deductive reasoning to reach their conclusions. They rely, if you will, on intuitive reasoning. Expert handicappers of the information age will want to do the same.

The Carnegie-Mellon researchers labeled the computer's thinking a "power" strategy, or the logical-deductive search of all possible options and their consequences. The experts' thinking was labeled a "knowledge" strategy, or inductive reasoning that moves relatively quickly from particular situations to general conclusions about events and their probable consequences. The knowledge strategy more closely resembles the mental process invariably employed by chess masters and other experts.

WHO ARE THE REAL EXPERTS?

Fortunately for thoroughbred handicappers as well as for chess players, one of the most prominent topics under scrutiny by cognitive psychologists today is mental performance and expertise. The idea is to discover what separates the experts from the novices, the superior performers from the inferior performers. Nowhere in the Carnegie-Mellon experiments can be found references to a study of racetrack handicappers—though that would be an especially fascinating population (they use real money to back their decisions) for the researchers to put under the microscope—but they have looked closely at chess players, securities traders, musicians, taxi drivers, lawyers, chemical engineers, cardiologists, and baseball fans.

What do the experts have in common? What distinguishes the very best performers in each field? Some tentative conclusions have been formulated and apply to regular practitioners of the great game of handicapping.

Memory experts were among the first to understand how the chess masters beat the computers. They pointed out that the key to the human success was knowledge. But it is knowledge of a special variety; not isolated facts or principles but knowledge that has been organized. The organization of prior knowledge was the secret of the experts' ability to retain in the working memory numerous chessboard positions similar to the new positions encountered during the computer challenge.

Novice chess players perceive positions on a board simply in terms of individual pieces sitting on individual squares.

Chess masters instead see positions as clusters of several pieces linked together. They automatically relate the clustered positions to a vast wealth of knowledge and information (experience) about chess positions, moves, and consequences of moves accumulated through years of playing chess.

By one Carnegie-Mellon estimate, the typical chess master has 50,000 board positions stored in his memory. When new or unfamiliar board positions present themselves as problems, experts can recall these "chunks" of stored information as easily as novices can recall single pieces. The experts integrate the new situations with the stored information and recognize or intuit swiftly what they should do. It's the relational nature of information on a single subject.

Other, more mundane experiments have demonstrated indisputably that memory improves dramatically when bits of data or pieces of information are clustered or chunked and associated with other information already stored in the memory. A spectacular illustration of the skill is the capacity of average people to remember as many as eighty-seven randomly ordered digits: 151867927484067-4562086736457203948574839609425. The trick is to cluster several digits into a group, such as "1518," and link each group of digits to meaningful information in the long-term memory—names, dates, images. A jogger who was trained to recall a string of digits remembered the "1518" cluster by linking it to the record final time for a certain marathon run, 15 minutes and 18 seconds. Meaningless sequences of words can be remembered when arranged into coherent sentences.

From these experiments a startling conclusion has been handed down to all of us. We can all become experts. Expertise is not a function of intelligence or aptitude but rather a result of accumulating a large store of "domain-specific" knowledge that can be accessed rapidly and efficiently. Experts have highly developed memory skills. The skills can be learned by just about anyone who cares. What really is involved, says a Carnegie-Mellon cognitive psychologist, is having an awful lot of *relevant* facts at one's disposal.

In studying expert mental performance in numerous fields, the researchers have also learned that superior performance does not relate to any kind of superior brain power but to *strategies for acquiring information*. Experts have a tremendous capacity for retaining information and experience that is new or unfamiliar by linking it in infinite ways to prior knowledge that has been stored in the memory base. They do it by transferring the new information and experience from the short-term memory to the long-term memory. The skill is called mnemonic association, the linking of new information to information already stored in the long-term memory.

The short-term memory is unrelated to expertise, as anyone who has soon forgotten the names of several new faces introduced at dinner parties will attest. Robert Bjork, professor of cognitive psychology at UCLA, says the typical short-term memory can retain only seven unrelated digits or letters, plus or minus two, for sixty seconds or less. That places quite a limitation on inductive reasoning, which relies on specifics as its raw data and proceeds from the particular to the general. It explains also why historically students have customarily been encouraged to learn general principles, rehearse them, store them in the memory, and apply them to new problems or situations that have familiar characteristics. But cognitive psychologists now demur. Once the digits, letters, data items, or information are successfully transferred to the long-term memory, they can be retrieved indefinitely. This is what experts do so well.

To make the connections, experts do not merely rehearse new information—repeat it to themselves—they reshape it, elaborate on it, organize it, and link it to other information that has some personal impact for them. The strategy works because the long-term memory operates through an almost infinitely complex system of interconnected information clusters and linkages. It's the experts' extensive accumulation of domain-specific knowledge, such as class handicapping, speed handicapping, or just handicapping, that allows them to form new linkages from new information sources and retrieve so much information so quickly.

CHUNKING

The memory skill of logically associating new information and new experiences with domain-specific knowledge is called "chunking," or expanding the knowledge domain. Memory chunks can become as large as experts decide. How do you improve expertise? Professor Bjork puts it simply: "Make bigger chunks." Experts do precisely that all the time, constantly relating new information and new experiences to what they already know. Experts consistently increase the size of their memory chunks by adding new information to them. The cognitive psychologists say nonexperts trying to retain new information—one thinks plaintively of novices and all that handicapping information—should do what the experts do; organize it into more comprehensive chunks and create retrieval cues by linking it to prior knowledge in as many ways as makes sense.

An outstanding example of the chunking phenomenon in handicapping is attached to the probability studies issued by Fred Davis and Bill Quirin in the seventies. These contributed numerous *impact values* (I.V.s) associated with several of the fundamental propositions of classical handicapping, or characteristics of handicapping as a general practice. No one remembers the impact values as isolated facts. Can handicapping experts list the ten strongest I.V.s in the game? They cannot. The probabilities of handicapping characteristics are not recalled in list form, not retrieved out of context.

Instead, handicappers use the data to *make larger chunks*. That is, experts have long since added the significant impact values to their prior knowledge domains about class, speed, form, distance, pace, trainers, jockeys, pedigree, mud, and on and on. When presented now with past performance information related to distance, for example, experts recall automatically the strongest impact values associated with the distance factor.

Suppose a horse is stretching out today from a sprint to a middle distance following a recent lengthy layoff. Experts look knowingly at the pattern of races preceding the stretch out. They *know* the probability of a horse stretching out successfully after a single sprint following a layoff has a decidedly poor chance of winning—though the actual I.V. may not be remembered at all.

Alternately, experts know that the stretch out following two sprints is the strongest probability pattern of all, far stronger than a single route and stronger than a sprint plus a route.

Likewise, when confronting the fall routes for two-year-olds, experts recall the probability data indicating two-year-old routes are the most unpredictable races of all. By the *chunking* technique, experts have associated probability values with knowledge of distance already stored in the long-term memory. In this manner experts can retrieve various probability data efficiently and use it to make better decisions.

When making class appraisals, though, they do not recall the actual impact values; experts remember easily that as a group horses dropping in claiming class win twice their rightful share of the races, while horses dropping by 30 percent or more win almost 300 percent their rightful share.

At the same time, experts can connect this chunk of the knowledge domain of class to other well-known information: that speed figures and back class are reliable indexes of whether claiming horses can stand class rises; that improving three-year-olds are more likely to move up in class successfully than any other thoroughbreds; that class barriers in nonclaiming races are usually tied tightly to the progressively stricter class demands of eligibility conditions; that two-year-old class is better evaluated by accurately adjusted speed or pace figures rather than by claiming prices or eligibility conditions. When observing demonstrations of class at the track, such as manner of performance or victory, or of troubled trips that interfere with expressions of real abilities, experts in turn link these experiences to basic principles of class appraisal and to the probability data. The "chunks" get bigger, as Professor Bjork has advised.

When new information about class is discovered in the future, those data are not rehearsed into memory to be recalled as isolated facts but rather are linked immediately to the existing knowledge domain and to other chunks.

As mentioned here before, recently Bill Quirin published a persuasive impact value for double-advantage horses, horses whose two best or recent speed figures are each superior to any figure of any other horse in the field and found by William L. Scott's ability times and Andrew Beyer's speed figures to represent exceptionally strong contenders. Quirin studied the assertions statistically. He reported the following results:

Double-advantage horses

NH	NW	WPCT	MPCT	I.V.	$Net
120	37	30.8%	55.0%	2.78	$2.74

Handicappers now realize that double-advantage horses represent one of the best statistical bets in the game. They win 278 percent their fair share of the races they contest, returning thirty-seven cents profit on each dollar wagered.

How is this information best remembered and used?

Experts integrate the statistics with other knowledge about class appraisal and speed figures, perhaps the greater importance of recent consistency in allowance races or the interpretation of ability times under various eligibility conditions or perhaps the interpretation of small numerical differences in speed ratings or the predictions of which horses are more likely to jump up in class successfully.

If the current concern is form analysis, experts have already integrated the most recent research with traditional principles and practices. As a result, experts have surely liberalized their form guidelines. Claiming horses can be out of action three weeks to thirty days now without being marked down as before, nonclaiming horses for six weeks or longer without being summarily dismissed. The highly specific form defects and form advantages discovered in 1984 by author Scott are chunked quickly onto the more liberal statistical guidelines on recency, and this revised knowledge domain is in turn linked to vital considerations of class appraisal and pace analysis, as experts intuit the critical relationships today.

As a result of the probability studies, too, the importance of early speed was recognized as not before, and this factor now represents a fundamental knowledge domain for handicappers of all stripes. Memory *chunks* of early speed information have grown larger fast and include information about track biases, daily variants, velocity ratings, and class drops. Many important decisions in handicapping today emerge from these *chunks* of early speed information.

To carry the thinking skills of mnemonic association and chunking to a logical conclusion, cognitive psychologists point out that the most sophisticated of experts rely on knowledge domains that have been organized into *hierarchical structures*. Expertise accumulates in stages, first by organizing a vast miscellany of facts into logical chunks and later by ordering the logical chunks into hierarchies.

A hierarchy is an ordered arrangement of information, such that the most important principles and ideas are positioned at the top, and the associated concepts, guidelines, practices, and facts descend in a line from them. Now new information and experience is not only related to prior knowledge, it is stored at specific levels of the hierarchy as experts determine.

The hierarchical arrangement of knowledge in handicapping is debatable in the abstract and situational in practice: Is class more important than speed; is speed more important than pace; does early speed belong above final time but below class ratings; or what? It depends on the situation. And because the relational nature of handicapping information is so complex and interconnected, forming chunks of knowledge, information, and experience makes more practical sense than does building hierarchies.

Handicappers become experts by organizing facts, data, and information into knowledge domains; by chunking knowledge, information, and experience; by increasing the magnitude of the chunks by linking new knowledge, new information, and new experiences to existing chunks in a variety of ways; and ultimately by sensing or intuiting the critical interrelationships in a context of particular racing situations. It's an unfamiliar but delightful process that might be referred to as intuitive handicapping. It's definitely a wave of the future, a brand of handicapping for getting ahead in the age of information.

The next chapter describes the guidelines and procedures of the intuitive approach in more detail and reports on its application during a day at the races at Del Mar racetrack in southern California.

The Chunking Test of Expertise

Handicappers who have battled this game for several years should submit themselves to the chunking test of expertise. It's an open-book test. There are no pat or correct answers to the pointed questions. The idea is to determine how much you know or don't know; how much knowledge, information, and experience you use or don't use. The best way to determine whether you are an expert is to determine to what extent you do what experts do. Experts make decisions by relating an extensive knowledge base, that is, all the information they can collect and as much experience as they can gather, to the situation (race) at hand.

If you are a systematic method player exclusively, for instance, such as a class handicapper or speed handicapper, in the purified sense, it's fair, if painful, to declare that you are not really an expert. If you are expert at what you do, whether it's speed handicapping, class handicapping, or neither, you are merely a highly proficient method player, but you are not an expert. If you restrict your play to races highly susceptible to the power of your methods, you may even be a big winner, but you still are not an expert. Sorry to take such a hard line on the point, but it's despairingly true and relatively important.

The test of expertise is simple. Write down your history as a handicapper. Emphasize change, evolution, insight, and growth. Describe how your knowledge base has grown, how your methods have broadened. Be as specific, but brief, as possible.

My own twelve-year evolution can serve as a model of how to take the test. In the early seventies I learned the principles and practices of classical, comprehensive handicapping promoted by the books of Tom Ainslie. In their application, I emphasized class, form, and pace. Class-form dynamics isolated my contenders, and pace ratings separated them. Distance and the other factors of handicapping were employed negatively; if potential winners did not measure up on what I called the secondary factors, I either passed the race or entertained second choices. I never employed speed figures, early speed, pedigree, track bias, trip notes, trainer data, or body language. At the windows, I practiced unit wagering of modest amounts, mostly, and rarely looked consciously for overlays. An early mentor had warned: Races are won on the track, not on the tote board. I believed that naive view a truism. As a result I supported many an underlay. Horses that figured to win were bet, prices notwithstanding. In consequence, my largest bets usually landed on short-priced horses. Many of these lost, of course.

Nonetheless, I emerged a consistent winner, even though profits for a season never exceeded $10,000, a reality that increasingly discomforted me, considering the time and effort I had committed to handicapping. In 1976 I read Beyer's *Picking Winners* but the speed techniques he promoted did not influence my thinking. They seemed to me, well, not exactly contrary to classical handicapping but superfluous. Much ado about something not altogether new. I considered brilliance, or speed, an attribute of class; still do. The crucial attribute, to be sure, but an attribute all the same. From this perspective I had determined the classiest of horses would also be the fastest of horses, and I already knew how to locate the class of the field.

What I had failed to understand was the significance of speed where other attributes of class were missing—endurance, willingness, competitive spirit, courage. In retrospect, it was dull of me. Speed horses certainly won a lot of races, particularly sprints, and all kinds of claiming and allowance events too. My personal peeve, widely proclaimed, was losing so many races to the "cheap speed horses." I blamed the glib southern California surfaces, not my handicapping. In fact, I had overestimated the role of class, underestimated the role of speed, in contemporary racing. I thought of myself as an expert handicapper at this point, but I was actually far removed from that status. Not until years later did I understand how my inability to recognize the role of speed in handicapping, and thus its several attendant factors, early speed, track bias, adjusted times, and trips, had prevented my real growth and development toward expertise.

A first breakthrough for me occurred when I read Bill Quirin's *Winning at the Races* in 1979. The book was statistical, a language I understood and respected, and it opened my eyes widely to the importance of early speed, enhanced my ability to analyze turf races, and contradicted with indisputable evidence a number of the principles and practices I had previously accepted as inviolate.

Quirin demolished as myth, for instance, the importance of the stretch gain; it was relatively insignificant. The book also suggested that the classical standards of form analysis had been much too rigid, too conservative.

As a result, I began to change my mind. I perceived at last that I had a lot more to learn about the art and science of handicapping. I decided to open my mind widely to change, accumulate more knowledge, broaden my methods, and vary my experience. I revisited Beyer closely, was absolutely persuaded, and determined to incorporate speed concepts into my methods, notably pars, daily variants, and adjusted final times. The next season I emphasized adjusted final speed in claiming sprints, relied on speed as a fundamental indicator of progress in the nonwinners allowance series, and demanded improving speed figures as a telltale sign of improving three-year-old claiming horses. Aspects of speed handicapping are now part and parcel of every race I analyze.

In the next years I continued to explore additional information sources. I decided that at the start of each season I would focus more intently on an aspect of handicapping I had previously undervalued. I also altered my daily practices in some basic way. In 1981 I emphasized trainer patterns and have since singled out a potentially profitable stable or two to follow in-depth with each new season.

In 1983 I chose trainer Pedro Marti, Laz Barrera's former assistant, who had impressed me briefly toward the end of 1982. It was a fortunate choice. Marti won with 20 percent of 105 starters, his winners including several generous overlays. When he started eventual Eclipse winner *Heartlight No. 1* in the Gr. 1 Hollywood Oaks following her maiden win, a maneuver that startled owner Burt Bacharach, I bet $1,000 on her nose. She won by 12 lengths and paid $8.40.

In 1984 I followed Neil Drysdale, an "A" trainer, unquestionably, who keeps a low profile in southern California and is therefore widely underestimated. Drysdale had just started a public stable after years of training privately for the Saron Stable of Corbin Robertson, an arrangement that produced the division leaders *Forceten, Hail Hilarious,* and *Bold N' Determined,* but in the main did not work well enough for Drysdale. The trainer had been performing far below his true level in the years preceding 1984—because Saron did not produce the live ones—and I anticipated backing a number of the public Drysdale's winners at nice prices. Drysdale did not disappoint. Early on, in January, he scored with the stakes filly *Bid for Bucks* at $35 down Santa Anita's hillside turf course under nonwinners allowance conditions. Later, in the Gr. 1 Vanity at Hollywood Park, the decisive test there for older fillies and mares, he beat Barrera's odds-on division leader *Adored* with *Princess Rooney,* at 5 to 1.

In following stables, the clever part is to identify outstanding trainers who are generally not perceived that way. It leads to very solid overlays.

In 1982 I attempted to become an expert on body language, but my aptitude for that specialty proved poor, my training worse, and I remain a dilettante on body language today—although I can occasionally spot a "sharp" horse I have favored on paper, which does sweeten the wagers.

In 1983 I concentrated on trip handicapping, and midway through the Santa Anita season spotted a troubled "trip" horse that was returning in seven days from Mike Mitchell's claiming barn with a drop from $25,000 to $16,000 and Chris McCarron. The horse won the ninth race at 8 to 1 and keyed a $1,160 Exacta. I reaped more than $5,000 on that race, my biggest score. I struggled with trip handicapping that season but was assisted in the fall with publication of Beyer's treatment of the subject in *The Winning Horseplayer.*

Coincidentally, Beyer played at Santa Anita that 1983 winter and it came to pass he alerted me to a $16.40 trip winner I otherwise would have ignored. I had not been a keen observer of truck biases, which are much less frequent and far less significant as a rule in southern California, and I did not have enough experience in 1983 to interpret them intelligently. In late February I bumped into Beyer along the mezzanine in the box section and we chatted briefly about an unattractive card.

Beyer asked if I had any strong opinions about the card. I did not.

Whereupon he suggested a horse in the fourth that had raced in the rail lane on February 11 against the strongest bad-rail bias of the meeting and had lasted until inside the eighth pole. I had no opinion about the fourth, other than it looked awful, a maiden claiming six-furlong sprint.

Beyer's tip was named *Es Mi Amigo,* a cheap speed type that had lost eight or nine maiden claiming races consecutively. I do not normally bet on previous maiden-claiming losers, one of my few remaining prejudices in handicapping. I therefore did not attach much personal significance to Beyer's information when first he shared it.

As the fourth approached, *Es Mi Amigo* was poised on the odds board at 7 to 1. I looked at its record again. It had the high speed points in the field by two, but had consistently quit. Suddenly, intuitively, I appreciated full force that I did not understand track bias, but Beyer did, and this pragmatic insight immediately translated into a fat overlay on the lone speed in a pitiful field. I had been eating lunch but removed myself from the table in time to bet $100 on a horse I did not imagine supporting an hour before. *Es Mi Amigo* went immediately to the lead and won handily, wire to wire, by five lengths.

In the two months I had known him Beyer had not touted me onto anything, in the fashion of the professional he is, so as I walked toward the press box to thank him for the information, I anticipated he had made another of his "killings," a $1,000 bet at least. Wrong. He had bet only $300 on *Es Mi Amigo,* Beyer said, because he feared another horse with a high speed figure.

Ironically, but unimportantly, the figure horse Beyer feared would not have bothered me at all, as I have found that high figures earned by maiden claiming losers are unreliable figures.

I mention the incident because it dramatizes not only how a good player can be a valuable element in a handicapper's information system but also how uncommon experience can enlarge the handicapper's knowledge domain. In this case my comprehension of early speed in relation to biases was improved. Following a $700 profit that was not anticipated and resulting from available information that had not been properly assimilated into memory *chunks,* a handicapper is likely to be more attuned to the prevalence of track biases, which I am now.

In 1984 I had intended to focus carefully on the foreign horses racing at Santa Anita—there were five hundred of them in the stalls—and prepared diligently for that chore months in advance by preparing an international catalogue of all the graded and listed stakes races programmed in seven countries, including grade designations, purse values, and eligible ages. But as the Santa Anita season approached I decided instead to alter my traditional methods drastically, embracing what this book has described as the information management approach to handicapping. I assembled and updated all sorts of information furiously and used all of it practically every day. The results were rewarding. Investing amounts ranging from $50 to $100 on most occasions, moderate amounts for full-time handicapping, profits at Santa Anita alone topped $15,000. Almost one-third of that was netted from information about foreign races in the stakes catalogue. A more ambitious betting strategy would have netted me a corporate executive's salary. The information management approach worked splendidly well at first tryout and will be repeated through the seasons.

So, my personal evolution as a handicapper has taken me from the relatively narrow confines of classical handicapping to the very broad strokes of contemporary practice; from strict adherence to systematic methods to a deeper reliance on information management; from making logical-deductive selections to making intuitive decisions. To my consolation, this kind of evolution simulates the directions usually followed in most fields by the experts. Forming bigger information *chunks* all the time. Linking all of them variously to one another as situations warrant.

Am I a handicapping expert, finally? Not yet. I lack a wide enough experience. In my modest opinion, in October 1984, I lost the World Series of Handicapping at Penn National Race Course due to inexperience with very small, highly volatile pari-mutuel pools. A key horse I passed at 3 to 1 at the seven-minute mark, when bets must be placed, won at 9 to 2. The maximum bet I had intended would have placed me third. Subsequent key horses all won. First prize was $90,000. A next phase for me is to broaden my experience to racetracks outside of southern California, especially to the medium-class and minor tracks. The wider and more diverse the experiences, the larger the information *chunks* will become. If chess masters have 50,000 board positions stored in long-term memory, handicapping experts should have a like number of race situations stored there. By that standard, I am not yet an expert.

INTUITIVE REASONING

Beyond memory structure, the truly exciting discovery of cognitive psychology tells us how experts use knowledge, information, and experience to solve problems and to make decisions. They do not use the formalistic deductive reasoning axioms and laws stressed in textbooks. They do not rely on hypothesis testing. They do not bother much with the application of general principles to particular situations. In fact, one of the more startling conclusions of the research on experts is that "deduction is a poor model of human thinking."

The illustration typically advanced is that of top medical specialists, the brightest of doctors, making their diagnoses. In Morton Hunt's classical book on contemporary cognitive psychology, *The Universe Within*, psychology professor Paul Johnson of the University of Minnesota asserts, "The expert finds logical thinking a pain in the neck and too slow." Johnson refers to the studies of doctors.

"So the medical specialist, for instance, doesn't do hypothetical deductive, step-by-step diagnosis, the way he was taught in medical school. Instead, by means of a wealth of experience, he recognizes some symptoms or syndrome, he quickly gets an idea, he suspects a possibility, and he starts looking for data that will confirm or disconfirm his guess."

UCLA's Bjork refers to this inductive process of reasoning as a reliance on "thinking schemes." By this he means again that experts have vast knowledge and experience so well organized that they can quickly relate new information and new situations to past learning, to past experience. Bjork's "thinking schemes" enjoy a formal name among cognitive psychologists—heuristics. Heuristics are broad, informal thinking principles that through trial-and-error methods provide demonstrably effective guidelines for discovering knowledge, processing information, and revealing relationships. The heuristic thinking of experts is characterized as *intuitive, inferential, nonaxiomatic,* and sometimes *illogical* or *creative.* I have summed up this kind of thinking for purposes of handicapping, fairly I think, as *intuitive reasoning.* The skill translates well as intuitive handicapping.

Effectiveness in intuitive reasoning depends greatly on the handicapper's range of experience. Experts get that way in large degree through the capacity to apply a vast amount of broad, theoretical, in-depth knowledge to wide-ranging practical experiences. From diversified experiences experts learn how to apply not only what they learn from books but also from prior applications of the knowledge.

Comparative studies of the problem-solving skills of experts and novices prove the point indisputably. When both experts and novices are provided with all the facts, information, and formulae needed to solve complex problems, experts invariably solve the problems faster and better. They know how to apply prior knowledge to problem situations that are new.

Consider the handicapper who notices that a horse in its latest race suddenly flashed early speed, where before its form looked dull. Last out the horse was fourth, beaten only three-quarters length at the first call of a fast six-furlong sprint. At the second call it raced fifth, beaten two lengths, and it lost more ground thereafter, finishing eighth of twelve, beaten ten lengths. Its last three lines look like this:

Points of Call				Fractions			
4¾	5^2	6^6	8^{10}	22	45^1	57^4	110
6^6	6^5	7^8	7^9	22^2	45^4	58^2	111
$7^{8½}$	5^7	5^7	7^{10}	22^3	45^4	58^3	111^2

Novices attach little significance to the latest running line. After all, classical handicapping instruction on form analysis defines a "good" race in sprints as (a) finishing in the money or (b) finishing within two lengths of the winner. An "acceptable" race has been defined as beating half the field and finishing within seven lengths of the winner.

William L. Scott's recent empirical research on form analysis considered the last running line a key indicator of positive or negative form, and at six furlongs assigned a form defect to any horse not within 2 3/4 lengths of the leader at the stretch call, with some exceptions.

So novice handicappers eliminate the horse that flashed impressive early speed for a quarter mile, but not much more. The novice sees a horse unacceptable on classical guidelines of form analysis—a "short" horse. In logical deductive terms, the novice is absolutely correct. But experts see much more.

The expert seizes on the flash of early speed as a sign of life, an indicator of improved form, and quickly begins to associate that signpost with a wealth of knowledge and experience organized in the long-term memory. Exactly as a top medical specialist does, the handicapping expert has recognized a telltale symptom, and he instinctively searches for other data to confirm or disconfirm his guess, or intuition, that this might be a meaningful sign.

How might handicapping experts proceed from this point? They proceed inductively from specifics to general conclusions. The process of inquiry might meander along these familiar lines: Is the horse dropping in class today? If so, the horse can be expected to move up dramatically, as improving form and a drop in class often return positive results.

Is there a favorable jockey switch, engineered by a winning or clever stable? Fundamental advantages, such as class drops and improving form cycles, are often intensified by incidental advantages, experts realize.

Will the horse be advantaged today by a better post position and trip? Experts know sprinters that break from the far outside at six furlongs and press a fast pace down the backside and around the far turn often get parked outside and are taking an extra wide trip that saps energy reserves. An inside post next out should help move that horse up, perhaps by lengths. Suppose the six-furlong trip last out looked like this:

Slo-1, 4B, 4T, 5E, 5S,

indicating that the horse was a length slow out of the gate, raced four-wide down the backstretch, four-wide around the far turn, and five-wide entering the stretch. How does that information connect to a flash of speed for a half mile against a fast pace?

Will today's field be paceless? Will the early pace be weakly contested? What of the class, form, and early lick of other frontrunners in the field? Presumably, the last race was hotly contested early—the horse was wide all the way against a fast clock. If today's probable pace should allow the improving horse to breathe easier, how much more might it move ahead?

Is there a bias favoring inside early speed today? Was there a bias against the outside speed last out? If the rail were advantaged while our improving hero were racing four-wide, its brief speed and improved form would be all the more impressive. If the bias today will assist the horse's front speed, move it up!

What about the horse's basic class or its back class? Can it beat this competition when in shape to do so? If yes, then the horse may figure strongly today, if the pace will be softer too. Maybe it can throttle this bunch, when ready?

What about the reverse sides of those possibilities? Is the improving speed horse up in class today? Is the trainer weak, the jockey ineffective? Will the early pace today also be contested, and by classier horses? Was the horse advantaged by a bad-rail bias last out, but probably disadvantaged today? Does the horse lack intrinsic class at this level, even when sizzling fit?

Finally, what are the odds? Is there attractive value in the potential payoff offered by the public? Is the horse a relative underlay or overlay? If the play were chanced one hundred times, how frequently could handicappers expect the horse to win? What to do? What to do?

Novices cannot be expected to know—they are best advised to throw the horse out, to be sure, and stick to the fundamentals until enough experience teaches them how to proceed differently. But experts already know what to do. They combine vast knowledge with vast experience. They are experts. They add up the situation accurately, decide, and act.

Experts proceed through this sort of inductive, intuitive reasoning process all the time. They do it quickly, sharply. They go beyond the logical deductive style of thinking that eliminates the early flash of speed out of hand as well as many similar handicapping characteristics that are untenable in classical, fundamental terms. Instead, they recognize a potentially vital symptom of impending success, add to that all the relevant facts and current information available to them, and make an elaborately informed and interconnected decision. They may toss the improving horse out in the end, but not at first glance.

The process of confirming or disconfirming original clues as meaningful or not is called *additive weighting*. It's the central concept of intuitive reasoning in handicapping. As each data relationship is clarified and understood, experts add or subtract some value from the whole. Values may be quantitative or qualitative, objective or subjective, no matter. As the *additive weighting* process continues, the original intuition grows stronger or weaker. Potential handicapping decisions begin to take on an undeniable force of their own.

In the illustration provided, if the improving speed horse will benefit by a real drop in class today, be mounted by a better jockey, face a relatively paceless field or at least a slower early pace, be advantaged by the track bias, and exhibit "back class" that can handle today's competition, obviously the original intuition will have been reinforced to the breaking point. If the odds will be delightful, favorable, or even acceptable, a bet will be placed, perhaps a sizable bet.

Most times the process of *additive weighting* is more uneven and inconclusive. Scenarios develop, not final selections. Races are far from indecipherable, but many remain unplayable in strict classical terms: the single selection at favorable odds. The intuitive thinking process continues. A critical final factor becomes the size of the odds. Will a horse enjoy a reasonably strong chance to win and return fair or good value for the wager invested? If both questions can be answered positively, the bet makes sense. Some would insist the horse cannot be ignored; it must be bet.

So intuitive reasoning at the track begins as often as not from sensing the potential importance of a small amount of information—a flash of improved early speed that may or may not prove significant—and proceeds through a process of uneven *additive weighting* in which the original clue or symptom is associated with all the other handicapping information and the full range of knowledge and experience organized in the long-term memory. As information and data relationships are added or subtracted from the equation, the intuitions grow stronger or weaker. Final decisions normally take on a force of their own; sometimes they cannot be resisted. Or they can depend variously on a cost-benefit analysis of the relative chances of several horses to win or place and the potential values of wagers on each.

That's how the experts do it. Broad knowledge. Information. Wide-ranging, in-depth experience. Memory structure. New information. Problem situations. Heuristic thinking schemes. Inductive reasoning. Information-processing skills. Additive weighting of relations. Intuitive decision-making.

A helpful first step for regular handicappers who want to become experts is to become explicit as to the information clues or symptoms they have found especially significant in leading to the kind of intuitive thinking and decision-making process outlined here, and resulting at times in some of their more memorable collections at the windows.

Expert speed handicappers will have little difficulty, for example, identifying with the reasoning processes described above, in relation to a recognition of much-improved speed for a quarter mile in a six-furlong sprint. They'll probably note that they've made some of their biggest scores in precisely that way. They are experts, however, and nobody should act surprised. But speed novices cannot do it. They usually do not have enough knowledge and experience. Even when they've read the books and collected the information, perhaps in a personal computerized database, as recommended here, novices still lack the practical experience. So they remain novices.

My own intuitions, not surpisingly, often involve a stable's manipulations of horses and eligibility conditions. Whenever I spot the following maneuvers, I take serious notice, as I know from vast practical experience that something interesting and rewarding might be afoot.

1. Class drops from advanced nonwinners allowance conditions (NM3x) to preliminary nonwinners allowance conditions (NW1x or NW2x).
2. In analyzing three-year-olds, a move from a powerful preliminary nonwinners allowance win to a restricted stakes *late in the year.*
3. A strong second or third place finish in a claiming race where the winner was dropping from higher claiming levels.
4. A high-priced claiming race win to a preliminary nonwinners allowance race, again *late in the season,* and assuming no late-blooming three-year-old is moving through its preliminary allowance conditions and looks conspicuous in this field.
5. A decent Gr. 1 or Gr. 2 stakes performance into any other stakes.
6. A minimally restricted classified allowance race into either a highly restricted classified race, a nonwinners allowance race, high-priced claiming races, or restricted stakes that bar former stakes winners. (You'll get juicy winners from this ploy.)

Other handicapping signposts I follow intuitively:

1. Improved early speed for a quarter mile in sprints, as described above, notably when accompanied by "back class."
2. A leading turf sire in grass racing, notably when young horses try the grass for the first or second time.
3. Foreign imports exiting "listed" stakes overseas and entered under allowances or minor stakes conditions here.
4. A strong positive speed bias that hurt a classy off-pace horse with acceptable front speed—not plodders.
5. Trainer pattern data at high odds, notably in weaker races.

When confronted by any of the handicapping information above, I anticipate the long-searching process of intuitive handicapping. But I know from practical experience it will often result in my best bet of the day or week.

For the same reason speed handicappers stand on alert whenever they find a figure horse whose adjusted times reveal it as superior to horses with apparently faster clockings; pace analysts react the same when horses with adjusted pace ratings shine in comparison with horses whose class ratings or unadjusted speed ratings might look fancier to the crowd; trip handicappers respond the same when they isolate competitive horses whose bad trips remain invisible to the public.

Cognitive psychologists are also busy investigating the creative thinking process, but haven't reported much evidence as to what separates geniuses like Picasso and Mozart from ordinary experts. They promote vague strategies only, such as forcing oneself to think in unconventional ways, so you might at least consider farfetched and improbable but potentially innovative alternatives. On this matter, so far the findings of cognitive psychology sound very much like the creative thinking of horseplayers landing on their longshots. Better handicappers should forsake the leap to genius, and try to become experts.

All handicappers can develop the expert's skills. All it takes is massive knowledge, all the information that's fit to collect, a vast practical experience, and highly polished intuitive reasoning skills. Of the latter, handicappers will be reassured to know that even high school students of the information age will be expected to develop intuitive thinking skills. These are thought by curriculum experts to be the critical thinking skills for dealing effectively with mass amounts of information that is new and unfamiliar. For instance, *futures* curriculum of reasoning skills specifies the following learning outcomes for all high school students:

Outcome Statement:

The student will exhibit intuitive and inductive thinking in problem-solving.

Competencies:

1. Know that mnemonic association is a memory function that recalls new information by relating it to dates, times, and images already stored in the memory.
2. Given a series of digits or random numbers of twenty or more, recall the digits through mnemonic association.
3. By "chunking," recall unfamiliar information items that are related variously to an already developed knowledge base.
4. Given facts, information, or knowledge, and practical situa-

tions/experiences that are poorly defined and characterized to an extent by irrelevant information, identify the specific facts that are relevant to solving the problems or coping with the situations at hand.

5. Given problem situations, make intuitive and inferential decisions supported by the given information.
6. Given practical problems (real life) and alternative solutions, estimate effectively the risks versus rewards or costs versus benefits of the proposed solutions.
7. Use cost-benefit analysis to select the most cost-effective of competing business problem solutions/decision alternatives.
8. Handicappers who can perform those thinking skills are far along the path to becoming expert information managers and intuitive problem solvers.

GENERALISTS BEAT SPECIALISTS

Specialists apply the principles and practices of systematic handicapping methods to all races. Generalists apply broad knowledge, current information, and practical experience to all races, such that alternative ideas and diverse methods are variously employed and decisive. Speed handicappers are specialists. Class handicappers are specialists. Disciples of this book will be generalists.

In the age of information, generalists will beat specialists. They will have access to more information and will collect the full-blown knowledge and experience by which to process it, not to mention that they will have microcomputers and databases to enhance their processing capability. This does not imply that specialists cannot also be expert handicappers and winners. That is altogether possible, but several conditions, interrelated, have conspired to thwart the contemporary specialist.

First, systematic method players uncover only a few potentially playable horses and wagering opportunities a day, usually at relatively low odds. The sport has been moving in other directions, providing more handicapping information and betting opportunities than could be imagined just a decade ago. Specialists might wait patiently for the appearances of key selections at acceptable odds, but simultaneously they will either sacrifice their motivation to boredom or wager indiscriminately in small ways on other races. This shrinks the profit margin, lowers the dollar return, and dampens the spirit.

Second, systematic method handicappers face stiffer competition from other bright, well-informed specialists all the time. As the literature of handicapping has grown and blossomed, more handicappers than ever are fully capable of identifying the class standouts and top figure horses. Increased competition for the key selections of systematic methods has depressed the odds on those horses steadily for years. Because it relies on past performance data exclusively, classical handicapping is practically passé on this point. Class handicapping and speed handicapping have suffered seriously from the same affliction. These trends can be expected to persevere.

Third, fresh information sources have bolstered the arsenals of handicappers who are willing to become information managers, and this has dealt a corresponding blow to the relative prosperity of the specialists. The trendy sources of handicapping information today include trips, pace, pedigree data, body language, biases, and trainer stats. Pace excepted, these data are not exactly fodder for conventional systematic methods. Information buffs will find the true overlays more frequently and, therefore, will receive greater value from their wagers.

The significant handicapping literature of the past two years alone supports the case for the generalists. In each of three major book publications, nationally reputed authors broadened methodologies that had been rather specifically defined in their previous works.

In *The Winning Horseplayer* (1983), Andrew Beyer urged the juxtaposition of speed figures and trip notes. Thus Beyer's advanced speed handicapping is broadened now to encompass trip handicapping. Beyer's key selections and big bets today depend variously on interplays among speed figures, trips, biases, and trainers.

In *How Will Your Horse Run Today?* (1984), William L. Scott broadened his ability times methodology to honor new developments in form analysis. Scott's preoccupations now combine speed, class, early speed, and form. Rumor has spread that Scott is presently studying exotic wagering techniques, an apparently radical departure for an author who has consistently concentrated the applications of his research on the two or three top betting choices.

In *Thoroughbred Handicapping: State of the Art* (1984), Bill

Quirin broadened his speed methodology to feature speed and pace figures in combination, tilted strongly in favor of using trip information alongside speed numbers and nodded approvingly in several directions littered with varying types of handicapping information.

Quirin emphasized more than once the contemporary handicapper's need for types of information not contained in the past performance tables, a theme this book has intended to expand on dramatically.

Most professionals of today have long since begun the inevitable process of broadening their ideas and methods. Within a few months of the book's publication, my friend and colleague Ron Cox, of San Francisco, a top-notch professional, had turned full bore toward Quirin's concept of "race shapes," which encompasses nine configurations of pace. Cox was even supplying that information for all races at Bay Meadows to subscribers of his weekly handicapping information service in northern California.

Steve Davidowitz, whose writings on handicapping have always been characterized by a highly divergent experience and eclectic points of view, continues to broaden his actual experience, now handicapping at the new Canterbury Downs in Minnesota for the *Minneapolis Star Tribune* and previous to that for Florida tracks during winter, Atlantic City during spring and summer, and independently at other times and circuits. Davidowitz more than most has accumulated the ingredient that prevents many excellent handicapping practitioners from joining the ranks of experts—abundant experience at a variety of racetracks.

For the same reason, Beyer's reputation among practitioners as a storied individualist, big bettor, and genuine expert is justified on all counts. It derives from wide and varied practical experience that has been well chronicled, by Beyer himself at first and increasingly by interested observers.

It is important to note that Beyer's reputation as a genuine expert does not derive from his writing two influential books. As an author Beyer imparts considerable knowledge of handicapping, but that leaves two criteria of expertise—information-processing skills and wide practical experience—unaccounted for. Yet his fellows will testify that no one collects and processes more updated information about handicapping than does Beyer (amazingly, he does not yet use a microcomputer), and no one who has played alongside the man for a time would dare to trifle about his range of experience. The reputation is deserved. Beyer fits the definition provided by the cognitive psychologists. He is a genuine expert.

Current trends in handicapping are clear. Handicappers who are specialists today are well-advised to become generalists by tomorrow. Those who do will beat the specialists to the bigger mutuels and will do it by relying more and more on intuitive reasoning skills. Those handicappers will be doing what experts in all fields do, according to the persuasive conclusions of social science.

CHAPTER 11

INTUITIVE HANDICAPPING

I keep reading the entries in the *Racing Form*, over and over. I don't make any notes of specifics. I just keep reading all the entries, again and again, and gradually I grow to despise a few horses in the race. I stop looking at them. After a few more reads I start to get all warm and glowing over two or three. Finally, I fall in love with one of them. That's my pick.

MILO MULCH,
Member, American Mensa,
the Handicapping Special Interest Group, explaining in Mensacap, June, 1982, how he came to pick seven winners for the members on their group trip to Churchill Downs

FOR REGULAR HANDICAPPERS who depend on the deductive logic of systematic methods for their contenders, selections, and most serious wagers, the truly revolutionary aspect of this book is not a high-tech information management approach to handicapping utilizing personal computers and databases. It is the book's promotion of new ways of weighing information and making decisions. It is intuitive handicapping.

When Andrew Beyer cashes ninety-five grand in the feature Exacta on Belmont Day, he doesn't do it by backing the logical deductive selections of speed handicapping. Instead, he has added up a multitude of facts and data relationships from various information sources in the specific context of two races and *intuited*—inferred, sensed—from the weight of all the guidance the highly probable outcomes at distinctly favorable odds. The handicapping situation achieves a force of its own, Beyer is sensitive to that, and he plays to it.

Among the several advantages of intuitive handicapping in the high-tech information age is the enhanced possibility of reaping huge rewards in the short run, the big kills that are not the province of systematic method play and logical deductive handicapping. But besides comprehensive knowledge and information, success at intuitive handicapping is dependent also on extensive experience, the practiced skills of applying the knowledge and information not only accurately and effectively in familiar situations but imaginatively in situations that are new, infrequent, or improbable. Few regular handicappers armed with the knowledge and information Beyer possessed on Belmont Day in 1984 would have approached a similar score. I could not have done it.

Most regular handicappers lack the intuitive experience to size up highly favorable handicapping situations in terms of the real possibilities. They are tied too tight to systematic methods and the logical selections they typically provide. The intuitive decisions depend on reasoning skills of different colors. These are highly advanced, artistic skills. They become refined only with practice, with repeated applications in a variety of situations. Handicappers who ply the trade at tracks on various circuits across the nation will benefit in ways denied to colleagues who stay close to home base.

INTUITIVE HANDICAPPING

To attempt a definition, intuitive handicapping refers to the *inductive, nonaxiomatic* judgment as to how knowledge, data, information, experience, and basic handicapping principles can be organized theoretically and linked empirically in specific situations or races. Intuitive handicappers rely on the *additive force* of knowledge, information, and experience. They combine the three sources to make the best decisions about the race situations at hand.

Earlier I mentioned that my handicapping friend Lee Rousso, at age twenty-five, beat the races by applying everything he knows to every race he plays. Rousso is not easily classified into a handicapping routine. He is not a class handicapper but uses class appraisals to make some decisions. He is not a speed handicapper but uses speed figures to make some decisions. He is not a pace analyst but uses pace ratings to make some decisions. He is not a trainer specialist but uses trainer data to make some decisions. He is not a trip handicapper but uses trip notes to make some decisions. He is not a bloodlines specialist but uses dosage data and turf I.V.s to make some decisions. He is not a body language expert but uses post parade inspections to make some decisions. He is not a devotee of form but uses evaluations of form cycles to make some decisions.

Rousso adds up everything he knows about the field of horses under scrutiny, forms linkages between that knowledge and vital information about racetracks and trainers, and makes final decisions that reconcile the best betting values with the most probable outcomes. Rousso is an intuitive handicapper. His distinction as a big winner is not accidental, and it might be usefully pondered by struggling handicappers wedded to the selections of systematic methods. The systematic methods are not working well enough. Among regular handicappers I know, Rousso boasts the highest average mutuel on his winners. He finds overlays consistently. The patterns are real, not random.

Handicappers should remember, for it is second nature to forget, that the essence of intuitive handicapping is a *reliance on knowledge, information, and experience in combination to make decisions* and *not* to identify contenders or make selections. The distinction is crucial and difficult to adhere to in practice. An approach that emphasizes information management and intuitive thinking eschews the reduction of a field to contenders and noncontenders, based on some model of handicapping, however broad and inclusive, and ultimately to final selections at acceptable prices. Intuitive handicapping instead looks at the same field of horses in multiple ways, sometimes with opposite logics, encompasses the full range of possibilities, and eventually reduces the handicapping problem to potential outcome scenarios and alternative decisions, from which final choices must be made. Conventional handicapping practice is selection-driven, but intuitive handicapping is decision-driven. Intuitive handicappers make decisions, not selections.

At the same time intuitive handicapping utilizing information management is an approach, not a methodology. A methodology consists of the systematic application of ideas in a sequence of steps or procedures, as with the procedures of class handicapping or speed handicapping.

An approach is a global, generalized orientation. Within a generalized information management approach handicappers can practice a favorite methodology or several methodologies. An analogy is the masking of multiple application programs against a database. In my personal routine I have hardly abandoned my class appraisal methods, which evaluate horses by comparing the full records against the typical class demands of today's eligibility conditions. Indeed, I incorporate this class evaluation method into every handicapping situation I confront. But my general orientation is now much broader, encompassing all sorts of information and relying on intuitive reasoning to arrive at final decisions.

Intuitive reasoning in the practice of handicapping proceeds through the five stages associated with the information management approach. The two correspond well, as intuitive thinking is the most efficient way to deal intelligently with relatively large amounts of information that must be processed in a problem-solving or decision-making situation. These stages were defined and elaborated during the earlier discussion of information management in Chapter 5. They are revived here in brief:

1. *The prerace information set.* This entails comprehending the class demands of the eligibility condition and recognizing the implications of the running surface and track biases for evaluating the past performances in terms of speed, pace, and trips.
2. *Information analysis* of the past performances. This phase intends to determine the weight of information each horse has in its favor today. Horses eliminated here practically eliminate themselves. They do so by default, by not revealing any information that supports their cause.
3. *Problem-solving.* This phase relies on the components of the management information system, that is, results charts, databases, applications programs, handicapping colleagues, to answer unresolved questions or explore data relationships that can strengthen or weaken the case for individual horses.
4. *Outcome scenarios.* This late phase identifies the probable outcomes of races, depending variously on how races might be run or which handicapping factors might prove dominant. Outcome scenarios reflect the alternative decisions to be evaluated.
5. *Making decisions.* Intuitive handicappers make handicapping and wagering decisions by relating the probable outcomes to the relative value of each, as indicated by the public odds. The best decisions will *intuit* the most likely outcomes at the best prices.

THE KNOWLEDGE CHECKLIST

If successful handicapping decisions depend on forming the linkages between knowledge, information, and experience in specific racing situations, the most important element in the equation is knowledge. The effective interaction of knowledge, information, and experience is dependent first of all on the handicapper's range and depth of knowledge. As we have seen, handicapping information is processed by linking it to domain-specific knowledge, such as knowledge about pace or about pedigree. Experience at the track is meaningless unless it can be interpreted in terms of the skillful application of knowledge, as when handicappers learn to appreciate how a GR+ track bias can propel cheap speed along the rail to defeat superior class running in the more tiring outside lanes.

Without a handicapping knowledge base, in fact, experience at the track can be the worst of teachers. That explains why so many veterans who have lagged so far behind the contemporary handicapping literature are numbered among the sorriest losers at any track. Perhaps it also explains in part why racing has not yet been able to attract the young adult market (learning to play the races sufficiently well to avoid the kind of financial shellacking young adults cannot afford normally requires more than occasional junkets to the track for entertaining features or slick giveaways), but that's another story.

So before information systems are built and vast experience is accumulated, a first step toward becoming a proficient intuitive handicapper is to acquire a massive academic knowledge of handicapping. Experience will turn academic knowledge into working knowledge. The information system will support the working knowledge by informing decisions to be made. If handicappers cannot respond positively to the items on the following checklist, they have fallen behind the contemporary knowledge boom in handicapping. Here are sixteen items:

1. Can you cite the general statistical impact of early speed on race outcomes, and do you know how to use *speed points* to estimate the early speed potential of horses?
2. Do you *know* the indicators of *form defects* and *form advantages*? What's the most advantageous use of *form defects?*

3. Do you understand the typical class demands of the various conditions of eligibility, and can you associate eligibility conditions with the kinds of past performance profiles that are best suited to each?
4. Can you calculate *ability times?* Do you know what a *double-advantage* horse is?
5. From raw final times can you produce adjusted final times and speed figures? Do you know how to calculate par times and daily variants?
6. Are you aware of the latest information on pace analysis? Can you apply an arithmetical method of expressing pace? What are velocity ratings? Can you interpret pace ratings?
7. Do you know the nine classic *race shapes?* Can you identify the kinds of horses advantaged and disadvantaged by each?
8. Do you know the recommended procedures of *trip* handicapping, using systematic observation skills and notation?
9. Can you recognize track biases? Can you describe the customary biases and can you explain how to interpret past performances in relation to them?
10. Do you know how to conduct paddock, walking ring, and post parade inspections? Can you describe the six basic profiles of body language? If horses are sweating, can you tell whether the lather is part of a *frightened* profile, or is it instead a *sharp* horse, primed to win?
11. Do you know the most significant statistical *impact values* associated with the numerous past performance characteristics?
12. Do you know what a dosage index is? Can you interpret and apply the dosage indexes of impressive three-year-olds? Two-year-olds?
13. Do you know how to interpret the grade designations of stakes races? Can you discuss the stakes hierarchy in this country and explain its ramifications for handicappers?
14. Can you tell which foreign horses have a class edge in United States allowance and stakes races?
15. Can you spot horses that are likely to win at juicy prices the first or second time they race on turf?
16. Can you evaluate trainer performance in various racing categories? Do you have a working knowledge of winning training patterns on your circuit?

If handicappers' responses to a handful of these questions are negative, they trail the knowledge boom in contemporary handicapping and cannot possibly benefit as will other more aware handicappers from high-tech information systems. They cannot mature as quickly into successful intuitive handicappers. Uninformed handicappers lack the knowledge base that personal growth demands.

The remedy is a close association with the literature in this book's bibliography. Method handicappers can transform themselves into intuitive handicappers only in the same manner. By acquiring the considerable knowledge that falls outside of personal methods. A broad, liberal dose of handicapping education is basic to success in intuitive handicapping. Modern handicappers need to know just about everything there is to know.

The Importance of the Handicapping Paradigm

This book has repeatedly promoted a reliance on multiple information sources for decision-making and not solely on systematic methods. Yet systematic methods remain part and parcel of an information management approach. The methods bring logical order to a mass of data items, and in skillful hands the most powerful of them are fully capable of yielding positive results, seasonal selections that are demonstrably effective in generating profits. After all, that is the goal of systematic method play and represents the traditional thrust of handicapping instruction.

In the intuitive, information management approach systematic methods serve as a centralizing filter of information and source of potential selections against which other information and alternative decisions can be balanced and weighed. Systematic methods deal in recognizing fundamental data items and in processing that data into information that is itself fundamental.

Fundamental handicapping information tends to be intrinsic to the abilities and preferences of horses. Trip handicapping aside, in almost all cases systematic methods reflect horses' status in relation to either speed or class, however modified by considerations of form, early speed, or pace. Other kinds of handicapping information are often incidental. The information is related not to horses' innate abilities and preferences but to situational factors surrounding today's races. Pace, because it reflects horses' relative speed capacities, might be the most significant of the incidental types of information. But also situationally important are track bias, body language, trips, trainer patterns, and jockey assignments, not to mention the relative odds lines.

Handicappers who become information managers will find that many final decisions in this decision approach to handicapping depend on the relative strength of relationships that combine the fundamental and the incidental. When Beyer proposes a juxtaposition of speed figures and trip information he is doing just that. Incidental information is not subordinate to fundamental information but rather merely associated with it. Horses that combine fundamental advantages and incidental advantages are especially difficult to defeat. Because incidental information that many racegoers do not have access to is involved in final decisions, the odds are often delightfully high.

Perhaps most important of all, the consistently successful application of systematic methods reflects basic handicapping proficiency. Method handicappers can be fairly sure of their effectiveness merely by tracking the method's win percentage and the average mutuel on winners. The two figures should be large enough to generate profits on a series of flat bets. Unsuccessful method players cannot expect to overcome basic shortcomings in proficiency by switching to information management. The information management approach is an extension and complication of the handicapping process designed to increase opportunities and propel profits, mainly in the short run, such as a single season. The approach is more challenging than systematic method play and the logical deductive reasoning that style entails. If fundamental handicapping prowess is missing, the complications of information management and intuitive reasoning should only serve to make matters worse.

A CASE-STUDY APPROACH

Because intuitive handicapping success depends ultimately on the skill of applying knowledge, information, and the lessons of experience in real situations that are new, its instructional method proceeds best in the real world and not so well in books. Once the handicapper's knowledge base has been accumulated and the information system has been developed, only the lessons taught by repeated applications in a variety of racetrack settings are left. The skills of intuitive, informational handicapping are applied skills. They develop best with applications, thousands of applications. A case-study approach is appropriate for demonstrating the applications of the information management approach. The instruction is similar to the technique used to teach applications of the law, where a multiplicity of principles and precedents must be applied anew to relatively unorganized ill-defined complex situations. Races can be viewed as unorganized and complex case studies susceptible to applications of intuitive handicapping.

This chapter's case study simulates a recreational handicapping approach to a day at Del Mar, the resort track near San Diego on the southern California circuit.

An information management approach is desired, complete with greatly intuitive decisions, but the author has not been playing the races since Santa Anita ended in late April, four months ago, and desires to issue a few disclaimers before proceeding. Regional computerized databases to which I might turn are not yet in place, although five years from now they will be. Certain types of information, therefore, will be missing, including speed figures, adjusted pace ratings, trip notes, and information about track biases. This thwarts intuitive handicapping and decision-making to no end. Yet in this kind of common practical situation, recreational handicappers must rely on whatever information system and sources remain at their disposal. Local, up-to-date information services and information sources outside of one's personal handicapping work station can be contacted. At Del Mar there are two first-rate handicappers conducting daily seminars, Lee Rousso and Jeff Siegel; Greg Lawlor appears with Siegel to identify spot plays from trainer data; and in 1984 there was for the first time the impressive high-tech outlet, the PDS sports computer.

So after perusing several days of Del Mar past performances, all of which proved despairingly unattractive to an information specialist, I finally landed on Saturday, September 1, which shaped up as an appetizing card, and decided to drive down. I say that with hesitation, as Del Mar has never been a personal oasis. The track is peculiar to the extreme, featuring extremely sharp turns that are improperly banked and an 832-foot stretch run. Until this very season, Del Mar lacked a back stretch chute that permitted sprints to be carded at six-and-a-half and seven furlongs. Because these two distances are new in 1984, handicappers will have no track time records, or par times for them, and many young horses with unknown dimensions will dominate the racing scene. Moreover, until 1983 the racing surface had decisively favored speed on the inside, but that season the footing changed dramatically enough to give latecomers a fair chance, notwithstanding the unbanked turns and short stretch.

Ironically, one reason I have never made hay at Del Mar is because that oval has favored precisely the kind of intuitive, informational handicapping I have been developing only lately and not systematic methods that emphasized class appraisal. Perhaps this tricky place is a suitable testing ground for these newfangled procedures?

So, on September 1, 1984, I contacted handicapping associates who store speed figures, trainer data, and track bias information for every conceivable purpose. Beyond that, I telephoned the Post Data Services (PDS) computer, located in Torrance, near TRW, the aerospace giant. By means of a *modem,* a cheap high-tech instrument, which attaches to a regular telephone and converts sound waves to binary digits, the PDS sports computer responds to voice-recognition input and is programmed to provide high-tech handicapping selections. Handicappers can dial the computer and learn of the day's picks by The Sports Judge, a high-powered regression model of handicapping that is commercially marketed as the PDS applications program. On cue from callers, the program supplies its information by voiced output. The Sports Judge had been picking 35 percent winners for thirty racing days at Del Mar, at a $6.60 average mutuel on winners. According to the Kelly optimal betting formula, The Sports Judge held a strong 7 percent advantage at Del Mar.

Armed now with key information I had lacked, I felt I might even enjoy a prosperous day should the odds on certain horses fall favorably in my direction. I particularly looked forward to the feature, the Gr. 2 Del Mar Oaks, as the highly favored *Lucky Lucky Lucky,* from the Wayne Lukas Barn, shaped up as a huge underlay and had an especially solid chance to lose.

The computerized selections of The Sports Judge for September 1 at Del Mar appear in Table 21.

Table 21
The Sports Judge: Computerized selections from Post Data Services for Del Mar, Saturday, September 1, 1984

The first and second selections for September 1, 1984, at Del Mar of *The Sports Judge,* a handicapping applications software program that employs a regression model of handicapping.

1st race	Cream Pocket Killora	6th race	Derby Dawning Image of Greatness
2nd race	Sit High Irish Guard	7th race	Embolden Stand Pat
3rd race	Emerald Cut Emperoroftheuniverse	8th race	Auntie Betty Patricia James
4th race	Positioned Cool Victoria	9th race	Fortune's Kingdom Gray Missile
5th race	Dare You II Quantum Leap		

Here's what happened:

1st Del Mar, 6f, 3-year-old fillies.
Claiming prices $25,000 to $22,500.

Pre-race information set. Open claiming races are often analyzed successfully by looking for three conditions in combination: satisfactory early speed, acceptable or improving form, and a drop in class. Three-year-olds are best supported on the rise, unless a one-level drop follows a solid performance.

Track bias: BR (bad rail). Early at Del Mar horses that had been favored by strong speed biases both on dirt and turf at Hollywood Park could be passed by in favor of off-pace types. Horses that had suffered from the speed biases at Hollywood could be expected to move up at Del Mar, notably on the grass.

Information analysis. As many as five fillies sported no information to bolster their case in this race. Moreover, four frontrunners figured to compromise one another's chances. The best of the speed horses, *Killora,* had not raced in three months and would be forced to handle fast early fractions plus a late challenge from *Cream*

Pocket, the computer choice. By my analysis, *Cream Pocket* and *Swiftly Mine* added up strongest. When *Swiftly Mine* scratched, a potential straight play had begun to loom up early.

Examine *Cream Pocket*'s record. The filly was being returned in six days after a horrible trip against much better. Despite that, it finished fourth of twelve. Its stretch loss was excusable; a wide, wide trip. The jockey switch to Kenny Black should be a plus. The big win August 11 indicated the class jump might be easy pickings, notably since *Cream Pocket* had beaten better at Santa Anita in April. This filly shaped up as the proverbial improving three-year-old that can negotiate Del Mar's turns.

Killora had seven speed points in its favor, acceptable form (workouts), and the best races versus the strongest competition, but the horse figured to tire and would probably go favored.

Problem-solving. No unanswered questions here. If *Killora* were a five-year-old, a search of the database would reveal any tendency to win or run big when fresh. No trainer patterns apparent here.

Outcome scenarios. Two possibilities for me. *Killora* would contest the early pace and draw clear to win. *Cream Pocket* would race in midpack, come on strongly in the late stages, and win.

Final decisions. At 4 to 1 *Cream Pocket* was backed to win ($50) and was coupled in the Daily Double with a colt that would be dropping in class and switching to the leading jockey ($20DD).

Comment. The winner Superstar Sunny, a shipper from Golden Gate and points north, showed blazing workouts and fast race times

Killora — Dk. b. or br. f. 3, by Killany River—Leal's Jewel, by Africanus

Own.—Leal Ranches **116** Br.—Leal Ranches (Cal) Tr.—Priddy Ray $25,000 — 1984 9 2 1 1 $23,850; 1983 3 1 0 0 $6,550

Lifetime 12 3 1 1 $30,400

24May84-3Hol 6f :221 :45 1:104ft 3½ 116 31½ 32½ 31½ 32½ Toro F3 (F) 40000 80-21 What Magic, Mill St. Lass, Killora 7

10May84-1Hol 6f :22 :45 1:104ft 7 116 1hd 1½ 12½ 21½ Toro F1 (F) 32000 81-16 Hot N Sandy, Killora, Wha: Magic 11

28Apr84-1Hol 6½f :22 :451 1:163ft 13 115 31 43 46 510¾ Toro F1 (F) 50000 76-19 La Fonteyn,TableTen,Gaeli:'sFavor 8

12Apr84-4SA 6f :214 :452 1:11 ft 9½ 115 56½ 53¾ 54½ 57 Toro F4 (F)Aw26000 76-10 Kyflaca, Luv's Force, Love Smitten 6

12Apr84—Broke in air

5Apr84-7SA 6½f :221 :461 1:183ft 5 114 21½ 42½ 63¾ 56½ ShmkrW3 (F)(S)Aw26000 71-21 Awkward Age, Quillummo, Zythum 6

5Apr84—Lacked clear racing room, steadied at intervals down backstretch,; checked sharply at 1/8

21Mar84-3SA 6½f :221 :452 1:174ft 5 114 31 22½ 41½ 4¾ ShoemakerW6 (F) 45000 80-15 Caucasia, LaFonteyn,BladeOfLuck 10

21Mar84—Lugged in stretch

29Feb84-1SA 6½f :212 :45 1:18 ft 7 115 21 2hd 1½ 1no Toro F3 (F) 32000 80-15 Killora, La Fonteyn, PullinStrings 11

25Jan84-3SA 6f :213 :45 1:112ft 8-5 115 11 12½ 14 13½ VlenzulPA2 (F)(S) 25000 81-19 Killora, Air Of Majesty, Bold Fancy 7

12Jan84-3SA 6f :213 :45 1:112ft 7½ 116 3½ 41½ 63½ 67 Fell J5 (F) 40000 74-21 HideAPenny,LFontyn,Sizzlin'Suzn 10

29Dec83-1SA 6f :213 :444 1:112ft 12 116 31 32½ 32 43 ValenzuelPA3 (F) 40000 78-23 MilliesMink,Sizzlin'Suezn,LFontyn 11

Aug 25 Dmr 5f ft :593 h Aug 19 Dmr 5f ft 1:001 h Aug 13 Dmr 5f ft 1:024 h Aug 7 Dmr 4f ft :53 h

Cream Pocket

Own.—Bernstein–Guiliano–Guiliano **116**

Dk. b. or br. f. 3, by Brents Creme—Pocket Lady, by Pocket Ruler
Br.—Giuliano & Sons (Cal) 1984 8 3 0 1 $26,385
Tr.—Bernstein David $25,000 1983 5 1 1 0 $6,880
Lifetime 13 4 1 1 $33,265

25Aug84-1Dmr 6½f :22^2 :45^2 1:17^{2}ft 24 114 6^{1¾} 4^3 3^{2½} 4^4 Garcia J A^{11} Ⓕ 35000 — — RomnticRomn,Dbi'sCourg,YttWng 12
25Aug84—Extremely wide 3/8 turn
11Aug84-1Dmr 6½f :22 :45^2 1:18^{3}ft *2 116 6^{7½} 6^7 2^2 1^{4½} ShoemakerW^4 Ⓕ 20000 — — CremPocket,TheSmelde,BlckFntsy 9
29Jly84-1Dmr 1 :46^3 1:12^2 1:39^{1}ft 6¼ 116 1^1 1^{hd} 2^{1½} 7^6 Pedroza M A^7 Ⓕ 2:000 66-17 Superfine,MissViMgnum,DistntJule 7
21Jly84-2Hol 6f :22 :45 1:11^{1}ft 9e 117 11^{8¾} 12^{15} 12^{13} 10^8 Pincay L Jr^{10} Ⓕ 40000 73-20 HttieWing,SpringBid,ShdyHostess 12
21Jly84—Wide 3/8 turn
16Jun84-9LaD 7f :22^2 :45 1:23^{3}ft 4¼ 114 8^{6½} 7^{8½} 6^{12} 5^{11} Kaenel J L^2 Ⓕ 50000 95-09 TylorsPromis,Erimo'sLdy,CollgCuti 9
9May84-7LaD 6f :22^4 :46^4 1:12^{1}ft 3 114 3^{2½} 4^{3½} 3^{6½} 3^{8½} FranklinRJ^2 ⒻAw12000 73-21 Moskee,Deborah'sSis,CreamPocket 6
20Apr84-1SA 6f :22 :45^2 1:11^{1}ft 5¼ 117 3^{1½} 3^2 3^{½} 1^3 Pincay L Jr^8 Ⓕ 32000 82-21 CreamPocket,SucyDncer,BoldDme 10
11Apr84-1SA 6f :22 :45^2 1:11^{4}ft 25 116 2^1 2^{2½} 2^2 1^{1½} Black K^8 Ⓕ 25000 79-20 CremPocket,SlightlyLced,Cli'sGirl 12
3Nov83-4LA 6f :22^1 :46^2 1:12^{2}ft 3¼ 117 4^{2½} 7^{5¾} 7^{10} 7^{11¾} Castaneda M^2 Ⓕ 25000 71-23 Avenger Mist, Awful, Pronto Miss 7
3Nov83—Bumped start
23Oct83-3SA 6½f :22 :45^3 1:19^{1}ft 12 118 4^{1½} 5^{6½} 5^{12} 5^{11½} Black K^4 Ⓕ 32000 62-19 GrandSlamBaby,SlightlyLced,Grbcz 7
Aug 20 Dmr 3f ft :35^4 h Aug 6 Dmr 3f ft :36^1 h Jly 14 Hol 6f ft 1:14 h Jly 7 Hol 4f ft :48^2 h

Superstar Sunny

Own.—Cass R A & R Jr **109^5**

Ch. f. 3, by Rising Market—Principle Lady, by Go Marching
Br.—Bingcang–Isaacs–Levy (Ky) 1984 10 2 1 2 $17,525
Tr.—Canani Julio C $22,500 1983 3 M 0 0 $1,125
Lifetime 13 2 1 2 $18,650

17Jly84-10Sol 6f :22^1 :44^3 1:09 ft 2¾ 114 3^2 3^2 3^{4½} 3^6 Chapman TM^5 Ⓕ 25000 92-06 TantaMr,NturlSquw,SuperstrSunny 7
17Jly84—Broke in a tangle, bobbled break
5Jly84-10Pln 6f :22 :44^2 1:09^{2}ft 3¼ 109^5 2^{½} 2^{½} 1^{hd} 1^3 Lozoya D A^6 Ⓕ 16000 97-08 SuprstrSunny,AnothrLlHllr,ThSmld 8
14Jun84-6GG 6f :22 :45 1:10^{2}ft 17 109^5 7^8 8^{9½} 9^{11} 9^{10½} Castro J M^7 Ⓕ 25000 76-13 Sleazy Looker, Sharlee, TantaMara 9
14Jun84—Wide
2Jun84-2GG 6f :22^3 :45^2 1:10^{4}ft *3¼ 107^5 3^{1½} 1^{hd} 1^1 1^{½} Lozoya D A^6 ⒻM25000 85-14 SprstrSnny,BlstGold,OlympcMmry 12
23May84-6GG 1 :47^2 1:12 1:38^{3}ft 3¾ 107^5 1^{hd} 1^{hd} 1^{½} 5^{5½} Garcia J J^3 ⒻMdn 70-19 Implosive,Chilaquack,GoldenBllerin 7
3May84-4Hol 1 :46^3 1:11^3 1:37^{4}ft 12 113 2^{½} 9^{21} 9^{30} 9^{34} Hawley S^9 ⒻM35000 43-19 PollyHigh,Meliss'sRiver,SeCommnd 9
3May84—Wide 7/8 turn
18Apr84-4SA 1$\frac{1}{16}$:47 1:12^1 1:46^{4}ft *2½ 118 1^1 2^{½} 2^{2½} 2^{1½} Hawley S^3 ⒻM28000 65-22 MnyASlp,SprstrSnny,Gmblr'sDghtr 9
22Mar84-3SA 6f :22 :45^3 1:11^{4}ft 2¼ 117 3^{½} 1^{hd} 2^{hd} 3^{4½} McCrronCJ^2 ⒻM32000 74-18 CrmelKey,Crownlet,SuprstrSunny 12
17Feb84-4SA 6½f :21^3 :45^2 1:18^{2}ft 11 117 3^2 3^1 3^1 6^{3½} Fell J^5 ⒻM40000 75-17 What Magic, Wine Kiss, Bulig Z. 9
20Jan84-4SA 6f :21^4 :45 1:11^{2}ft *2e 113 6^{2¾} 4^6 4^{7½} 5^{8¾} McCrrnCJ^{11} ⒻM40000 72-19 MusicalBall,AlotaGlitter,PollyHigh 11
●Aug 25 Dmr 5f ft :58^2 h Aug 14 Dmr 5f ft :59^2 h Jly 31 BM 5f ft 1:04^3 h

2298—FIRST RACE. 6 furlongs. 3 year olds. Fillies. Claiming prices $25,000-$22,500. Purse $10,000.

Index	Horse and Jockey	Wt.	PP	ST	¼	½	¾	Str.	Fin.	To $1
1088	Superstar Sunny, Lozoya	x109	7	5	5^2	4^2	-	2^{hd}	1^2	4.70
2244	Cream Pocket, Black	117	6	2	4^1	3^{1½}	-	3^{hd}	2^{¾}	4.30
2037	Distant Jule, Pincay	121	2	10	10^3	6^{hd}	-	5^4	3^2	3.80
2163	Bob Hutson Esquire, Meza	118	11	1	1^{2½}	1^5	-	1^3	4^{½}	4.70
1193	Killora, Delahoussaye	116	5	4	3^{1½}	2^{hd}	-	4^{3½}	5^{1¾}	3.20
(2157)	Winning Gold, McCarron	118	8	6	9^1	10^1	-	6^{hd}	6^2	10.00
2208	Vedalia, Delgadillo	114	1	9	8^{hd}	7^{½}	-	7^{1½}	7^{nk}	37.60
2047	Waikoloa, Lamance	114	9	11	11	11	-	11	8^{2½}	109.7
2084	Indian Art, Garcia	118	3	7	6^{hd}	9^{hd}	-	9^1	9^{1½}	57.80
2136	The Same Idea, Dominguez	x111	4	3	2^1	5^4	-	8^{1½}	10^2	25.10
2163	Minor League, Ortega	118	10	8	7^{1½}	8^{hd}	-	10^{½}	11	40.10

Scratched—Miss Via Magnum, Swiftly Mine Claimed—Killora by Round Meadow Farm (trained by D. Vienna), for $25,000.

7-SUPERSTAR SUNNY	11.40	6.80	4.20
6-CREAM POCKET		6.00	3.80
2-DISTANT JULE			3.20

Time—.22, .44 4/5, .57 3/5, 1.11 1/5. Clear & fast. Winner—ch f 3. Rising Market—Principle Lady. Trained by J. C. Canani. Mutuel pool—$175,821.

at Pleasanton and Solano on the California fair circuit. Without pars, variants, and adjusted times—information—its raw final times were hard to evaluate, and the colt looked questionable at the higher class level. Would handicappers having the key information have taken this winner? Maybe; it won impressively. Cream Pocket finished a strong second.

2nd Del Mar, 6f, 3-year-olds.
Claiming prices $25,000, to $22,500.

Pattern Match had much in its favor, including five speed points, *up close* form last out at every call (a plus factor), a drop in class from $32,000 to $25,000, ordinarily a real change in the order of competition on this circuit, and a switch to leading rider Chris McCarron. The class drop-jockey switch system, for which this horse qualified, has shown fantastic profits on this circuit almost continuously for two seasons. I realized the wire-to-wire wins at Hollywood, once against better May 27, had occurred on a biased strip, but I succumbed to the class-jockey maneuver at 7 to 2 and bet *Pattern Match* to win ($50).

I did not like the favorite, *Sittin' High,* which had the top speed figure. The colt was up in class after losing three lengths of an uncontested lead versus softer August 22 and should be pressed today by a better horse. The stretch loss amounted to a form defect, and form defects represent solid reasons to eliminate low-priced horses and favorites. With jockey Laffit Pincay and the horse's high figures, this colt was overbet.

3rd Del Mar, IM, 3-year-olds.
Claiming prices $25,000 to $22,500.

A dandy illustration of procedure for high-tech handicappers using information management and intuitive reasoning follows.

Examine the past performances for *Emerald Cut, Triumphantly,* and *Emperoroftheuniverse.* Which do you prefer? How will you bet? *Emerald Cut* had seven speed points, figured to control the early pace, was trained by leading claiming trainer Mike Mitchell, was dropping in class significantly, and was returning to action only seven days after beating better. That covers speed, pace, trainer, class, and form. This colt had taken to Del Mar, besides, and had

Sittin' High
Dk. b. or br. c. 3, by Rise High—Smart Folly, by Smart Bug
Br.—Stedmand II & Wolcott Rose (Cal) 1984 3 2 0 0 $10,175
Own.—Cameron Margaret A 118 Tr.—Tinsley J E Jr $25,000 1983 0 M 0 0
Lifetime 3 2 0 0 $10,175

22Aug84-5Dmr 6f :22^{1} :45^{3} 1:10^{1}ft *2½ 118 1^{hd} 1^{2} 1^{3} 1^{nk} Pincay L Jr^{5} 20000 87-14 Sittin'High,OvrlndJourny,Willodon 11
6Aug84-2Dmr 6f :22^{3} :46^{2} 1:11 ft 3½ 117 $1^{1\frac{1}{2}}$ $1^{2\frac{1}{2}}$ 1^{5} 1^{5} Pincay L Jr^{4} Ⓢ M32000 83-17 Sittin' High,Kejolie,KeepCharging 12
13Jly84-4Hol 6f :22 :45^{1} 1:10^{2}ft 24 116 3^{1} $4^{5\frac{1}{2}}$ 5^{13} 6^{12} Garcia J A^{4} Ⓢ M32000 73-19 BoldPldg,PhiBtKpp,SpcilInvntory 12
Aug 29 Dmr 4f ft :48 h Aug 20 Dmr 4f ft :47^{2} h Aug 13 Dmr 3f ft :37^{2} h Aug 2 Dmr 5f ft 1:00 h

Pattern Match
Gr. c. 3, by Jumping Hill—Vanilla Nose, by Kinsman Hope
Br.—Mishkin R & Chase (Cal) 1984 8 2 1 0 $17,375
Own.—Mishkin R & Chase 116 Tr.—Mulhall Richard W $25,000 1983 13 2 2 0 $26,450
Lifetime 21 4 3 0 $43,825

16Aug84-1Dmr 6f :22^{1} :46 1:11 ft 10 114 $4^{2\frac{1}{2}}$ 2^{hd} 3^{nk} $5^{2\frac{3}{4}}$ Pedroza M A^{8} 30000 80-16 MostDetermind,EmrldCut,StblEys 12
16Aug84—Lost whip start
7Jun84-7Hol 6f :22^{1} :45^{1} 1:10^{1}ft *2½ 117 $2^{\frac{1}{2}}$ $2^{1\frac{1}{2}}$ $4^{4\frac{1}{2}}$ 8^{5} Pincay L Jr^{3} 40000 81-16 ThreeForTwo,OfficeSkr,Af'ciondo 10
7Jun84—Bobbled start
27May84-1Hol 6f :22^{1} :45^{2} 1:11^{2}ft *9-5 117 1^{hd} 1^{2} 1^{2} 1^{2} Pincay L Jr^{4} 32000 80-18 PtternMtch,BrgingAhed,OfficeSekr 7
6May84-3Hol 6f :22^{1} :45 1:11 ft *7-5 117 1^{hd} 1^{hd} 1^{hd} 2no Pincay L Jr^{9} 32000 82-17 JohnJrry,PttrnMtch,TrmphntBnnr 10
26Apr84-1Hol 6f :22 :44^{4} 1:10^{3}ft 4¼ 117 $1^{\frac{1}{2}}$ $1^{2\frac{1}{2}}$ $1^{3\frac{1}{2}}$ 1^{4} Pincay L Jr^{7} Ⓢ 25000 84-19 PtternMtch,TostMster,BoverSong 9
13Apr84-3SA 6f :21^{3} :45 1:10^{3}ft 8¼ 114 2^{1} $6^{4\frac{1}{2}}$ 6^{8} 6^{10} Pedroza M A^{1} Ⓢ 20000 75-20 Tonopah Low, Nosfos, John Jerry 7
13Apr84—Bumped at 5 1/2
24Mar84-2SA 6f :21^{4} :45 1:10^{4}ft 8¼ 116 $1^{1\frac{1}{2}}$ 2^{1} $5^{5\frac{1}{2}}$ $6^{7\frac{1}{2}}$ Pedroza M A^{2} 32000 76-18 MistkenIdentity,SpeedyTlkr,Nosfos 8
1Feb84-1SA 6½f :21^{2} :44^{3} 1:17^{2}ft 5 116 6^{6} 10^{11} 10^{12} $10^{14\frac{1}{4}}$ DelahoussayeE^{4} 50000 69-17 HilTheEgle,TostMster,TonophLow 10
1Feb84—Bumped, jostled after start, again at 5 1/2
21Dec83-4Hol 6f :22 :44^{4} 1:09^{4}ft *9-5 114 $1^{\frac{1}{2}}$ 1^{5} 1^{7} 1^{6} Pedroza M A^{5} 25000 88-16 PatternMtch,TrgicToy,ProudYnkee 7
3Dec83-2Hol 6f :22 :45^{1} 1:11^{3}sy 8¼ 114 2^{hd} 1^{hd} 1^{hd} $4^{3\frac{1}{2}}$ Pedroza M A^{9} 25000 75-22 FleetJoey,ProudYnkee,MikesClico 12
Aug 24 Dmr 5f ft 1:00 h Aug 9 Dmr 6f ft 1:13 h Aug 4 Dmr 3f ft :35^{4} h Jly 30 Dmr 5f ft 1:01^{4} h

2299—SECOND RACE. 6 furlongs. 3 year olds. Claiming prices $25,000-$22,500. Purse $10,000.

Index	Horse and Jockey	Wt.	PP	ST	¼	½	¾	Str	Fin.	To $1
2172	Contequos, Lipham	116	8	6	8^{1}	6^{1}	-	3^{3}	1^{2}	15.60
1389	King Of The Green, Garcia	114	5	5	2^{hd}	$1^{2\frac{1}{2}}$	-	1^{3}	$2^{1\frac{1}{4}}$	28.00
2194	Rejected Suitor, Hawley	116	9	3	3^{hd}	2^{hd}	-	$2^{1\frac{1}{2}}$	3^{hd}	11.10
2059	Jolly Josh, Delgadillo	116	4	12	12	9^{hd}	-	6^{hd}	$4^{1\frac{1}{2}}$	23.90
——	Irish Guard, Ortega	118	10	9	9^{1}	11^{2}	-	7^{hd}	5^{hd}	5.80
2104	Phi Beta Kappa, Lozoya	x113	3	11	$10^{\frac{1}{2}}$	8^{2}	-	5^{2}	$6^{\frac{1}{2}}$	16.30
2172	Baron Lucky, Sibille	116	1	10	11^{1}	10^{hd}	-	9^{3}	$7^{1\frac{1}{2}}$	17.60
2108	Golden Watch, Meza	116	7	4	5^{1}	$4^{\frac{1}{2}}$	-	$8^{\frac{1}{2}}$	8^{2}	24.00
2257	Magic Memo, Fernandez	116	11	1	1^{hd}	3^{2}	-	4^{hd}	9^{hd}	34.50
2172	Pattern Match, McCarron	116	6	8	6^{1}	$7^{2\frac{1}{2}}$	-	10^{2}	10^{4}	3.80
2172	Lucky Saros, Black	117	12	2	7^{1}	12	-	12	11^{10}	6.50
(2221)	Sittin' High, Pincay	118	2	7	$4^{2\frac{1}{2}}$	5^{1}	-	11^{2}	12	2.30

Scratched—Rosalie's Choice. Claimed—Rejected Suitor by Mrs. T. L. Daniels (trained by R. S. Cofer), for $25,000 Pattern Match by Royal T Stable (trained by D. Cross Jr), for $25,000

8-CONTEQUOS 33.20 13.80 14.80
5-KING OF THE GREEN 29.20 17.40
9-REJECTED SUITOR 11.20

Time—.22 2/5, .45 1/5, .57 4/5, 1.11. Clear & fast. Winner—b c 3. Nantequos—Stoop To Conquer Trained by Mel Stute Mutuel pool—$292,305. Daily double pool—$313,308.

$2 DAILY DOUBLE (7-8) PAID $252.40

benefited from the Mitchell claim of July 29. The trainer was obviously getting *Emerald Cut* back into an action spot, even though that meant taking an unwarranted dip in class. No wonder the colt was sent postward at 3 to 5. *The Sports Judge* computer program had identified *Emerald Cut* as one of its two top picks of the afternoon.

However, high-tech handicappers using information management and intuitive decision-making would have forsaken this colt at the price immediately. The class drop is suspect at least, and trainer Mitchell's stats show a far more proficient performance with horses on the rise as opposed to horses dropped in claiming price.

Emerald Cut should be pestered again on the early pace and could conveniently tire and lose. Besides, two other colts deserved more attention here. *Triumphantly* was starting for the second time since May 30. Favored, the colt had been close to $32,000 horses after six furlongs on August 1, a race in which it encountered considerable trouble. It was dropping down today and was being sent out by a high percentage trainer. More significantly, below are the unadjusted pace figures for the latest races (leaders) of *Emerald Cut* and *Triumphantly*, and the finish figure earned by each horse.

	4f	6f	1 1/16M
Emerald Cut	86	79	83
Triumphantly	85	81	79

Triumphantly might be challenging at the six-furlong mark today and should improve its final figure, perhaps by several lengths. The computer picked as its second choice *Emperoroftheuniverse*, a long number at 10 to 1. This horse was being raised in class following a wide trip and a recent claim and would be benefiting today by a jockey shift from the nondescript Hoogie Drexler to journeyman Fernando Toro.

I decided to box the $5 Exacta between *Triumphantly* and *Emperoroftheuniverse* two times ($20) and use each horse on top twice, with *Emerald Cut* in the two hole only ($20). Handicappers should understand that odds-on favorites often become overlays when used on the underside of the Exacta. They are usually conspicuous underlays when used on top, just as in the straight pools.

Comment. Systematic methods pick an underlay here, and traditionally handicappers would choose to pass. But the high-tech

Triumphantly
Own.—El Peco Ranch 116
Gr. c. 3, by Decidedly—Hill Flag, by Hillary
Br.—Pope G A Jr (Cal) 1984 5 1 0 0 $14,700
Tr.—Gregson Edwin $25,000 Turf 1 0 0 0
Lifetime 5 1 0 0 $14,700

Entered 31Aug84- 9 DMR
1Aug84-5Dmr 1 1/16 :47 1:114 1:44 ft *8-5 116 76 1/2 45 57 76 McCarron C J8 32000 74-18 Ward C., Bite TheBuck,CrimeFree 12
1Aug84—Bumped at 3 1/2, wide final 1/4
30May84-5Hol 1 1/16 (T):4711:1121:421fm 64 120 68 1/2 74 3/4 86 76 1/2 Fell J9 Aw22000 82-12 MjsticShor,Promontory,HonorMdl 10
11May84-7Hol 1 :464 1:113 1:363ft 9 1/2 120 73 3/4 66 1/4 45 1/2 48 1/4 Fell J1 Aw22000 75-18 TnnssRit,QuiroDinro,CountThHous 8
23Apr84-6SA 1 1/16 :471 1:12 1:442ft 3 1/2 114 63 1/4 1 1/2 1 1/2 1 1/2 Fell J4 Mdn 79-18 Triumphantly, SoloOrbit,Muztagata 8
22Mar84-6SA 6f :214 :444 1:10 ft 6 118 10 12 812 59 1/4 59 DelhoussyeE10 M50000 79-18 BenRedeemed,BargingAhe1,Lothr 11
22Mar84—Broke slowly, wide stretch
Aug 28 Dmr 4f ft :47 h Aug 23 Dmr 5f ft 1:004 h Aug 18 Dmr 4f ft :503 h Jly 26 Dmr 5f ft 1:003 h

***Emerald Cut**
Own.—Gould-Heers-Heers 118
B. c. 3, by First Amendment—County Emerald, by County Clare
Br.—Achar V (Mex) 1984 9 2 2 3 $25,600
Tr.—Mitchell Mike $25,000 1983 0 M 0 0
Lifetime 9 2 2 3 $25,600

23Aug84-9Dmr 1 1/16 :464 1:121 1:442ft *3-2 116 3 1/2 2hd 1 1/2 1 1/2 McCarron C J9 32000 78-19 EmerldCut,TrdyChoic,LimstonLoui 9
16Aug84-1Dmr 6f :221 :46 1:11 ft 4 116 3 2 31 2hd 2 1/2 McCarron C J5 32000 82-16 MostDetermind,EmrldCut,StblEys 12
29Jly84-3Dmr 6f :223 :454 1:104ft 4 116 3nk 3nk 3nk 3 1/2 McCarron C J5 c25000 83-17 IrishGuard,KingsJester,EmeraldCut 6
15Jun84-5Hol 1 1/16 :464 1:112 1:442ft 9 117 21 1/2 33 98 915 1/4 Pincay L Jr3 40000 58-21 OfficeSeekr,JohnThTough,InAMssg 9
28May84-2Hol 6 1/2 f :22 :444 1:17 ft 2 1/2 115 41 1/4 32 1/2 33 11 1/2 Hawley S2 M40000 85-19 EmeraldCut,WhiteCloud,IrishGuard 9
25Apr84-6Hol 7f :223 :46 1:232ft 4 1/2 113 51 1/2 51 3/4 43 31 1/2 Hawley S7 M45000 78-17 EmprdorAlNort,MstroMo,EmrldCt 12
12Apr84-3SA 6f :214 :453 1:112ft 6 1/2 116 3 1/2 41 1/2 42 3/4 31 1/4 Delgadillo C4 M45000 79-18 Lothar, Teddy's Love,EmeraldCut 12
12Apr84—Bumped at 1/16
16Feb84-6SA 6f :221 :46 1:114ft 3 1/2 116 32 32 1hd 2hd ValenzuelaPA3 M45000 79-23 Stress, EmeraldCut,NuclearAttack 12
16Feb84—Veered in, bumped after start, lacked room, steadied at 3/8
1Feb84-2SA 6 1/2 f :214 :442 1:172ft 5 1/2 116 42 1/2 34 1/2 46 43 1/2 Cauthen S3 M45000 79-17 Dust To-Riches, Travet,IrishGuard 11
Jly 24 Dmr 5f ft :594 h Jly 18 Hol 5f ft 1:011 h Jly 13 Hol 5f ft 1:00 h Jly 8 Hol 5f ft 1:00 h

Emperoftheuniverse
Own.—Carr M S 116
B. g. 3, by Universal—Imperialistic, by Colorado King
Br.—Rosenblatt Farms (Cal) 1984 12 2 2 2 $22,563
Tr.—Feld Jude T $25,000 1983 0 M 0 0
Lifetime 12 2 2 2 $22,563

19Aug84-2Dmr 1 1/16 :464 1:114 1:443ft 7 1/2 115 86 53 64 1/4 42 1/2 Drexler H2 20000 75-17 Edi'sRunnr,Kvn'sTrcks,Lynnwould 12
19Aug84—Wide final 3/8
10Aug84-3Dmr 1 1/16 :454 1:12 1:443ft *2 116 612 41 3/4 1hd 12 Drexler H7 c16000 77-20 Emperoftheunivrs,ProprDlvry,Ondl 8
27Jly84-2Dmr 6f :223 :461 1:111ft 15 116 78 78 42 3/4 31 1/2 Drexler H2 [S] 16000 81-19 HvymnClnt,Holm'sBst,Emprfthnvrs 7
17Jun84-4Hol 1 1/16 :463 1:12 1:46 ft 6 1/2 118 58 98 1/4 815 913 DelahoussayeE1 20000 52-23 Lynnwold,SonOfChf,LovrByMcky 12
26May84-9Hol 1 1/16 :471 1:113 1:434ft 18 116 21 42 58 1/4 613 1/4 Gilbert C J6 40000 63-16 MikesClico,PtrioticPledge,InAMssg 6
26May84—Wide 7/8 turn
2May84-9Hol 1 1/16 :461 1:11 1:442ft 5 1/2 119 58 713 717 718 Pincay L Jr4 32000 55-18 KnghtSkng,PlntyConscos,MksClco 8
2May84—Bumped, jostled off stride after start
20Apr84-4SA 1 1/16 :462 1:114 1:453ft *1 118 612 44 21 1/2 1 1/2 Pincay L Jr11 [S]M32000 73-21 Emperofthnvrs,PssportPhot,Algbr 12
20Apr84—Fanned wide 1st turn
7Mar84-2SA 1 1/16 :462 1:112 1:442ft 3 1/2 118 710 27 22 1/2 23 3/4 Garcia J A2 M40000 76-16 SkyPrnc,Emprfthnvrs,RnThGlxy 12
7Mar84—Bumped start
1Mar84-6SA 7f :223 :453 1:243ft 16 118 42 1/4 44 912 1013 1/4 Sibille R3 [S]Mdn 64-21 Philosopher,Stickette,Sri'sDelight 12
16Feb84-2SA 1 1/16 :464 1:123 1:451ft 4 1/2 118 41 1/2 2hd 2 1/2 24 1/2 Sibille R6 M40000 78-23 Tabare,Emprrfthunivrs,HppyGrtn 12
Aug 3 Dmr 1 ft 1:412 h Jly 21 SLR tr.t 5f ft 1:02 h ● Jly 14 SLR tr.t 4f ft :483 h

2300—THIRD RACE. 1 mile. 3 year olds. Claiming prices $25,000-$22,500. Purse $10,500.

Index	Horse and Jockey	Wt.	PP	ST	1/4	1/2	3/4	Str.	Fin.	To $1
2200	Emperoftheuniverse, Toro	116	4	7	6 3 1/2	5 1	3 1	2 2 1/2	1 nk	10.20
(2234)	Emerald Cut, McCarron	118	3	1	1 hd	1 hd	1 1/2	1 hd	2 1/2	.60
——	Triumphantly, Meza	116	1	8	7 1/2	7 1 1/2	6 2	4 hd	3 1	7.80
2124	North Seaway, Delahoussaye	116	6	5	3 1 1/2	4 1/2	5 hd	6 2	4 3	7.10
2059	In Natural Form, Sibille	116	2	3	5 1/2	6 2 1/2	4 1 1/2	3 hd	5 1/2	16.60
——	Doctor Foote, Pincay	117	5	2	2 2	2 5	2 1/2	5 1 1/2	6 1/2	14.80
——	Limestone Louie, Dominguez	x111	8	6	8	8	8	7 3	7 6 1/2	12.10
2234	Jolly Writer, Pedroza	116	7	4	4 3	3 hd	7 1 1/2	8	8	14.70

Scratched—A Sparkle. Claimed—Emerald Cut by I. S. Longo (trained by D. Velasquez), for $25,000.

4-EMPEROFTHEUNIVERSE	22.40	5.20	4.20
3-EMERALD CUT		2.80	2.40
1-TRIUMPHANTLY			4.20

Time—.22 4/5, .46 1/5, 1.11 3/5, 1.38. Clear & fast. Winner—b g 3. Universal—Imperialistic. Trained by Jude Feld. Mutuel pool—$182,216. Exacta pool—$297,033

$5 EXACTA (4-3) PAID $123.00

selection wins and pays big, and a $10 Exacta with the favored underlay on the bottom yields $246. If *Emerald Cut* tired just a bit more, the payoff between the two horses isolated by the information management approach would have been splendid. Notice that a straight bet on *Triumphantly* loses all. Exotic wagering was appropriate here. If *Triumphantly* had been offered at 12 to 1 or 15 to 1, I would have bet the horse to win besides.

The net on the investment was $206, pushing the intuitive approach ahead by $86 for the day.

4th Del Mar, 6f, 3-year-olds up.
Maiden fillies.

With seven first starters and no genuine early speed visible in this maiden field, trainer Neil Drysdale's *Positioned* figured to battle it out with *Cool Victoria*. Both fillies figured to handle the Lukas-trained first starter by *Seattle Slew, Le Slew*. Why? Because first-timing three-year-olds win only 50 percent their fair share of the races they enter. They are therefore automatic eliminations, except when trainer pattern data commands otherwise, and the price is appealing. Neither condition prevailed here. Positioned not only experienced a trouble trip not of its own making August 19 but the filly was improving, flashing better speed with each outing. It already possessed the high figure, as its selection by the computer indicated. *Le Slew* was getting no hot toteboard action.

Cool Victoria went postward at 5 to 1. Though I did not fancy its chances as much as *Positioned*'s, I might have bet the filly, except that John Gosden, a class A trainer overall, wins approximately 8 percent with sprinters. It's Gosden's main weakness, and maiden sprinters are not his game in the least. When *Positioned* was bet down to 2 to 1, I passed the race, fortunately.

Le Slew — Dk. b. or br. f. 3, by Seattle Slew—Le Moulin, by Hawaii
Own.—Ashment A L — **117** — Br.—Northwest Farms (Ky) — 1984 0 M 0 0
Tr.—Lukas D Wayne — 1983 0 M 0 0
Lifetime 0 0 0 0
Aug 25 Dmr 5f ft 1:02^3 h — Aug 18 Dmr 4f ft :47^4 h — Aug 11 Dmr 5f ft 1:00^4 h — Aug 3 Dmr 5f ft 1:02 h

Cool Victoria — B. f. 3, by Icecapade—Caregien Jo, by Victoria Park
Own.—Kuppin H Jr — 117 — Br.—Johnson & Caves (Ky) — 1984 2 M 1 0 $3,000
Tr.—Gosden John H M — 1983 0 M 0 0
Lifetime 2 0 1 0 $3,000

10Aug84-6Dmr 6f $:22^{4}$ $:46^{3}$ $1:11^{3}$ft 15 117 1^{1} 2^{hd} $2^{1\frac{1}{2}}$ $2^{1\frac{1}{2}}$ Lipham T^{12} ⓕMdn 79-20 NturlyNtl,CoolVctor,SchoolPrncss 12
22Jly84-3Hol 6f :22 $:44^{4}$ $1:10^{4}$ft 15 116 $4^{5\frac{1}{2}}$ 5^{7} 6^{12} $6^{9\frac{1}{2}}$ Lipham T^{7} ⓕMdn 74-15 Aunt Stel, Tabula Rasa, Incubus 7
Aug 26 Dmr 5f ft $1:00^{4}$ h   Aug 20 Dmr 4f ft $:50^{3}$ h   Aug 5 Dmr 6f ft 1:12 h   Jly 15 Hol 6f ft $1:15^{2}$ h

Positioned — B. f. 3, by Cannonade—Logistic, by Crimson Satan
Own.—Summa Stable (Lessee) — 117 — Br.—Hunt N B (Ky) — 1984 3 M 0 1 $2,805
Tr.—Drysdale Neil. — 1983 0 M 0 0
Lifetime 3 0 0 1 $2,805

19Aug84-4Dmr $6\frac{1}{2}$f $:22^{2}$ $:45^{4}$ $1:17^{4}$ft $5\frac{3}{4}$ 117 4^{nk} $5^{1\frac{1}{2}}$ $5^{\frac{3}{4}}$ $4^{1\frac{1}{2}}$ DelahoussyeE[9] ⓕMdn — — QucksDrlin',LOfStrs,SchoolPrncss 11
19Aug84—Wide 3/8 turn, bumped at 1/8
4Aug84-2Dmr $6\frac{1}{2}$f $:22^{1}$ $:45^{1}$ $1:17^{1}$ft 12 112 1^{hd} 1^{hd} $2^{2\frac{1}{2}}$ 8^{11} Meza R Q^{6} Mdn — — Matafao, Chanago's Son, Fitzallen 10
24Jun84-4GG 6f :22 $:45^{2}$ $1:11^{2}$ft 3 114 7^{2} $7^{8\frac{1}{2}}$ $6^{4\frac{1}{2}}$ $3^{3\frac{1}{2}}$ Chapman T M[9] ⓕMdn 78-10 SetForOne,AlwaysAHit,Positioned 12
Aug 24 Dmr 6f ft $1:13^{1}$ h   Aug 14 Dmr 6f ft 1:16 h   Jly 21 Hol 1 ft $1:41^{2}$ h   Jly 15 Hol 6f ft $1:14^{4}$ h

2301—FOURTH RACE. 6 furlongs. Maidens. Fillies & mares. 3 year olds & up. Purse $15,000.

Index	Horse and Jockey	Wt.	PP	ST	1/4	1/2	3/4	Str.	Fin.	To $1
——	Le Slew, Pincay	117	3	5	4^{1}	3^{3}	-	$2^{2\frac{1}{2}}$	$1^{2\frac{1}{2}}$	3.40
——	Missy Binkie, Shoemaker	117	4	4	1^{hd}	1^{1}	-	1^{hd}	$2^{\frac{1}{2}}$	7.60
2202	Positioned, Delahoussaye	117	9	9	7^{hd}	4^{1}	-	$3^{2\frac{1}{2}}$	$3^{4\frac{1}{2}}$	2.10
——	Singing Snow, Sibille	117	5	7	$8^{3\frac{1}{2}}$	$6^{1\frac{1}{2}}$	-	5^{2}	4^{hd}	28.90
2132	Cool Victoria, Lipham	117	8	3	5^{1}	$5^{1\frac{1}{2}}$	-	$6^{3\frac{1}{2}}$	$5^{3\frac{1}{2}}$	5.20
——	Palm Reader, Dominguez	x112	10	8	12	10^{6}	-	8^{2}	6^{1}	32.60
——	Dear Josie, McGurn	117	2	11	3^{1}	2^{1}	-	4^{hd}	7^{2}	40.30
2069	Perfect Joy, Delgadillo	117	11	2	9^{1}	9^{hd}	-	$7^{\frac{1}{2}}$	$8^{1\frac{1}{2}}$	96.30
——	Tangata, Lamance	122	1	12	$10^{\frac{1}{2}}$	8^{hd}	-	10^{6}	9^{hd}	80.00
——	Cyane's Button, Fernandez	117	12	1	$6^{1\frac{1}{2}}$	7^{2}	-	$9^{\frac{1}{2}}$	10^{10}	18.60
——	Dershan, Lozoya	x112	6	10	$11^{\frac{1}{2}}$	11^{6}	-	11^{4}	$11^{4\frac{1}{2}}$	29.60
——	Half Angel, Pedroza	117	7	6	2^{1}	12	-	12	12	5.00

Scratched—Aerturas, Dr. Anne, Naskarion, T.V. Spot.

3-LE SLEW 8.80 6.20 3.60
4-MISSY BINKIE 8.80 4.20
9-POSITIONED 3.00

Time—.22 3/5, .45 4/5, .58 4/5, 1.11. Clear & fast. Winner—br f 3. Seattle Slew—Le Moulin. Trained by D. Wayne Lukas. Mutuel pool $336,566.

5th Del Mar, 1 1/8M (turf), 3-year-olds up. (Exacta)
Claiming prices $80,000 to $70,000.

Fundamental problems nagged at every horse in this high-priced claiming field. Another opportunity perhaps for the information management approach to reach to the bottom of mysterious matters?

During the problem-solving phase of the handicapping the crucial questions concerned the six-year-old John Gosden import from England, *Indian Trail,* unraced since October 1982 and entered today for a claim. Had the horses raced well in listed stakes in England? If so, might he not trample a claiming band?

The database responded affirmatively. The Rose of York stakes, which it had won at York, was unlisted, but *Indian Trail* had also run well in three listed stakes, the Chesterfield Cup, the John Smith Magnet Cup, and the Royal Hunt Cup. As regular handicappers should appreciate by now, listed stakes are black-type races judged prestigious enough to be *listed* on the pages of international sales catalogues. They are therefore superior in quality to unlisted stakes and outstandingly superior to claiming races.

Furthermore, trainer Gosden specializes in revitalizing these overseas shippers. His performance data indicated he wins with them consistently at first crack. The workout line supported this intriguing case. The works were regular, characterized by longer slower bottom works, suggesting the horse was long since fit and ready. Gosden also relies on jockey Terry Lipham in just these situations. The odds on *Indian Trail* hovered at 19 to 1. In these unpredictable circumstances, with a formless field and a dubious pace confronting me, I decided to wheel an Exacta box ($80), keying top and bottom on the longshot *Indian Trail.*

Comment. Indian Trail looked a threatening second at the eighth pole, beaten a head there, but tired and finished fourth of nine. I can assure handicappers who possess vital information about the stakes performances of foreign horses and know how to interpret them smartly that they will be prepared to tap one of the most generous sources of overlays in major United States racing today. But not today.

Indian Trail — B. h. 6, by Apalachee—Majestic Street, by Majestic Prince
Own.—Sangster R E — 113 — Br.—Brady W T (Ky) — Tr.—Gosden John H M — $70,000
1982 8 1 2 2 $37,44
1981 9 3 2 1 $52,58
Lifetime 20 4 4 5 $90,674 — Turf 20 4 4 5 $90,67

30Oct82♦4Newmarket(Eng) 1¼ 2:102gd *5 123 Ⓣ 6 5½ CuthnS Tia Mri Atmn H Cannon King,FineSun,LadyJustice 17
2Oct82♦3Newmarket(Eng) 1⅛ 1:564gd 12 124 Ⓣ 2nk CuthenS Cambridge H Century City, Indian Trail,Oratavo 29
19Aug82♦5York(Eng) 1 1:384gd *4 123 Ⓣ 1 2½ CuthnS Rose of York H IndinTrl,CorditeSper,SilverSeson 11
31Jly82♦3Goodwood(Eng) 1¼ 2:133fm *2¼ 123 Ⓣ 3nk CuthenS Chstrfld Cp H Criterion, Oratavo, Indian Trail 7
10Jly82♦4York(Eng) 1⅜ 2:083fm 4 124 Ⓣ 4 2½ CthS Jhn Smt Mgnt Cp H BuzzardsBay,CannonKing,Aberfield 6
16Jun82♦3Ascot(Eng) 1 1:421gd 7 124 Ⓣ 7 3½ CuthenS Ryl Hunt Cp H Buzzards Bay, Paterno, Tower Joy 20
31May82♦3Redcar(Eng) 1¼ 2:06 fm 9-5 136 Ⓣ 33 CuthenS Ztld Gld Cp H Say Primula,MeekaGold,IndianTrail 7
13May82♦6York(Eng) 1⅜ 2:114gd *2 130 Ⓣ 23 CuthnS York Gn Ansty H Say Primula, Indian Trail, Ski Run 8
Aug 30 Dmr Ⓣ 4f fm :51 h (d) — Aug 23 Dmr 6f ft 1:161 h — Aug 17 SLR tr.t 1 ft 1:41 h — Aug 7 SLR tr.t 1 ft 1:413 h

Quantum Leap * — Dk. b. or br. h. 6, by Windy Sands—Walking in Space, by Victory Morn
Own.—Mountcastle Stable — 117 — Br.—Crowl Mr-Mrs F (Cal) — Tr.—Glauburg Louis — $80,000
1984 11 2 3 1 $61,825
1983 13 3 4 1 $65,525
Lifetime 48 9 12 6 $245,850 — Turf 35 7 8 6 $199,125

25Aug84-5Dmr 7½f Ⓣ:232 :4641:29 fm 21 111 3 2½ 5 2¼ 55 6 6¾ PdrMA3 Ⓢ Escndido H 87-08 Go Dancer,Acquisition,ColorBearer 8
25Aug84—Rank 1st turn; Run in divisions
4Aug84-5Dmr 1 1/16 Ⓣ:4811:1121:43 fm 9¼ 113 3 3½ 4 2½ 55 8 5½ Pedroza M A11 90000 88-08 Bengeo, Match Winner, Sagamore 11
4Aug84—Bobbled at start
7Jly84-9Hol 1 1/16 Ⓣ:46 1:0911:402fm *1 119 3 1½ 2 1½ 2 2½ 2 3½ McCarron C J1 80000 93-06 Amorous II, QuantumLeap,Berengo 8
10Jun84-5Hol 1 1/16 Ⓣ:4731:11 1:411fm 5½ 116 2½ 2½ 11 1hd McCarron C J3 80000 93-06 QuantumLeap,Bengeo,RoyalCptive 8
20May84-5Hol 1 Ⓣ:4621:1011:344fm *2½ 116 21 21 3 1½ 2 1½ McCarron C J5 80000 92-09 Mr.Rector,QuntumLp,FlthcrpMrinr 8
6May84-9Hol 1 1/16 Ⓣ:48 1:1111:421fm 7½ 116 32 1hd 2hd 5 1¼ Black K4 90000 87-11 Royal Captive, Valais, Dare You II 7
8Apr84-5SA a6½f Ⓣ:22 :4441:152fm 5 121 32 3 2½ 32 34 Pedroza M A2 80000 78-18 Prosperous,EbonyBronz,QuntumLp 8
25Mar84-5SA a6½f Ⓣ:213 :4411:141fm 7½ 118 43 5 4½ 713 812 Pedroza M A7 90000 76-11 Steelinctive,Prosprous,Nighthwkr 10
4Mar84-3SA a6½f Ⓣ:213 :4421:153fm 5¾ 116 4 1½ 2hd 1hd 1nk Pedroza M A10 80000 81-19 QuntumLp,Prosprous,EboryBronz 11
5Feb84-6SA 7f :23 :454 1:23 ft 12 116 2hd 2hd 5 2½ 7 6½ Castaneda M4 100000 78-17 Norbet,RoyalCaptive,ChiefCornstlk 7
Aug 24 Dmr 3f ft :37 hg — Aug 18 Dmr 6f ft 1:142 h — Aug 12 Dmr 5f ft 1:002 h — Aug 2 Dmr 3f ft :354 h

***Dare You II** B. g. 6, by Daring Display—Sanga, by Charlottesville
Br.—Boussac M (Fra) 1984 8 2 3 1 $62,580
Own.—Rimrock Stable **117** Tr.—Hutchinson Kathy $80,000 1983 10 3 0 1 $98,475
Lifetime 41 11 6 4 $288,731 Turf 38 10 6 3 $203,481

Date-Race	Dist/Times	Odds	Wt	Calls	Jockey	Claim	Speed	Company	Field
4Aug84-5Dmr	1 1/16 Ⓣ:48^{1} 1:11^{2} 1:43 fm	3¼	119	6$^{6\frac{1}{2}}$ 10^{11} 10^{13} 11$^{12\frac{1}{2}}$	Garcia J A^{8}	100000	81-08	Bengeo, Match Winner, Sagamore	11
22Jun84-5Hol	1 1/16 Ⓣ:46^{4} 1:10^{1} 1:40^{4} fm	3-2	119	5^{13} 3$^{7\frac{1}{2}}$ 2$^{3\frac{1}{2}}$ 1^{hd}	Pincay L Jr^{1}	125000	95-09	DareYouII,ChmpgneBid,RoylCptive	6
25May84-9Hol	1 1/16 Ⓣ:47 1:10^{4} 1:41 fm	*2¾	117	6$^{4\frac{1}{2}}$ 4$^{3\frac{1}{2}}$ 2^{1} 2$^{\frac{1}{2}}$	Pincay L Jr^{4}	140000	93-04	Champagne Bid, Dare You II,Valais	7
6May84-9Hol	1 1/16 Ⓣ:48 1:11^{1} 1:42^{1} fm	*4-5	120	4$^{2\frac{1}{2}}$ 6$^{2\frac{3}{4}}$ 6$^{3\frac{1}{2}}$ 3^{1}	Pincay L Jr^{2}	100000	87-11	Royal Captive, Valais, Dare You II	7
29Apr84-9Hol	1 Ⓣ:46^{2} 1:10^{1} 1:34^{2} fm	*2-3	117	6^{7} 5^{2} 1^{hd} 1^{3}	Pincay L Jr^{4}	80000	96-06	Dare You II, Phosphurian, Mufti	7
12Apr84-8SA	1⅛ Ⓣ:46^{4} 1:11 1:47^{1} fm	6¼	116	7^{3} 3$^{1\frac{1}{2}}$ 3$^{\frac{1}{2}}$ 2^{nk}	Garcia J A^{4}	80000	91-12	Thorndown,DareYouII,RustyCnyon	9
15Mar84-5SA	1⅛ Ⓣ:46^{4} 1:13^{1} 1:49^{1} fm	5¾	116	3$^{2\frac{1}{2}}$ 1^{hd} 1^{2} 2$^{\frac{1}{2}}$	Garcia J A^{1}	80000	88-19	Phosphurian, Dare You II,ItemTwo	8
6Jan84-5SA	1⅛ Ⓣ:47^{1} 1:13^{1} 1:49^{2} fm	6¼	118	7$^{4\frac{1}{2}}$ 8^{12} 9^{26} 9$^{36\frac{3}{4}}$	Hawley S^{3}	100000	43-19	Lithan, Item Two, Broadly	9
6Jan84—Lugged out 7/8									
21Dec83-8Hol	1⅛ Ⓣ:47^{3} 1:12 1:49 fm	8¾	115	1^{hd} 2$^{\frac{1}{2}}$ 6$^{9\frac{1}{2}}$ 6^{10}	ValenzuelPA^{5}	Aw40000	75-17	Valais, Lithan, Lucullus	7
10Jly83-9Hol	1⅛ Ⓣ:46^{1} 1:10^{1} 1:47^{4} fm	6¾	122	3$^{4\frac{1}{2}}$ 2$^{2\frac{1}{2}}$ 3$^{2\frac{1}{2}}$ 6^{8}	Pedroza M A^{2}	100000	83-06	RedCrescent,FabulousDd,Perengo	10

● Aug 30 Dmr Ⓣ 4f fm :48^{4} h (d) Aug 23 Dmr Ⓣ 6f fm 1:16^{1} h (d) ● Aug 13 Dmr Ⓣ 5f fm 1:00^{3} h (d) ● Aug 2 Dmr Ⓣ 4f fm :48^{3} h (d)

2302—FIFTH RACE. 1-1/8 miles on turf. 3 year olds & up. Claiming prices $80,000-$70,000. Purse $26,000.

Index	Horse and Jockey	Wt	PP	ST	¼	½	¾	Str	Fin.	To $1
2086	Dare You, Pincay	117	9	8	7^{2}	6^{hd}	6^{hd}	4^{1}	1^{1}	2.00
2169	Match Winner, Lozoya	x112	3	1	1^{1}	1^{1}	1$^{\frac{1}{2}}$	1^{hd}	2$^{\frac{1}{2}}$	8.60
(1551)	Rusty Canyon, McCarron	113	6	2	3^{1}	3$^{1\frac{1}{2}}$	3$^{1\frac{1}{2}}$	3$^{\frac{1}{2}}$	3$^{2\frac{1}{2}}$	2.80
——	Indian Trail, Lipham	113	1	4	4^{hd}	4$^{\frac{1}{2}}$	4^{1}	2$^{\frac{1}{2}}$	4^{1}	19.30
2068	Chancey Bidder, Shoemaker	113	8	7	6^{hd}	7^{2}	7$^{2\frac{1}{2}}$	6$^{1\frac{1}{2}}$	5^{nk}	16.90
1497	Felthorpe Mariner, Garcia	117	5	6	8^{2}	8^{hd}	8$^{1\frac{1}{2}}$	8$^{1\frac{1}{2}}$	6^{nk}	24.00
2080	Timberjack, Delahoussaye	117	2	9	9	9	9	9	7^{1}	4.20
9194	Lithan, Sibille	115	4	3	2$^{\frac{1}{2}}$	2^{1}	2^{hd}	5$^{\frac{1}{2}}$	8^{hd}	14.60
2248	Quantum Leap, Pedroza	117	7	5	5$^{2\frac{1}{2}}$	5$^{2\frac{1}{2}}$	6^{2}	7^{2}	9	10.30

6th, Del Mar, 1M, 2-year-olds.
Maidens.

The information analysis and problem-solving phases of the handicapping here were typically constrained by the realization that two-year-olds at the route represent the most unpredictable of all races. I ended with three outcome scenarios, but three low-priced horses go with them. I passed. The computer picked the winner.

7th, Del Mar, 6½f, 3-year-olds up. (Exacta)
Claiming prices $50,000 to $45,000.

Presuming a turf speedball named *Sheriff Muir* that had never raced on dirt and was returning from a three-month respite did not blast out fast from the bad-rail one-hole today, this sprint shaped up as a classic confrontation between two running styles—the frontrunner against the closer. The two outcome scenarios were as brightly clear as the sun's shine. As handicappers generally agree, in these classic confrontations, stick with the frontrunners. So be it. But today that prescription was intuitively unsettling.

Despite its wire-to-wire romp August 15 under Bill Shoemaker, the frontrunning *Record Catch* was definitely a cheapster. He had sped unmolested around the Del Mar turns following consecutive losing races under preliminary nonwinners allowance conditions at Golden Gate, barely surviving that kind of competition when transferred to Pleasanton on the weaker fair circuit.

The late runner, Embolden, alternatively, is perhaps the best late-running claiming sprinter on the southern California circuit. It is handled as well by the remarkable Bobby Frankel. *Embolden* was exiting fast, competitive races at a classier level than today's. The computer picked *Embolden,* to its everlasting credit, as computers favor early speed as much as humans do. The Frankel colt figured big, to be sure, having everything in its favor except the early pace. If *Record Catch* should be pressed earlier not only would *Embolden* win but the frontrunner might finish down the course. The anti-speed bias today also supported *Embolden.* On the other hand, *Record Catch* and Shoemaker might again grab an easy lead, scamper around the sharp Del Mar turns, and endure to the wire of the 832-foot lane.

```
Embolden                                 Ch. g. 4, by Plotting—Future Era, by Envoy
                                         Br.—Frankel & Roberts (Cal)                1984 11 3 3 2    $61,375
Own.—Alsdorf-Frankel-Roberts       116   Tr.—Frankel Robert              $45,000    1983  9 3 2 1    $38,900
                                       Lifetime   20 6 5 3   $100,275
6Aug84-7Dmr   6½f :221 :452 1:16 ft    7½ 115  67¾ 66 41½ 21½   Ortega L E2      50000 — — LaughingBoy,Embolden,GHatSpecil 7
    6Aug84—Wide 3/8 turn
23Jly84-5Hol  1 :463 1:111 1:364ft      3 116  54½ 68 510 59     Ortega L E2      55000 73-18 Mandato, Cyril's Choice, Ice Caper 6
6Jly84-9Hol   7f :22 :443 1:213ft       3 113  79½ 76¾ 2½ 1½     Ortega L E5      55000 89-15 Embolden, Laughing Boy, IceCaper 8
27Jun84-5Hol  6f :22 :45 1:094ft       *2¾ 116  75¾ 54 43 1hd     Ortega L E7      40000 88-19 Embolden, Insearchof, World Ruler 7
    27Jun84—Wide 3/8 turn
7Jun84-9Hol   7f :221 :45 1:22 ft     *8-5 116  67¾ 54¾ 1½ 14½    Ortega L E2      32000 87-16 Embolden, L'Natty, Shantin 6
20May84-3Hol  6f :212 :44 1:091ft       7½ 116  915 613 69 53¾    Ortega L E8      32000 87-19 Mr.Unbelievbl,Clbong,HndromPckg 9
    20May84—Bumped start
18Mar84-1SA   6f :22 :444 1:103ft      *3 118  1118 1120 911 32¾ DelahoussayeE5   40000 82-18 Socrates, Okubo, Embolden 11
    18Mar84—Extremely wide into stretch
11Mar84-2SA   6½f :221 :45 1:161ft     *1 118  76¾ 77½ 48 44½    DelahoussayeE3   50000 84-18 LondonCross,ChuliaStreet,StandPt 8
    11Mar84—Bumped, checked at 3/16
19Feb84-3SA   6½f :222 :452 1:16 ft     2¾ 116  77 78½ 66½ 32     DelahoussayeE1   60000 88-14 Ice Caper, Item Two, Embolden 7
    19Feb84—Extremely wide
28Jan84-4SA   7f :23 :46 1:222ft        3 117  33½ 34 32½ 2no    DelahoussayeE4   60000 88-12 Estate, Embolden, Ice Caper 6
  Aug 25 Dmr 5f ft 1:003 h       Aug 20 Dmr 4f ft :493 h       Aug 3 Dmr 4f ft :501 h       Jly 15 Hol 4f ft :482 h
```

```
Record Catch                             Ch. c. 4, by What Luck—Salmon Lake, by Tom Fool
                                         Br.—Hunter Barbara (Ky)                    1984  8 2 1 2    $29,175
Own.—House J & C N                 118   Tr.—Greenman Walter             $50,000    1983 11 1 3 0    $34,075
                                       Lifetime   23 4 5 3   $77,000                Turf  8 0 2 1    $18,795
15Aug84-5Dmr  6f :213 :443 1:091ft      16 118   13 13 12½ 14½     Shoemaker W4     50000 92-19 RcordCtch,LondonCross,AstnsC.W. 9
27Jun84-11Pln 6f :223 :443 1:101ft      7½ 1105  3½ 12½ 12½ 1hd    Garcia J J10   Aw13000 93-12 RcordCtch,DimondGorg,RcidRogu 10
14Jun84-8GG   7½f (T):233 :47 1:292fm   9½ 114   55 56½ 49½ 413½   Munoz E3       Aw16000 79-07 French's Luck, Viron, Motivity 5
    14Jun84—Lugged out early
30May84-8GG   1 :461 1:104 1:361ft      3½ 115   1hd 1½ 31½ 48     Diaz A L6      Aw17000 79-18 Splendid Tab, Viron, L'Marquis 6
11May84-8GG   7½f (T):224 :4621:313fm  8-5 114   24 23 21 32½      Diaz A L4      Aw17000 79-18 Wishbone, L'Marquis, RecordCatch 5
6May84-4GG    6f :223 :452 1:093ft      2¾ 113   2hd 21 22 23½     Diaz A L1      Aw19000 87-14 BulLibr,RecordCtch,DimondGeorge 6
26Apr84-8GG   6f :223 :452 1:101ft       2 113   2hd 2hd 21½ 31¾   Diaz A L1      Aw18000 86-17 BluRmRock,DmondGorg,RcordCtch 6
4Feb84-7SA    1⅛ (T):4711:1211:503fm    14 114   32 52¾ 713 821½   Meza R Q5      Aw28000 53-30 Airfield, Pair Of Aces, Negundo 8
    4Feb84—Lugged in, bumped shortly after start
```

29Dec83-7SA 6f :221 :444 1:10 ft 5½ 115 41½ 32½ 43 44 Shoemaker W3 c50000 84-23 BoldRuddy,GoodbyeJ.Y.,GreyBrbrin 7
29Dec83—Bumped start
24Nov83-9Hol 1 1/16 ⓣ:473 1:114 1:42 fm 2¾ 116 3½ 31½ 86½ 95 Shoemaker W8 80000 84-11 GoodFinish,RondsDJmb,DucLDuc 10
24Nov83—Lugged in and steadied through stretch
Aug 29 Dmr 3f ft :36 h Aug 12 Dmr 3f ft :364 h Jly 14 GG 3f ft :373 h

London Cross
Own.—Burke G W **116**
Dk. b. or br. h. 5, by London Company—Something Gay, by Sir Gaylord
Br.—Lin-Drake Farm (Fla) 1984 14 2 1 3 $45,650
Tr.—Lage Armando $50,000 1983 10 2 4 0 $34,350
Lifetime 27 4 5 4 $83,375 Turf 4 0 0 0

15Aug84-5Dmr 6f :213 :443 1:091ft 31 116 77½ 76¾ 66¾ 24½ Pedroza M A7 50000 87-19 RcordCtch,LondonCross,AstnsC.W. 9
15Aug84—Bumped start, wide 3/8 turn
12Jly84-7Hol 6f :22 :444 1:094ft 3 116 77½ 66½ 610 69¾ Meza R Q6 40000 78-20 ToghEnvoy,Jojohnck,HndsomPckg 7
15Jun84-1Hol 6f :222 :452 1:094ft 5½ 116 62½ 61½ 32 32½ Black K7 50000 85-21 LghngBoy,AMomntInTm,LndnCrss 7
26May84-7Hol 6f :221 :442 1:094ft 4½ 116 75½ 76½ 54 31½ Black K7 50000 87-16 MorActon,AstnsC.W.,LondonCross 8
20May84-5Hol 1 ⓣ:462 1:101 1:344fm 24 114 88 86 88½ 86½ McGurn C6 70000 80-09 Mr.Rector,QuntumLp,FlthorpMrinr 8
2May84-6Hol 6½f :221 :453 1:154ft 4 116 2hd 11 1hd 33½ Black K1 62500 87-18 Stand Pat,Prosperous,LondonCross 5
23Apr84-1SA a6½f ⓣ:21 :441 1:15 fm 5½ 116 65½ 55 53½ 73½ Pedroza M A6 62500 80-19 Pellin, Great Eastern, Travelguard 8
14Apr84-7SA a6½f ⓣ:22 :451 1:152fm 27 114 41¾ 42 2½ 62¾ Pedroza MA2 Aw33000 79-16 Prosperous, Fenny Rough, Ansuan 8
21Mar84-7SA 6½f :214 :443 1:151ft 4 116 53 43½ 35 46½ Pedroza M A7 62500 80-15 Whelp, Dedicata, Chief Cornstalk 8
21Mar84—Lugged in down backstretch and 3/8 turn
11Mar84-2SA 6½f :221 :45 1:161ft 8¼ 116 2hd 13 12 11½ Pedroza M A5 50000 89-18 LondonCross,ChuliaStreet,StandPt 8
Jly 9 Hol 4f ft :472 h

Austines C. W.
Own.—Delisle-Harada-Zamora **116**
Dk. b. or br. g. 4, by Austin Mittler—Lotta Talk, by Top Conference
Br.—Crisafi & Warren (Cal) 1984 6 1 2 1 $23,550
Tr.—Wilmot William B $50,000 1983 21 6 1 3 $43,630
Lifetime 33 8 3 6 $69,632 Turf 1 0 0 0

15Aug84-5Dmr 6f :213 :443 1:091ft 4½ 117 34½ 33 34 34¾ Pincay L Jr5 50000 87-19 RcordCtch,LondonCross,AstnsC.W. 9
15Aug84—Veered out, bumped start
19Jly84-5Hol 6f :22 :442 1:10 ft 9½ 117 55½ 57 54½ 21 Pincay L Jr3 50000 86-24 Inserchof,AustinsC.W.,EbonyBronz 6
6Jly84-9Hol 7f :22 :443 1:213ft 13 113 63½ 64½ 43½ 55½ Hawley S3 55000 83-15 Embolden, Laughing Boy, IceCaper 8
15Jun84-1Hol 6f :222 :452 1:094ft 2½ 116 75½ 73¾ 52½ 44½ Hawley S3 50000 83-21 LghngBoy,AMomntInTm,LndnCrss 7
26May84-7Hol 6f :221 :442 1:094ft 6 114 2½ 2hd 1½ 2¾ Hawley S3 45000 87-16 MorActon,AstnsC.W.,LondonCross 8
19May84-7Hol 6f :222 :453 1:101ft 15 117 63¾ 64½ 33½ 12 Hawley S3 25000 86-17 AustnsC.W.,FrnchMjsty,ToghEnvoy 9
28Dec83-1SA 6f :22 :452 1:114ft 4 116 2hd 2hd 5¾ 127¾ Tejeira J8 Ⓢ 16000 71-19 PerfectCover,Socrtes,BostonMgic 12
16Dec83-7BM 6f :23 :47 1:14 sl 8¼ 114 1011 78 89 86¾ Dillenbeck B D6 25000 63-49 CandyStore,TempestWys,MorgnD. 12
13Nov83-2SA 6f :214 :451 1:102m 11 115 31½ 22½ 23½ 46½ Tejeira J1 20000 80-26 HndsomePckge,Grenoble,NewsFlsh 9
6Nov83-1SA 6f :214 :444 1:104ft 3¾ 116 21½ 34 35 34 Tejeira J1 25000 80-18 LghtnngSwft,CndyStor,AustnsC.W. 7
Aug 25 Dmr 5f ft :592 h Aug 13 Dmr 3f ft :362 h Aug 8 Dmr 5f ft :593 h Aug 1 Dmr 4f ft :491 h

2304—SEVENTH RACE. 6 1/2 furlongs. 3 year olds & up. Claiming prices $50,000-$45,000. Purse $19,000.

Index	Horse and Jockey	Wt.	PP	ST	¼	½	¾	Str.	Fin.	To $1
2106	Embolden, Ortega	116	2	11	10 1½	7 1	-	4hd	1½	3.00
(2167)	Record Catch, Shoemaker	116	4	5	1 1	1 2	-	1 4	2 3	1.20
2106	Stand Pat, Hawley	116	9	9	9½	8½	-	7½	3½	6.20
9304	Polly's Ruler, Lamance	116	8	4	3½	2hd	-	3 3	4½	22.50
——	Ackermann, Delahoussaye	116	7	3	2hd	3 3	-	2½	5 1½	16.00
1215	Homto, Dominguez	x109	3	10	8 1	9 6	-	8 2½	6½	109.8
2167	Austines C.W., Pincay	117	11	2	5hd	4½	-	5½	7½	8.20
2167	London Cross, Pedroza	116	10	1	6hd	5 3½	-	6 1	8½	11.80
——	Siglo Diaz, Sibille	116	6	8	11	11	-	10 6	9 2½	124.2
——	Mocito Pilucho, Toro	116	5	7	7 6	6 2	-	9 5	10 17	36.00
1045	Sheriff Muir, Meza	116	1	6	4 2½	10 1½	-	11	11	55.40

Scratched—Laughing Boy. Claimed—Record Catch by H. & K. Alter (trained by W. Spawr), for $50,000.

2-EMBOLDEN	8.00	3.20	2.80
4-RECORD CATCH		2.80	2.40
9-STAND PAT			3.20

Time—.22, .44 3/5, 1.09 3/5, 1.16 2/5. Clear & fast. Winner—ch g 4. Plotting—Future Envoy. Trained by Robert Frankel. Mutuel pool—$218,398. Exacta pool—$340,743.

Pick Six pool—$263,980. Winning numbers (8-4-3-9-8-2) paid $105,486.40 to 1 ticket with 6 winners. Consolation pool paid $1,883.60 to 56 tickets with 5 winners.

$5 EXACTA (2-4) PAID $48.80

I therefore boxed the $5 Exacta ten times, coupling *Record Catch* and *Embolden* ($100). I next played $20 Exactas with *Embolden* on top to each of the two runners-up of *Record Catch*'s August 15 victory ($40). At 3 to 1 straight odds I chose not to bother with *Embolden* to win, which it did.

8th Del Mar, 1⅛M (turf), 3-year-old fillies.
The Del Mar Oaks, Gr. 2, $125,000-added.

Prerace information set. The class demands of Gr. 2 stakes favor two distinguished types of horses: Gr. 1 winners and Gr. 2 winners. When these races are restricted to three-year-olds, handicappers also accept the impressively improving horses that have won open stakes or have finished close in a graded stakes, preferably a Gr. 1 or Gr. 2 stakes.

Turf. On the southern California circuit, handicappers rightly discount three-year-olds that are frontrunners and have not yet raced on grass.

Information analysis. Examine the past performances of *Lucky Lucky Lucky,* the Lukas-trained multiple stakes winner and heavy prerace favorite here. Interestingly, neither the computer nor the author selected *Lucky Lucky Lucky* one-two today. Both preferred *Auntie Betty,* which I had anticipated might be my best opportunity on the card.

Lucky Lucky Lucky had just annexed the Arlington Oaks handily. This is a $100,000-added Gr. 3 stakes open to older horses. The filly's final time had been ordinary, a characteristic of her season-long adjusted times, with two exceptions.

Gr. 3 races are not comparable to Gr. 2 races. Yet *Lucky Lucky Lucky* had already won the Gr. 2 Black-Eyed Susan plus the Gr. 1 Kentucky Oaks, and she finished a short head away in the definitive Gr. 1 Alabama Stakes at Saratoga as recently as August 1. The filly's class credentials were undeniable. She was sharp, too, returning to competition seven days following the easy Arlington win. Lukas strikes repeatedly in this manner with stronger young horses.

The single question mark was the grass. But at even money that was enough. Neither the filly's sire Chieftain nor its broodmare sire Raise A Native had earned a winning impact value in turf racing. Experience has alerted southern California handicappers to be wary of faster three-year-olds when they first try turf. The grass slows them as the hard dirt ovals here cannot. Inexperienced on grass, younger horses typically race too fast and too hard early, get tired, and regularly lose, often to lesser horses. I was quick to forsake *Lucky Lucky Lucky* to this scenario today.

Outcome Scenarios. The alternatives were *Auntie Betty* and *Fashionably Late.* After pressing slow fractions, *Fashionably Late* had won the restricted San Clemente Stakes here two weeks back, but the stretch run had been uninspiring. *Auntie Betty* had been flying by, and in fact should have won. Shoemaker did not arrive in time. I discounted *Fashionably Late* as probably uncomfortable at nine furlongs. She also presented inconclusive class credentials. The Royal Native Stakes she won sprinting at Aqueduct March 10 was unlisted. The Rare Perfume and Prioress were both listed, but *Fashionably Late* finished next-to-last in the Prioress and finished fourth in slow time in the Rare Perfume.

Auntie Betty's best efforts had been her previous two, a telltale signal among high-potential, improving three-year-olds. This filly had attracted lots of money on her first starts at Calder and had even been tried in Hollywood's Gr. 1 Oaks, although without raising much commotion. In her two previous races at Del Mar she had run the late fractions in a smashing $:23\frac{3}{5}$ and $:30\frac{1}{5}$ seconds.

The five-furlong $:59\frac{4}{5}$ workout August 29 was the quickest this filly had trained in the morning. Final decisions. All systems appeared set. Today's Gr. 2 objective looked like the end result of a typical Charlie Whittingham prep job. The other big fillies had been knocked down on the tote, and I was eager to make a $250 investment to win, but wanted 5 to 2. When the crowd knocked *Auntie Betty* down to a surprising 8 to 5, I bet just $50 instead. The Exactas were not appealing.

Comment. Lucky Lucky Lucky tired in the stretch, as predicted, lugging out and bumping the winner twice. But *Auntie Betty* flattened late too. She lost; just didn't have it. The intuitive handicapping had been sound, but wrong. Next race.

Fashionably Late

Dk. b. or br. f. 3, by My Gallant—Two-Thirty Girl, by Songman
Br.—Newquist F R (Fla) 1984 10 3 1 2 $87,671
Own.—Firestone Mrs B R (Lessee) 119 Tr.--Frankel Robert 1983 2 2 0 0 $18,600
Lifetime 12 5 1 2 $106,271 Turf 2 1 1 0 $41,800

18Aug84-8Dmr 1 1/16 (T):48 1:12 1:431fm *6-5 114 3½ 2½ 2½ 1½ McCrrCJ2 (F)[R]SnClmt 93-08 FshionblyLte,AuntieBtty,PtriciJms 8
20Jly84-8Hol 1 (T):461 1:10 1:352fm 7½ 116 3nk 2hd 2hd 2hd McCrrC5 (F)[R]TrsTrsrs 91-13 SligoTown,FashionablyLate,Shelbin 9
16May84-6Bel 6f :232 :462 1:102ft 3½ 115 11½ 12 21½ 31 CordroAJr2 (F)Aw26000 89-17 TringMilli,Prd'sCrossng,FshonblyLt 8
25Apr84-8Aqu 6f :23 :461 1:102ft 4½ 115 22 33½ 33 53 CordroAJr5 (F)Prioress 86-22 ProudClrionss,Dumddumddum,Suvt 6
25Apr84—Wide
1Apr84-8Aqu 170[•]:483 1:124 1:432ft 12 115 2½ 2hd 2½ 41¾ VlsJ5 (F)Rare Perfume 81-19 Given, Maharadoon, Recharged 9
10Mar84-7Aqu 6f [•]:224 :464 1:122ft *3-2 118 2½ 1½ 1hd 12 VlszJ3 (F)Royal Native 82-17 FshionblyLte,WithTrdition,Din'sKin 7
29Feb84-7Aqu 6f [•]:221 :462 1:122ft *2½ 116 42 3½ 1hd 11½ VelsquezJ5 (F)Aw22000 82-24 FshionblyLte,TheDell,Jnnifr'sChoic 9
18Feb84-6Aqu 6f [•]:223 :462 1:112ft *2-3 114 21 31 42 54 VlsJ5 (F)[R]Chou Croute 83-12 SinghHony,WithTrdition,GottDoubl 7
11Feb84-8Key 6f :213 :442 1:104ft *2 116 811 55½ 42½ 3½ LopzCC3 (F)New Hope 86-21 SecondGlnce,WhizQuiz,FshionblyLt 8
11Feb84—Bobbled start
15Jan84-8Aqu 170[•]:492 1:153 1:47 ft *7-5 112 1hd 65½ 710 712½ VelsquezJ6 (F)Busanda 53-30 Mile High Lady, Given, Shavie 8
Aug 27 Dmr (T) 6f fm 1:16 h (d) Aug 13 Dmr (T) 6f fm 1:134 h (d) ●Aug 6 Dmr (T) 6f fm 1:162 h (d) Jly 14 Hol (T) 7f fm 1:283 h (d)

Auntie Betty

Dk. b. or br. f. 3, by Faraway Son—Manchester Miss, by Groton
Br.—Headley & Penna (Ky) 1984 7 3 2 1 $52,350
Own.—Green Thumb Farm Stable 114 Tr.—Whittingham Charles 1983 4 1 2 1 $12,000
Lifetime 11 4 4 2 $64,350 Turf 5 2 2 1 $40,200

18Aug84-8Dmr 1 1/16 (T):48 1:12 1:431fm 2¾ 117 811 85½ 54 2½ ShrW8 (F)[R]Sn Clmnte 92-08 FshionblyLte,AuntieBtty,PtriciJms 8
4Aug84-7Dmr 1 (T):48 1:122 1:363fm 6½ 1045 43 43 2½ 11 Lozoya DA3 (F)Aw24000 90-08 AuntieBetty,Lyphrd'sPrincss,ClrTlk 9
8Jly84-8Hol 1⅛ :451 1:094 1:491ft 18 121 810 89 67½ 58¾ McCrrCJ3 (F)Hol Oaks 76-16 MomntToBy,Mttrnd,LckyLckyLcky 9
24Jun84-3Hol 1 (T):464 1:111 1:361fm 4½ 114 34 31 1hd 31½ McCrrCJ1 (F)[R]Glenris 86-11 Leslie'sDeb,WineTster,AuntieBetty 7
23Mar84-8GP 1 1/16 :48 1:131 1:461sy 2¾ 119 54½ 32 2½ 21 Bailey J D3 (F)Aw14000 69-22 ‡Illaka, Auntie Betty, Proud Nova 8
23Mar84—Placed first through disqualification; Bumped, steadied
2Mar84-7Hia a1⅛ (T) 1:491fm *8-5 120 621 32 1hd 2hd Bailey J D1 (F)Aw20000 85-15 PurpleGlory,AuntieBtty,Rxson'sRos 8
2Mar84—Bumped
7Jan84-4Crc 1 1/16 (T):481 1:131 1:451fm 3¾ 117 1111 65¾ 43½ 1no Vasquez J10 (F)Aw12000 79-18 Auntie Betty, T. V. Siren,JayParee 12
19Dec83-5Crc 1 1/16 :491 1:152 1:482ft 2½ 116 33½ 32 21½ 2¾ Pennisi FA5 (F)Aw12000 76-18 Jigevr,AuntieBetty,OurBrightLight 7
10Dec83-6Crc 1 1/16 :50 1:153 1:493ft *1 118 33½ 32 21 12 Vasquez J9 (F)Mdn 71-18 AuntieBetty,BayouBirdie,VagueIde 9
21Nov83-6Crc 1 :49 1:16 1:433ft *3-5 118 52 41½ 68½ 38½ Pennisi F A3 (F)Mdn 61-18 His Ex, Commended, Auntie Betty 8
Aug 29 Dmr 5f ft :594 h Aug 16 Dmr 3f ft :354 h Aug 9 Dmr 3f ft :37 h Aug 2 Dmr 3f ft :353 h

Lucky Lucky Lucky *

B. f. 3, by Chieftain—Just One More Time, by Raise A Native
Br.—Spendthrift Farm (Ky) 1984 10 3 1 3 $388,016
Own.—Combs LII&EquitesStb(Lse) 124 Tr.—Lukas D Wayne 1983 8 3 2 2 $376,140
Lifetime 18 6 3 5 $764,156

25Aug84-8AP 1⅛ :46 1:10 1:494ft *3-5 121 11 11½ 14 13½ Davis R G2 (F)Arl Oaks 82-19 LuckyLuckyLucky,Mrs.Revere,Bsie 7
11Aug84-8Sar 1¼ :48 1:363 2:023ft *4-5e 121 11 1½ 1½ 2hd Davis R G5 (F)Alabama 87-15 Lif'sMgc,LuckyLuckyLucky,ClssPly 5
2Aug84-8Sar 7f :214 :443 1:223ft 2e 124 43 32½ 31½ 33½ Cordero A Jr6 (F)Test 86-19 Sintr,WldApplus,LuckyLuckyLucky 9
8Jly84-8Hol 1⅛ :451 1:094 1:491ft *8-5 121 32 2½ 21½ 33½ CordrAJr2 (F)Hol Oaks 80-16 MomntToBy,Mttrnd,LckyLckyLcky 9
6Jun84-8Hol 7f :212 :44 1:221ft 2½e 122 83¾ 76¾ 55 35½ Hawley S4 (F)Railbird 80-20 Mttrnd,Gn'sLdy,LuckyLuckyLcky 10
26May84-8Bel 1 :453 1:102 1:354ft 2¾e 121 21½ 42 66 68¾ Cordero A Jr6 (F)Acorn 77-15 MissOcen,Life'sMgic,ProudClrionss 9
18May84-9Pim 1 1/16 :463 1:103 1:411ft *4-5 121 11 12 12 1½ CrdrAJr3 (F)BEydSusn 99-13 Lucky LuckyLucky,Sintra,DuoDisco 7
4May84-9CD 1⅛ :482 1:132 1:514gd 6 121 11½ 11 2hd 1no CordroAJr4 (F)Ky Oaks 83-16 LckyLckyLcky,MssOcn,MyDrlngOn 6
4May84—Broke slowly
21Apr84-7Kee 1 1/16 :48 1:132 1:491sy 3-2 121 32½ 37 313 419½ VelasquezJ3 (F)Ashland 41-37 Enumerting,MissOcen,RoseofAshes 4
1Apr84-9Lat 1 1/16 :46 1:101 1:424ft 2¾ 116 33 33 54½ 53 CordroAJr1 Jim Beam 93-18 AtThThrshld,BldSthrnr,ThWddGst 12
Aug 19 Sar 5f ft 1:02 h ●Jly 26 Dmr 6f ft 1:102 h Jly 17 Hol 5f ft 1:00 h Jly 2 Hol 5f ft 1:002 h

2305—EIGHTH RACE. "Del Mar Oaks." 1-1/8 miles on turf. 3 year olds. Fillies. Purse $125,000-added. Total purse $160,150.

Index	Horse and Jockey	Wt.	PP	ST	¼	½	¾	Str.	Fin.	To $1
(2197)	Fashionably Late, McCarron ..	119	3	5	$5^{2\frac{1}{2}}$	$5^{3\frac{1}{2}}$	$5^{1\frac{1}{2}}$	$2^{2\frac{1}{2}}$	1^{hd}	3.70
1505	Lucky Lucky Lucky, Pincay ...	124	6	1	$1^{1\frac{1}{2}}$	$1^{1\frac{1}{2}}$	1^{1}	$1^{1\frac{1}{2}}$	2^{2}	1.40
2197	Auntie Betty, Shoemaker ...	114	4	6	6^{3}	6^{5}	6^{4}	3^{hd}	$3^{2\frac{1}{2}}$	1.60
2197	Lassie's Deb, Toro	117	7	7	7	7	7	7	$4^{\frac{1}{2}}$	15.50
2197	Patricia James, Delahoussaye	116	5	4	$3^{\frac{1}{2}}$	$4^{2\frac{1}{2}}$	$4^{1\frac{1}{2}}$	6^{1}	5^{hd}	18.10
2197	Shy Bride, Garcia	114	2	2	2^{1}	$2^{1\frac{1}{2}}$	3^{1}	$4^{\frac{1}{2}}$	$6^{2\frac{1}{2}}$	46.20
1505	Gene's Lady, Sibille	119	1	3	$4^{1\frac{1}{2}}$	3^{hd}	2^{hd}	5^{hd}	7	18.40

Scratched—none.

3-FASHIONABLY LATE	9.40	3.60	2.40
6-LUCKY LUCKY LUCKY		3.00	2.40
4-AUNTIE BETTY			2.20

Time—.23 2/5, .46 4/5, 1.11 1/5, 1.36 2/5, 1.49 2/5. Clear & firm. Winner—br f 3. My Gallant—Two-Thirty Girl. Trained by Robert Frankel. Mutuel pool—$254,759. Exacta pool—$222,616.

$5 EXACTA (3-6) PAID $63.50

9th Del Mar, 1M, 3-year-olds up. (Exacta)
Nonwinners of two races, allowance.

Prerace information set. The class demands of these tough nonwinners allowance races favor lightly raced three-year-olds of any real potential. Horses four and older and still eligible to these preliminary nonwinners conditions in September should be lightly raced themselves, nicely bred, and reveal no starts in claiming races. They usually lose anyhow. Interestingly, if the potentially good three-year-olds are absent, this late in the season the impressive claiming race winners can move into these preliminary nonwinners allowance contests and win.

The outcome scenarios here became quickly apparent, either *Fortune's Kingdom* winning from near the front or *Grey Missile* coming from farther behind. *Fortune's Kingdom* had been highly regarded on this circuit since its January 8 close call with division leader *Fali Time,* but its record intimated that the colt preferred sprinting to routing, and its class credentials in longer races remained suspect. Yet the probable pace here might favor *Fortune's Kingdom,* as no real early speed could be reliably detected.

The lack of speed might penalize *Grey Missile,* though its move between the second and third calls last out proved this was no footless laggard. Trainer Willard Proctor was changing jockeys today as well, from apprentice Lozoya to champion McCarron, which meant mainly that odds of 29 to 1 last out would also be changed, to approximately 2 to 1.

Intuitively, if you will, I began to search for other meaningful handicapping information about this field.

I found only one interesting possibility. Following a layoff, the two-sprint stretch-out pattern is best; by comparison, the one-sprint pattern is the worst—statistically.

The Graustark colt, *Appetite*, from the hot Jerry Fanning stable, a successful outfit with young horses, was trying the mile after two sprints that followed a four-and-a-half-month vacation. I liked the pattern. The latest sprint had been much better than the first, with *Appetite* finishing fast, completing the final five-sixteenths in :30⅕ seconds. That second sprint had been a fast race, the field passing the six furlongs in 1:08⅘ seconds. *Appetite*'s full record lacked signs of genuine class, but Fanning was switching interestingly to apprentice Dario Lozoya here. This colt could readily shoot to the lead in this paceless contest and steal the race.

I keyed Appetite in $20 Exacta boxes with Fortune's Kingdom and Grey Missile ($80). The pari-mutuel action suggested I had zeroed in on a runner, but that intuition did not pay off at the windows today.

Without a winner in the straight pools, I departed Del Mar with $221 in profits, a not unusual circumstance resulting from intuitive handicapping utilizing information management. Also not unusual is that profits accumulated from just two winning races, interspersed among six losing plays and two passes. ROI was .43.

Table 22
Intuitive Handicapping: Del Mar, Sept. 1, 1984

	Summary *Bet*	*Net*
1st	$70	($70)
2nd	50	(50)
3rd	40	206
4th	0	0
5th	80	(80)
6th	0	0
7th	140	345
8th	50	(50)
9th	80	(80)
Total	$510	$221

Grey Missile
Own.—Hawn W R 113
Ro. c. 3, by Vigors—Cornish Genie, by Cornish Prince
Br.—Hawn W R (Ky) 1984 9 1 2 1 $23,250
Tr.—Proctor Willard L Turf 3 0 0 0 $1,650
Lifetime 9 1 2 1 $23,250

19Aug84-7Dmr 1 1/16 :45³ 1:10⁴ 1:42³ft 29 108⁵ 8⁷ 3½ 1½ 2½ Lozoya D A³ Aw18000 86-17 BenBg,GryMissil,Fortun'sKingdom 8
29Jly84-9Dmr 1 1/16 ⓣ:48³ 1:13 1:45¹fm 12 116 8¹⁵ 9¹¹ 9¹² 8⁷ DelhoussyE¹⁰ Aw18000 76-13 RfldII,‡AtCmmndr,Agnstthkngdm 10
16Jun84-7Hol 1⅛ ⓣ:48 1:12³ 1:49¹fm 21 114 97½ 118½ 10⁶ 9⁶ Fires E⁴ Aw22000 78-16 StrMtril,BlushngGust,Promontory 11
3May84-7Hol 1 1/16 ⓣ:47² 1:11⁴ 1:41¹fm 5½ 120 7¹³ 6⁸ 59½ 410½ Sibille R⁷ Aw22000 82-07 M.DoubleM.,HeavenlyPlain,Bronzed 7
28Mar84-8SA 1⅛ :46³ 1:10⁴ 1:48³ft 19 118 9²⁰ 9¹⁶ 8¹¹ 710½ Sibille R⁴ Ⓢ Bradbury 76-18 TsnmSlw,MghtyAdvrsry,LotsHony 10
28Mar84—Bobbled start
11Mar84-6SA 1 1/16 :46¹ 1:11¹ 1:44 ft 3 118 7¹³ 57½ 23½ 1ⁿᵒ Sibille R⁴ Mdn 81-18 Grey Missile, Ocean View, Koshare 7
26Feb84-2SA 1 :46³ 1:12 1:37⁴ft 5 118 75¾ 64½ 65¾ 22½ Sibille R¹⁰ Mdn 76-15 Riva Riva,GreyMissile,CoopersHill 10
26Feb84—Wide into stretch
12Feb84-6SA 7f :22³ :45⁴ 1:24⁴ft 54 118 88½ 8¹² 5⁸ 31½ Sibille R⁴ Mdn 75-18 Barcelona, Guards, Grey Missile 8
12Feb84—Wide into stretch
22Jan84-6SA 6½f :22² :45⁴ 1:18 ft 56 118 54¾ 8¹² 9¹⁶ 914¾ Steiner J J³ Mdn 65-21 Carrizzo, Coopers Hill, River Yang 12
Aug 30 Dmr 4f ft :47⁴ h Aug 16 Dmr 5f ft 1:04¹ h Jly 22 Dmr 7f ft 1:27⁴ h Jly 16 Dmr 6f ft 1:15² h

Appetite
Own.—Gentry T E 108⁵
B. c. 3, by Graustark—Dinner Partner, by Tom Fool
Br.—Gentry T E (Ky) 1984 6 1 0 2 $16,775
Tr.—Fanning Jerry 1983 0 M 0 0
Lifetime 6 1 0 2 $16,775

22Aug84-7Dmr 6½f :22² :44³ 1:15¹ft 2 113 4⁶ 4⁹ 49½ 3³ ShoemkerW⁵ Aw18000 — — Music Master, Bunker, Appetite 5
22Aug84—Wide into stretch
2Aug84-7Dmr 6f :22¹ :45² 1:09⁴ft 18 108⁵ 98½ 9¹¹ 78¾ 34½ Lozoya D A³ Aw17000 84-18 ToughEnvoy,Sari'sDelight,Appetite 9
2Aug84—Wide into stretch
17Mar84-7SA 6f :21⁴ :44³ 1:09¹ft 9 120 7¹¹ 69½ 5¹⁰ 58½ Lipham T⁸ Aw25000 83-17 DebonirJunior,PrincPninsul,Artificr 8
4Mar84-6SA 1 1/16 :46⁴ 1:11 1:43¹ft 4e 115 78¾ 7¹¹ 6¹¹ 69¾ DelhoussyeE⁴ Aw23000 75-17 MightyAdversry,HevenlyPlin,Czdor 8
4Feb84-2SA 6½f :21³ :44³ 1:16⁴ft 7½ 118 7¹⁰ 7⁷ 64½ 11¾ Delahoussaye E⁷ Mdn 86-14 Appetite, Ocean View,SavorySauce 7
22Jan84-6SA 6½f :22² :45⁴ 1:18 ft 6¾ 118 85¼ 7¹² 5⁸ 55½ Delahoussaye E⁹ Mdn 74-21 Carrizzo, Coopers Hill, River Yang 12
Aug 29 Dmr 5f ft 1:01³ h Aug 19 Dmr 5f ft :59³ h Jly 26 Dmr 6f ft 1:14⁴ h Jly 17 Hol 5f ft 1:00² h

Fortune's Kingdom
Own.—Sarkowsky-Spehler-Wygod 113
B. c. 3, by King Pellinore—Fortune Teller, by Gummo
Br.—Wygod-Sarkowsky-Spehler (Cal) 1984 5 0 2 1 $28,575
Tr.—Mandella Richard 1983 6 1 1 4 $36,200
Lifetime 11 1 3 5 $64,775 Turf 2 0 0 2 $18,000

19Aug84-7Dmr 1 1/16 :45³ 1:10⁴ 1:42³ft *2½ 117 2² 5³ 3¹ 32¾ Pincay L Jr¹ Aw18000 84-17 BenBg,GryMissil,Fortun'sKingdom 8
2Aug84-7Dmr 6f :22¹ :45² 1:09⁴ft 2 117 88½ 8¹⁰ 46½ 45¾ Pincay L Jr² Aw17000 83-18 ToughEnvoy,Sari'sDelight,Appetite 9
14Mar84-8SA 1 1/16 :47¹ 1:11¹ 1:43³gd 28 115 43¾ 45½ 58½ 610½ Toro F¹⁰ Ⓢ Sta Ctlna 72-26 Tights, Prince True, Gate Dancer 10
14Mar84—Fanned extremely wide 7/8 turn
26Feb84-7SA 1 :46³ 1:11¹ 1:36⁴ft *2-3 117 4² 31½ 1ʰᵈ 2¹ Pincay LJr⁶ Ⓢ Aw23000 83-15 SarosSaros,Fortune'sKingdom,Tbre 8
8Jan84-8SA 7f :22² :45² 1:23¹ft 5¾ 117 3½ 1ʰᵈ 2ʰᵈ 2ⁿᵏ PincyLJr¹ Ⓢ Cal Brdrs 84-20 FliTime,Fortune'sKingdom,BuenJef 6
17Dec83-7Hol 1 1/16 ⓣ:47 1:11 1:43³fm 3 117 68¾ 59½ 5¹⁰ 33½ PncyLJr³ Ⓢ Cougar I I 77-16 FrnchLgonr,Tghts,Fortn'sKngdom 11
25Nov83-6Hol 1 ⓣ:48³ 1:13⁴ 1:38²gd 13 115 4² 5³ 41½ 3² DlhssyE⁶ Hst The Flg 74-22 Artichok,Ngurski,Fortun'sKingdom 8
25Nov83—Run in divisions; Bumped 1/4
28Oct83-8SA 6½f :21⁴ :44² 1:16¹ft 7¾ 117 57½ 5¹¹ 45½ 3⁸ DelhoussyeE³ Aw22000 81-19 Prcsonst,DonnrPrty,Fortn'sKngdm 7
16Oct83-6SA 6½f :22² :45⁴ 1:18 ft *6-5 117 22½ 2³ 1⁴ 16½ McCarron C J¹⁰ Mdn 88-20 Fortn'sKngdm,MntM.,KpOnTlkng 10
6Oct83-6SA 6f :21³ :45¹ 1:11¹ft *6-5 117 23½ 21½ 32½ 3⁴ McCarron C J⁸ Ⓢ Mdn 78-21 FftySxInRw,Rsl'sChc,Frtn'sKngdm 9
●Aug 29 Dmr 4f ft :46 h Aug 13 Dmr 1 ft 1:41¹ h Aug 10 Dmr 3f ft :36³ h Jly 29 Dmr 5f ft 1:00 h

2306—NINTH RACE. 1 mile. Allowance. 3 year olds & up. Purse $18,000.

Index	Horse and Jockey	Wt.	PP	ST	¼	½	¾	Str.	Fin.	To $1
2205	dq-Fortune's Kingdom, Hwly	113	7	4	3ʰᵈ	5⁴½	1½	11½	1ⁿᵒ	2.00
2206	Grey Missile, McCarron	113	3	7	7	6⁴	4³	21½	2⁵	2.70
2223	Appetite, Lozoya	x108	5	1	5½	3ʰᵈ	2ʰᵈ	3½	3ʰᵈ	2.90
2147	Trovista, Garcia	113	4	3	21½	1ʰᵈ	3²	4⁶	45½	34.10
——	Mangione, Sibille	118	6	2	42½	4½	52½	5²	5²	4.20
2205	White Cloud, Shoemaker	118	1	6	1ʰᵈ	2¹	6³	62½	63½	5.80
2005	Emergency Fund, Ortega	118	2	5	63½	7	7	7	7	56.70

dq-disqualified & placed second.
Scratched—Prando, Purple, Nitro, Take A Rest.

3-GREY MISSILE	7.40	2.80	2.80
7-dq-FORTUNE'S KINGDOM		3.20	2.80
5-APPETITE			3.00

Time—:22 3/5, :46, 1:10 1/5, 1:35 4/5. Clear & fast. Winner—ro c 3. Vigors—Cornish Genie. Trained by W. L. Proctor. Mutuel pool—$202,032. Exacta pool—$299,810. Attendance—23,529. Total handle—$4,359,406.

$5 EXACTA (3-7) PAID $49.50

The PDS sports computer, which tabbed three winners, helped me beat the third race. I would not have used *Emperoroftheuniverse* in the Exacta had it not been the computer's second choice to a heavy underlay and proposed at 10 to 1 odds. In applying the information management approach, it's foolhardy not to integrate the solid information of other practiced handicappers—or even computers implementing powerful regression models of handicapping—when the odds beckon you. Those computer models were erected by highly brainy handicappers.

The day's key investment decision embraced two opposing outcome scenarios combined in exotic wagering. It's practically an everyday tactic with the information management approach. Will the solid come-from-behind horse catch the cheaper but much-advantaged frontrunner? Why worry? Accept the opposite logics and entangle the two scenarios in exotic wagering.

The database of stakes data supplied essential information in the fifth race only today, but the resulting 19-to-1 horse almost hammered home the point of having handicapping databases with a mighty generous upset.

When backing this kind of information, promising horses at high odds, high-tech handicappers will most often rely on exotic betting and in these instances should be sure to cover all the real possibilities. Exacta wheels are hardly preferred types of handicapping investments but must be resorted to at times to protect the sizeable capital invested. If I had not hooked *Indian Trail* to *all* the other horses, by the way, I would have bet at least $50 to win on the horse. These well-recommended, carefully searched-out longshots cannot be allowed to pound home on top at 19 to 1 but return nothing to the handicapper's bankroll.

TYPES OF INTUITIVE DECISIONS

As handicapping information supporting the case for a horse accumulates and becomes concentrated, a *feeling* that the animal has an unusually good chance to win strengthens. Inductive, intuitive reasoning is additive, piling data relationships and information on top of each other, and as the rational cases develop so do the emotional counterparts. Handicappers can imagine intuitive decisions as positioned on a scale whose polar points are labeled rational and visceral. Each decision point on the scale has both rational and visceral elements. Paradoxically, the stronger the rational case for a horse, the more likely an intuitive decision will become more visceral than rational. That is to say, at some point the intuition takes over for the intellect. Rational handicappers need not be chagrined by this thought.

Regardless of the intuitive force of handicapping decisions, they all are based on rational considerations and reflect varying degrees of rationality. This is not a guessing game. Intuitive decisions do not derive from whim, whisper, or whimsy. Intuitive decisions are not hunches; they are not guesses.

All intuitive decisions have a rational basis. In fact, in practice the rational case ordinarily does not become so forceful that its visceral elements take over. Expert handicappers may intuit, sense, or infer the correctness of their choices, but they also can explain the rational bases for them. Regardless, neither computer applications nor systematic methods perform in this intuitive way. Both are heavily calculated and proceed *rationally* in straightforward manners. A useful way to classify intuitive decisions is therefore to consider the visceral force they are likely to develop, depending on the amount and weight of information supporting each case.

I have attempted to classify information management decisions at the track in this way by identifying six types of intuitive handicapping decisions, according to the intuitive force that is characteristic of each. The stronger the intuition, the more reliable the decision. I label the six: the overwhelming intuition, combining the fundamental and the incidental, the rational force horse, multiple keys, opposite logics, and creative handicapping.

Below is a succinct characterization of each type. Many handicappers will be able to recognize several of these decision types in relation to their personal decision-making behavior at the races.

THE OVERWHELMING INTUITION

The most powerful of all intuitive decisions, the overwhelming intuition attaches to that rare horse that handicappers *feel* will just not lose today. The nature of the intuition is overwhelmingly visceral, so much so that its rational elements appear vague or lost. Yet they are unmistakably present and are the root cause of the strong feelings. If asked to present the intellectual case for the horse, handicappers might sound tentative; they might appear stuck. That is because the intuition has already become irrational or at least distracting.

The rational case usually can be characterized variously by an intense comprehension of a few significant handicapping factors in combination, by a *Eureka!* gut reaction based on a single decisive factor, or by a highly generalized understanding of the total racing situation, in which the prime choice looks outstanding in all respects and all the circumstances of the race appear to be ideal.

Handicappers should trust the overwhelming intuition. Most have learned to do just that. The horses win and often win big. Moreover, they regularly return generous prices. Wager size should be escalated, perhaps pushed to the limits. If exotic wagering is offered, the key horse should be coupled repeatedly with all the real possibilities.

The overwhelming intuition, to be sure, is rare. It presents itself a few times a season. It should not be mistaken as a strong feeling. And it is not represented by a brilliant analysis or intricate calculation that suddenly renders a complex race completely clear. It is rather an overwhelmingly irresistible urge whose intellectual end points strike sharply and quickly, and then almost disappear. Handicappers who learn to recognize the overwhelming intuition must also learn to exploit it. That means betting personal limits and maximizing the potential gains of a one-shot case study.

To fly closer to reality, overwhelming intuitive decisions sometimes misfire, and the maximum wagers have been lost against all odds. When this happens, handicappers should proceed to the next race without feeling painstakingly sorry for themselves or attempting to analyze what went wrong. The only explanation is that all horses lose. When the overwhelming intuition appears again, handicappers who lost the last time should remember that these horses enjoy a tremendous advantage. Thus, they have an unusually good chance to win and deserve correspondingly strong betting support.

COMBINING THE FUNDAMENTAL AND THE INCIDENTAL

The great majority of intuitive handicapping decisions are not nearly so emotionally loaded as the overwhelming intuition. They reflect relatively equal components of rationality and affect, and fall into a wide classification best described as combining the fundamental and the incidental.

Fundamental information, to recall, relates to the intrinsic abilities and preferences of horses. Incidental information relates to the situational factors that influence the running of races.

A successful trainer maneuver is being repeated today with a horse rounding into the peak condition of its form cycle.

An impressively improving claiming horse suddenly on a steep rise has beaten this class before, the pace today should be favorable, and last out the horse raced extra wide while accelerating against fast fractions between calls.

The two horses described above combine fundamental and incidental advantages.

Whenever matters of class, speed, early speed, form, distance, and pedigree can be linked positively to matters of pace, bias, trips, trainer, jockey, body language, or even post position, but not weight, the combined weight of the relationships makes for a relatively sturdy case. Handicappers sense the situation for exactly what it is. The horses have a lot going for them. The greater the number and strength of the information relationships supporting a horse, the greater the intuition that it should win.

It is important to note that the information management approach treats incidental types of handicapping information differently from what has been traditionally true of systematic method play.

Method handicappers have customarily related situational information to method selections in negative ways. After all, the method horse has been accepted *a priori* as the probable winner.

Incidental information is therefore secondary to the method's selection. If method handicappers do not like the jockey, for instance, they pass. If the trainer is a proven flop, pass. If the post position and weight seem problematic, method handicappers pass again. If the probable pace should be uncomfortable or the footing uncertain, method handicappers once again pass or sometimes switch to second choices.

The information management approach relates incidental information to fundamental information in positive ways. It forms information linkages or information chains. By this approach handicappers are not testing out a prime selection on all handicapping points but rather building information cases for different horses, solving tricky problems with various information, identifying potential outcome scenarios, and making the best possible decision in specific situations. Incidental information is not subordinate to fundamental information. Instead, it complements and supplements it. Where fundamental factors are inconclusive, incidental information tells the tale. At times incidental information can be far more decisive than are fundamental factors, as when a track bias is powerful.

Bettable horses that combine fundamental and incidental advantages can be isolated two or three times a day on major racing cards. The handicapping intuition often depends on the negative aspects of opponents as well as the positive aspects of choices, as the case for Auntie Betty at Del Mar certainly did. The rationale for Auntie Betty was equally dependent on the shortcomings attached to Lucky Lucky Lucky and Fashionably Late as on the attractions of her own past performances.

Horses that combine strong fundamental and incidental advantages generally can be used as reliable keys in exotic wagers, regardless of whether they are supported in the straight pools. If the eighth at Del Mar had offered Exacta wagering, for example, I would have thrown out Lucky Lucky Lucky and keyed Auntie Betty in multiple Exacta boxes with Fashionably Late and perhaps one or two other outsiders. The reason: Should *Lucky Lucky Lucky*'s class advantage collapse, its disadvantages on incidental factors—strange footing or faster, tougher pace—should take on heightened importance. The heavily bet filly might run out.

Handicappers who accept the information management approach advocated in this book should begin today to stand on alert for horses that are unusually attractive on both fundamental factors and incidental factors. The stronger the relationships, the better. Horses that qualify for this category are among the best of plays at racetracks, representing as they so often do solid choices in advantageous situations at middling to generous odds.

THE RATIONAL FORCE HORSE

Rational force horses figure to win on straightforward fundamental handicapping and are usually the logical deductive selections of conventional thinking and systematic methods. The horses have earned better class ratings or higher speed figures or faster pace ratings and look acceptable on form, distance, and trainer-jockey patterns besides. They are fundamentally solid and can be accepted as probable winners. The logical reasoning that selects the horses proceeds from general principles, which have worked repeatedly when applied properly, and not so much from the additive force of various data relationships or diverse sources of information. These types were not meant to be abandoned by the information management approach, which merely recognizes that most races and horses remain unplayable by the systematic, logical-deductive approach, either because no horse stands apart on the fundamental principles, or the few a day that do are often overbet.

At Del Mar, on September 1, I supported two horses in this logical-deductive category, *Cream Pocket* in the first and *Embolden* in the seventh.

The intuitive aspect of rational force horses is relatively low. Intuition has been sublimated to the strict and orderly intellectual character of the reasoning that spots the horses. Highly rational racetrack decisions are stepwise and routine for experienced handicappers. But rational force horses can accumulate their own intuitive momentum. The stronger the rational case, the more confidence handicappers should feel. Handicappers should experience that confidence at the emotional level. Where that kind of confidence is not felt, the rational force horse is probably not very compelling at all.

MULTIPLE KEYS

A crucial assumption of the information management approach holds that most races cannot be reduced to a single selection at an acceptable price. The probable outcomes are instead various, with each scenario featuring a different horse or combinations of horses. Traditionally, the definition of a playable race has promoted its predictability in terms of solid fundamental selections. The concept of a fair price has often been conveniently buried. Yet two of three races are misplayed by the general public, and many of its winners become unplayable underlays. The majority of races also remain unpredictable by the methods of the most skillful handicappers, who face a ceiling on handicapping proficiency at 40 percent winners. Handicappers can benefit by replacing the ideal of a prime selection with the concept of multiple keys.

Two contemporary advances, increased exotic wagering opportunities and numerous new sources of handicapping information, have rendered the abstraction of the prime selection at an acceptable price practically obsolete. In concert with the information explosion in handicapping, nowadays if a horse stands out on fundamentals, too many informed, skillful handicappers will know it. The key selection becomes an outstanding underlay, the race unbeatable. Modern handicappers seeking meaningful profits from a season's play must rely instead on multiple keys in combination betting.

Multiple keys represent the intuitive handicapping tactic for handling unpredictable yet beatable races. If the probable outcome scenarios have been carefully constructed, handicappers have identified the multiple keys to unlocking the race. Combination bets can be underlays too, but more frequently they provide overlays, and they allow handicappers who are information managers unprecedented flexibility of coverage. When the odds warrant, of course, multiple keys can be supported separately in the straight pools.

Intuitive handicappers will not imagine that multiple keys can be combined only with one another. Outcome scenarios may differ in detailed ways other than merely predicting potentially different winners. The contending horses in each scenario often will be different. If scenarios feature overlapping horses, some of the same horses will be anticipated by each. The probable winners of the various scenarios will therefore be combined in intuitive ways with other horses in the pictures. The possibilities can be numerous and varied.

The intuitive elements of handicapping decisions that involve multiple keys will likely be stronger than their rational elements. During the information analysis and problem-solving phases of the handicapping process neither the logical-deductive nor the rational-informational data relationships have coalesced into a final decision. The information has been either inconclusive or indicative of alternative possibilities. Neither systematic methods nor information management has solved the handicapping puzzle, or perhaps both have by specifying different horses.

In either case, several horses have been judged as potential winners. Depending on how the race will actually be run, other horses that may not actually figure to win will be viewed as contenders for the runners-up prizes. The combinations of horses that might survive can appear exasperating, yet the probable payoffs and betting values will often be remarkably appealing.

In many cases the values offered handicappers by multiple keys in exotic combinations justify the rather ample investments in seemingly unpredictable races and rightly so. In this way real value accrues to information that has never been promoted as being fundamental or meaningful of itself, but which often is.

The key horses in many of these situations will be supported by what this book has referred to as incidental information—pace, biases, trips, body language, trainer patterns, dosage, or whatever. Because the information is "incidental," fewer handicappers will possess it or use it, and the odds will be greater.

When relying on multiple keys, handicappers should expect to support two or three key horses in multiple exotic combinations with other horses, provided the odds warrant each investment. The investment is steep, so value must be present. One of the keys might be a fundamental method selection. A common example is a method horse that has become an exaggerated underlay in the straight pools. Its true chances hardly warrant the picayune odds, but the horse is definitely a factor. This key is often used advantageously on the bottom side of Exactas. Where the underlay has been exaggerated, it's a waste of time to use these keys on top. The underlay situation persists.

OPPOSITE LOGICS

A specific variation of multiple keys to a single race can be termed *opposite logics,* which is merely a paired association of opposites. The key bet September 1 at Del Mar, in the seventh race Exacta, paired the frontrunner and the latecomer, an especially familiar illustration of the concept. Other common pairs of logical opposites combine class rises and class drops (notably with three-year-olds), favorites and longshots, and horses having higher class but questionable form with horses having especially sharp form but lower class.

I first encountered the idea of *opposite logics* in a manuscript entitled *Positive Expectations Handicapping,* developed in 1985 and prepared by southern California handicappers Mark Cramer and Dick Mitchell. The authors were discussing the *dialectics* of continuity and change in racing, and scolding regular handicappers for catering to expectations of continuity, while misapprehending expectations of change. They recommended an increased alertness to several specific patterns of change, noting that several spot play systems demonstrably capable of tossing considerable profits for a time appeared to be poles apart when viewed logically from the accepted fundamentals of handicapping. The authors' point was that handicapping information is so richly complex and diverse that any small sample of races, at times the same races, can be viewed in multiple ways, even contradictory ways.

The concept fits snugly with the information management approach to handicapping; of the outcome scenarios potentially playable in any short series of races, several are bound to seem poles apart in logic. Yet the opposite poles each represent a real possibility.

When confronting polar logics that appear to conflict or contradict one another in the analysis of the same race, handicappers should link the logical extremes in combination wagering. Unlike multiple keys, which might be combined variously with other horses toward optimal effects, opposite logics are best combined with one another. If the handicapping logics are indeed that polar, presumably certain logical elements of each position must be greatly persuasive. That is, both logical positions are likely to deserve relatively equal merit. The conflict is more logical than real, and handicappers should resign themselves to the possibility that both horses are apt to run well.

CREATIVE HANDICAPPING

Creative handicapping decisions will obviously be far more intuitive than rational, but even so the intuition will be relatively weak. The outcomes predicated on creative intuitions are always improbable, sometimes illogical, and occasionally farfetched. Years ago James Selvidge and his associates at Longacres perceived that horses adding tongue-ties to their equipment were winning lots of races. Extensive subsequent investigations suggested that whenever a power group of forty trainers tried this maneuver, handicappers could have made money by betting their horses. A less farfetched yet creative intuition might support a horse propped up by a single piece of incidental information. Or perhaps by a training pattern from the distant past, repeated now. Not surprisingly, the rational basis for considering these horses is tenuous and weak. The logic connects perhaps to obscure information or to data relationships that are at least unconventional if not actually contradicted by normal experiences. The case gathers no more logical or positive force as the handicapping proceeds than it had at the first, yet handicappers believe the horse has a legitimate shot to win.

A circumstance that *always* attaches to implementing creative handicapping decisions is long odds. Should the odds be low or moderate, the horses should be forsaken, no matter how ingenious the handicapping. These scenarios portray logical longshots, and the horses had better be longshots in the real world, too. Creative selections are also more inspiring if the race is unpredictable by more orthodox means, such as handicapping, but this is not necessary. Upsets do occur, many of them. Should the favorite and another shaky possibility or two be underlays, the creative handicapping decision becomes more digestible. The single necessary criterion is super odds.

In the fifth at Del Mar on September 1, the author boxed the 19-to-1 import *Indian Trail* with all the other horses in an Exacta wheel costing eighty bucks; I was hoping for much bigger bucks should the outsider finish first or second. That eventuality was improbable. The six-year-old horse had not raced in two years. Its workout pattern was unimpressive. A minor journeyman rider would be aboard.

Yet Indian Trail had once upon a time placed or finished close in three listed stakes in England, and trainer Gosden not only wins frequently enough with this kind but he consistently fires at first asking. The horse was 19 to 1, the race otherwise indecipherable. Indian Trail ran well but lost, a regular result of creative handicapping. So the price must be right—long.

In sum, intuitive handicapping decisions cover a wide range of possibilities, as the logical deductive selections of systematic methods do not. Through a process of additive weighing of information, handicapping is transformed into a flexible art of inductive reasoning and intuitive decision-making. With so many sources of meaningful information now available, however, handicappers should insist on a firm rational basis for each of their decisions.

Hunches, guesses, and stabs-in-the-dark do not qualify. As the rational case for each horse solidifies, the corresponding intuitive feeling will intensify. Rational problem-solving and intuitive decision-making in that way go hand in hand.

The bottom line is this. The intuitive approach affords handicappers who are fully armed with knowledge, information, and experience many more opportunities for success and far greater profits in the short run as well.

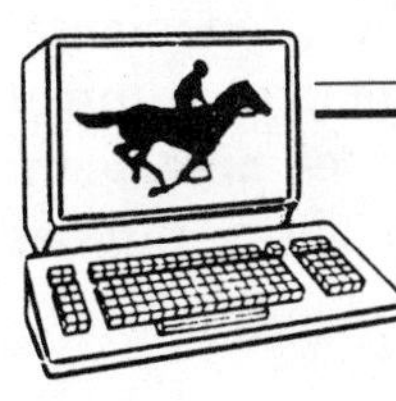

Chapter 12

INTUITIVE WAGERING

> In the history of thoroughbred racing, there have been two great revolutions in the way horseplayers bet. One occurred in the early part of the century, when pari-mutuel machines began displacing on-course bookmakers. . . .
>
> The other reached full flower in the late 1970s, as tracks offered increasing numbers of what are called multiple or gimmick or exotic bets—wagers that involve more than one horse. The proliferation of bets like the Exacta and the Triple has fundamentally altered the nature of the game. . . .
>
> In no area of the game does published theory conflict so sharply with reality. I know few if any professional horseplayers who would claim that they beat the game with a slow, steady accumulation of profits. They win with occasional windfalls.
>
> ANDREW BEYER
> *The Winning Horseplayer,*
> Houghton Mifflin, 1984

ALONGSIDE HIGH-TECH SYSTEMS, information management, and intuitive reasoning, the contemporary practice of handicapping can hail as its conquering hero the age of exotic wagering. If it hasn't already, exotic wagering will supersede straight wagering in the age of information for three compelling reasons at least:

1. Only combination betting permits the intelligent, efficient use of the abundant information on handicapping now available and widely distributed. Exotic wagering opportunities allow handicappers who are information managers to apply everything they know to every race they play. They therefore can benefit by covering all the real possibilities, by taking advantage of the best opportunities.
2. As information becomes increasingly available to more handicappers, an unhappy consequence is the oversupply of underlays in the straight pools, notably the win pools. Well-

informed, skillful handicappers can now convert straight-pool underlays to overlays through combination betting.

3. Exotic wagering is more compatible with the fundamental money-management principle of maximizing gain at minimal risk. Exotic wagering facilitates the making of significant profits at relatively low investment in the short run, as straight wagering does not.

The first point is crucial. With access to simulcasting, inter-track wagering, off-site betting systems, and information management systems, high-tech handicappers of the information age will also have access to more bonanza betting opportunities than ever, and they need to learn how to exploit them. Racetracks might also attend to the point, especially the bustling new marketing departments that are still struggling to market their product to the young adults they have never been able to attract and still are not doing a very enlightened job of it. A dramatic illustration should help. It happened to me just months ago at Hollywood Park.

Read the conditions for the *A Gleam Handicap,* the feature at Hollywood, June 27, 1984, and examine the records of the three horses I focused on.

This seven-furlong handicap was marked by the return to competition of Kentucky Derby favorite *Althea,* champion two-year-old filly of 1983. The class of the field for sure, *Althea* figured to lose this sprint just as surely, for reasons that should not need much explanation to practiced handicappers. Humbled in the Derby, when favored, *Althea* had not worked sharply for this comeback race and had not sprinted in nine months. Far more significantly, it does not require a wealth of racetrack worldliness to appreciate that no champion racehorse would be hard pushed for this kind of unimportant sprint and should not be well intended to win. *Althea* shaped up as a fantastic underlay in the *A Gleam* as soon as it appeared in the entries.

This creates other opportunities for handicappers. There were two alternatives to *Althea.*

Lass Trump is a Gr. 3 stakes winner (Test) that has placed twice in Gr. 2 stakes (Milady and Gazelle) worth $100,000-added or more. She had won six of seven sprints in her past performances and looked particularly terrifying at seven furlongs. She was in peak form today, having finished a close third behind division leaders

8th Hollywood

7 FURLONGS. (1.19⅘) 6th Running of A GLEAM HANDICAP. $60,000 added. Fillies and mares. 3-year-olds and upward. By subscription of $50 each, which shall accompany the nomination, $600 additional to start, with $60,000 added, of which $12,000 to second. $9,000 to third, $4,500 to fourth and $1,500 to fifth. Weights Friday, June 22. Starters to be named through the entry box by closing time of entries. A trophy will be presented to the owner of the winner. Closed Wednesday, June 20, 1984, with 20 nominations.

Lass Trump

Own.—Samford A G 116

Ch. f. 4, by Timeless Moment—Chief Nefertari, by One For All
Br.—Ward J (Ky) 1984 4 1 1 2 $53,767
Tr.—McGaughey Claude III 1983 8 6 1 1 $124,444
Lifetime 12 7 2 3 $178,211

16Jun84-8Hol 1 1/16 :46 1:091 1:41 ft 12 117 32 24 2¾ 3¾ Day P6 ⒻMilady H 86-16 Adored,PrincessRooney,LassTrump 7
16Jun84—Wide 7/8 turn
12May84-8CD 1 :464 1:112 1:37 ft *6-5 111 1hd 1hd 11 1¾ PttrsnG2 ⒻMint Julep 84-15 LassTrump,LadyHwthorn,Delhousie 7
18Apr84-7Kee 6f :22 :454 1:113m 9-5 123 63 33 3¾ 24 PttrsG7 ⒻThrobrd Clb 88-20 BidsAndBids,LssTrump,GrcnComdy 8
18Apr84—Wide
3Apr84-9OP 6f :221 :453 1:103ft *1 118 67½ 67½ 55 3¾ Day P3 ⒻAw25000 89-20 FrostyTail,HereAndGone,LssTrump 8
28Aug83-8Bel 1⅛:46 1:101 1:481sy *3-2 120 2½ 2½ 22 25 Fell J5 ⒻGazelle H 81-16 HighSchems,LssTrump,LdyNorcliff 6
13Aug83-5Sar 7f :214 :442 1:232m *2-3 118 65 53 1hd 1¾ Day P2 ⒻRevidere 85-16 LssTrump,OnTheBench,MdivlMoon 7
13Aug83—Brushed
28Jly83-8Sar 7f :212 :443 1:221ft 6½ 114 3¾ 32 1hd 15 Day P6 ⒻTest 91-11 LassTrump,MedievlMoon,ChicBelle 9
25Jun83-4CD 6f :21 :441 1:094ft *4-5 115 1hd 1½ 1¾ 14 Day P2 ⒻAw17300 97-11 LassTrump,HonestGlow,ClssiclBnd 5
10Jun83-7CD 6½f:23 :46 1:173ft *3-5 118 3½ 21 11 1¾ Day P5 ⒻAw13700 92-16 LassTrump,Slashing,StrightEdition 9
4Apr83-9OP 6f :214 :453 1:112ft 2½ 112 36 44 45 33 Nemeti W3 ⒻAw30000 85-23 Quixotic Lady,Ahsoloud,LassTrump 6
15Mar83-8OP 6f :22 :454 1:12 ft *6-5 109 3¾ 34 2¾ 1¾ Day P8 ⒻAw17500 85-22 LassTrump,DoubleDidit,GleoFolly 11
21Feb83-4OP 5½f:223 :474 1:071sy *8-5 112 3nk 2½ 1½ 1¾ Day P1 ⒻMdn 82-27 Lass Trump, Workin Girl, GreyBid 12
Jun 26 Hol 3f ft :354 h Jun 15 Hol 3f ft :352 h Jun 10 CD 5f ft 1:011 b Jun 6 CD 4f ft :484 b

Althea

Own.—Alexander-Aykroyd-Groves 116

Ch. f. 3, by Alydar—Courtly Dee, by Never Bend
Br.—Groves-Alexander-Aykroyd (Ky) 1984 5 3 1 0 $582,630
Tr.—Lukas D Wayne 1983 9 5 3 0 $692,625
Lifetime 14 8 4 0 $1,275,255

5May84-8CD 1¼:472 1:363 2:022ft *2½e121 11 55 1922 1930¾ McCrronCJ1 Ky Derby 54-19 Swale,CoxMeChd,AtTheThreshold 20
21Apr84-9OP 1⅛:463 1:101 1:464ft 3 121 11 1½ 16 17 VlnzulPA10 Ark Derby 109-04 Althea, Pine Circle, Gate Dancer 11
14Apr84-9OP 1 1/16:463 1:104 1:411ft *3-5 121 52 3¾ 1hd 2½ Pincay LJr1 ⒻFantasy 101-10 MyDarlingOne,Althea,PersonbleLdy 6
14Apr84—Poor st., drifted
11Mar84-8SA 1 1/16:472 1:113 1:433ft *3-5 117 32 1½ 11 1¾ PincyLJr2 ⒻSta Ssana 83-18 Althea,PersonableLady,Life'sMagic 5
25Feb84-8SA 1 :462 1:11 1:37 ft *1 124 51¾ 42 4¾ 1no PincyLJr6 ⒻLs Vrgnes 83-18 Althea,VagabondGal,MyDarlingOne 6
25Feb84—Broke poorly, wide 7/8 turn, crowded 3/8 turn to 1/8
18Dec83-8Hol 1 1/16:454 1:101 1:413ft 2½ 118 2½ 1hd 34 611 Pincay L Jr4 Hol Fut 76-14 Fali Time, Bold T. Jay,Life'sMagic 12
18Dec83—Veered out and bumped soon after start
27Nov83-8Hol 1 1/16:464 1:112 1:43 ft *6-5 120 11 11 1½ 1½ PincayLJr6 ⒻHol Strlt 80-20 Althea, Life's Magic, Spring Loose 6
12Nov83-8SA 1 1/16:463 1:113 1:442sy *4-5 117 11 1½ 1½ 2½ PincyLJr1 ⒻOak Leaf 78-24 Life's Magic, Althea, Percipient 9
26Oct83-8SA 7f :222 :451 1:231ft *1-9 123 2hd 2hd 2½ 2¾ Pincay LJr2 ⒻAnoakia 83-20 Percipient, Althea, PersonableLady 4
26Oct83—Wide backside,3/8
14Sep83-8Dmr 1 :454 1:102 1:344ft *1-3 117 11 1¾ 16 16½ Pincay L Jr4 Dmr Fut 94-14 Althea, Juliet's Pride, Gumboy 5
Jun 20 Hol 6f ft 1:143 h Jun 14 Hol 5f ft 1:003 h Jun 7 Hol 4f ft :482 b May 30 Hol 4f ft :504 b

Pleasure Cay

Own.—Farish W S III 115

B. f. 4, by Foolish Pleasure—Unfurled, by Hoist The Flag
Br.—Cox E A (Ky) 1984 8 3 1 1 $155,694
Tr.—Drysdale Neil 1983 5 2 2 0 $58,890
Lifetime 14 6 3 1 $221,640

9Jun84-4Bel 6f :223 :453 1:093ft 8 122 3¾ 5¾ 5¾ 6¾ DayP4 ⒻGenuine Risk 86-10 On The Bench,Nany,GratefulFriend 6
9Jun84—Wide
19May84-8Bel 1 1/16:46 1:104 1:43 ft 6 117 2hd 2hd 912 913¾ Davis RG3 ⒻShuvee H 73-26 QuenofSong,TrySomthingNw,Nrrt 10
6May84-8Aqu 7f :223 :452 1:24 ft *4-5 118 2½ 1½ 2hd 2½ DvisRG4 ⒻVagrancy H 88-19 GrtefulFriend,PlesureCy,SwtLughtr 7
7Apr84-7Aqu 170⊡:47 11:12 11:42 ft 8 115 15 1¾ 14 1¾ DsRG2 ⒻBed OrosesH 98-16 PlesureCy,SweetMissus,Sintrillium 8
24Mar84-8Aqu 6f ⊡:221 :452 1:11 ft 2½ 116 31 7¾ 7¾ 59 VelsquzJ4 ⒻDistaff H 81-24 Am Capable, Sweet Missus, Fissure 7
24Mar84—Steadied
3Mar84-7Bow 7f :222 :443 1:224ft 3 115 1½ 13 1¾ 12 MllrDAJr5 ⒻBFritchiH 88-19 PleasureCay,Kattegat'sPride,Amnti 9
3Mar84—Run in divisions
16Feb84-8Aqu 170⊡:47 1:12 1:41 m 3 115 1¾ 1½ 1¾ 1¾ Davis R G2 ⒻAw30000 96-10 Pleasure Cay, Ticketed, Bernadine 7
6Feb84-8Aqu 6f ⊡:221 :452 1:104m *2½ 117 1hd 12 1¾ 3¾ MacBethD1 ⒻAw27000 88-19 GiftOfTheMgi,StrictlyRisd,PlswrCy 7
30Dec83-8Aqu 6f ⊡:223 :463 1:13 ft 4½ 113 13 12 1hd 6¾ Samyn J L1 ⒻHcpO 76-26 GrtefulFrind,ClosToM,Micky'sEcho 9
19Dec83-8Aqu 6f ⊡:232 :471 1:13 ft *1 117 2¾ 2¾ 2¾ 2¾ McHrgDG5 ⒻAw25000 77-26 Fissure,PleasureCay,PrettySensible 8
Jun 21 Bel 4f ft :481 b ●Jun 7 Bel ⓣ 5f gd 1:002 b (d) Jun 3 Bel 6f sy 1:184 b May 18 Bel 3f ft :36 b

Adored and *Princess Rooney* on June 9 and would be handled by leading jockey Chris McCarron.

Now examine the interesting multiple stakes winner *Pleasure Cay*. A Gr. 3 winner as well (Bed of Roses), she had won six of fourteen races and $221,649. A shipper from New York, and therefore unknown to the crowd, she had been wide in the Genuine Risk on June 9 before traveling to Neil Drysdale's barn at Hollywood. On May 19, at Belmont Park in the Gr. 2 Shuvee stakes, *Pleasure Cay* had contested a fast pace for six furlongs and prior to that had been nipped in the Gr. 3 Vagrancy at odds-on. This filly clearly outstripped the conditions of the *A Gleam*, and handicappers should readily forgive the wide trip performance last out.

The odds on the three just before post looked like this:

Althea	6–5
Lass Trump	2–1
Pleasure Cay	12–1

I bet $100 on Pleasure Cay to win and lost it to a better horse in blazing time, as the result chart shows. I knew *Lass Trump* was superior, but what else could handicappers do? Well, I might have played the Exacta, but on this day, for some mystical reason only the executives at racetracks can comprehend, Hollywood Park offered its patrons Exacta wagering on the third, fourth, fifth, sixth, seventh, and ninth races, but not on the eighth. To rub it in, Hollywood Park offered Exacta wagering on the eighth race too every other day that week.

The reason this illogic becomes unsettling to racing's loyal customers should be plain. Had Exacta wagering been available on the feature June 27, I would have bought a $50 Exacta box coupling *Lass Trump* and *Pleasure Cay*. With the strong underlay *Althea* in the offering, the *Lass Trump-Pleasure Cay* combination might have been an overlay. At odds of 2 to 1 and 12 to 1, respectively, the fair value, break-even payoff for the $5 Exacta combination would have been $200. I'll take it, even less. Instead of losing $100 on a win investment that was correct at the odds but a high risk, I could have chosen to play the race as I understood it and preferred to play it, and thereby would have netted approximately $1,900 for the same investment. Do handicappers think that difference is meaningful? Can racetrack executives see it? Do they care?

The operating axiom of pari-mutuel wagering in the information

HOLLYWOOD PARK

(Continued from Page Thirty Nine)

EIGHTH RACE
Hollywood
JUNE 27, 1984

7 FURLONGS. (1.19⅘) 6th Running of A GLEAM HANDICAP. $60,000 added. Fillies and mares. 3-year-olds and upward. By subscription of $50 each, which shall accompany the nomination, $600 additional to start, with $60,000 added, of which $12,000 to second. $9,000 to third, $4,500 to fourth and $1,500 to fifth. Weights Friday, June 22. Starters to be named through the entry box by closing time of entries. A trophy will be presented to the owner of the winner. Closed Wednesday, June 20, 1984, with 20 nominations.
Value of race $66,400, value to winner $39,400, second $12,000, third $9,000, fourth $4,500, fifth $1,500. Mutuel pool $549,548.

Last Raced	Horse	Eqt.	A.	Wt	PP	St	¼	½	Str	Fin	Jockey	Odds $1
16Jun84 ^{8}Hol^{3}	Lass Trump		4	116	5	3	$3^{2\frac{1}{2}}$	$3^{2\frac{1}{2}}$	$2^{3\frac{1}{2}}$	1^{1}	McCarron C J	2.30
9Jun84 ^{4}Bel^{6}	Pleasure Cay		4	116	9	1	1^{1}	1^{1}	1^{1}	$2^{5\frac{1}{2}}$	Delahoussaye E	12.60
30May84 ^{8}Hol^{4}	Angel Savage	b	4	112	6	6	5^{1}	$4^{2\frac{1}{2}}$	4^{2}	3^{2}	Shoemaker W	8.10
3Jun84 ^{8}GG^{3}	My Native Princess		5	114	3	9	$8^{\frac{1}{2}}$	6^{3}	5^{3}	4^{hd}	Meza R Q	43.00
3Jun84 ^{5}Hol^{3}	Madam Forbes		4	117	1	7	$6^{\frac{1}{2}}$	$7^{\frac{1}{2}}$	6^{1}	$5^{3\frac{1}{2}}$	Pincay L Jr	5.60
5May84 ^{8}CD^{19}	Althea		3	116	7	2	2^{hd}	2^{1}	3^{hd}	$6^{\frac{1}{2}}$	Valenzuela P A	1.20
23May84 ^{8}Hol^{4}	Filomena Galea		5	116	4	5	7^{1}	8^{2}	7^{2}	7^{2}	Pierce D	26.60
19May84 ^{5}Hol^{6}	Skillful Joy	b	5	114	2	8	9	9	9	8^{3}	Lipham T	53.70
1Jun84 ^{8}Hol^{1}	Tyrosong		6	108	8	4	4^{1}	5^{hd}	$8^{\frac{1}{2}}$	9	Garcia J A	36.50

OFF AT 6:02. Start good. Won driving. Time, :21⅘, :44⅘, 1:08⅘, 1:21⅕ Track fast.

$2 Mutuel Prices:			
6–LASS TRUMP	6.60	3.80	3.00
11–PLEASURE CAY		9.40	5.40
7–ANGEL SAVAGE			4.40

Ch. f, by Timeless Moment—Chief Nefertari, by One For All. Trainer McGaughey Claude III. Bred by Ward J (Ky).

LASS TRUMP, in close contention from the outset, saved ground around the turn, rallied gamely outside PLEASURE CAY in the final furlong and was up in the closing yards. The latter outsprinted ALTHEA for the early lead, disposed of that one on the turn but could not hold the winner safe. ANGEL SAVAGE also saved ground around the turn but could gain little in the stretch drive. MY NATIVE PRINCESS was never dangerous. MADAM FORBES showed little. ALTHEA forced the early pace outside PLEASURE CAY and tired after five furlongs. NAUGTHY MADAM AND ROYAL DONNA WERE WITHDRAWN AND ALL WAGERS ON THEM WERE ORDERED REFUNDED.

Owners— 1, Samford A G; 2, Kilroy W S; 3, Garcia H; 4, Parks R Charlene; 5, Jones A U; 6, Alexander-Aykroyd-Groves; 7, Galea & Krentz; 8, Hooper F W; 9, Ashmore D R.

Trainers— 1, McGaughey Claude III; 2, Drysdale Neil; 3, King Hal; 4, Moreno Henry; 5, Barrera Lazaro S; 6, Lukas D Wayne; 7, Millerick M E; 8, Fenstermaker L R; 9, Stauffer Joe.

Overweight: Pleasure Cay 1 pound; Madam Forbes 2; Filomena Galea 1; Skillful Joy 1.

Scratched—Hi Yu Lulu (15Jun84 ^{8}Hol^{1}); Naughty Madam (30May84 ^{8}Hol^{2}); Royal Donna (20Oct83 ^{8}SA^{3}).

age should be freedom of choice, the opportunity to play any race as the race itself dictates or as the handicapper prefers. If handicappers align themselves with this point of view, they will frequently be frustrated while waiting on several racetracks to fall into step, but only for a time. In the meantime, they will find opportunities galore.

The lessons transported to this chapter are the result of long, difficult, but instructive experience. They mesh money-management theory and betting realities, with emphasis on the latter. Handicappers can beat the races in two ways: by grinding it out long term or by speculating on handicapping values short term. After years of floating, I believe I have learned how to proceed properly on both fronts, and I intend to be highly specific about procedure.

The specificity is deliberate. It is intended to guide unaware handicappers away from the major pratfalls of pari-mutuel wagering by handing them a carefully defined program that works. At the same time it is intended to be practical, the kind of advice handicappers can implement. It is likewise intended to offer the best chance for the greatest number of recreational players to succeed. Several of the specifics may seem strange or discomforting to practitioners, but they are carefully thought out, rooted in experience, and I urge handicappers to accept them for now as maxims.

Let's begin with a directive to regular handicappers in search of meaningful profits that may seem the most impractical, most unrealistic of all, yet it is absolutely necessary.

The money used for racetrack betting must be capital or investment money, not income and not discretionary income or savings, and the amount must be $10,000 and no less.

A large amount of capital is fundamental to success, especially in the age of information and exotic wagering. It controls for two enduring problems at the windows. First, it rids handicappers of the anxieties attached to "short" money. As the punters put it, short money is scared money, and scared money loses.

Sizable capital also allows moderately large investments at the outset of play without threat of running on empty. This is only expedient. All wagers are independent. They must be treated in that way. The amounts bet cannot depend on what happened beforehand or on the opportunities coming up later. Handicappers benefit by affording themselves a chance to win significant profits early in the season, not just later.

Second, a large capital investment permits the separation of capital into two piles, one for straight wagering, the other for exotic wagering. This book recommends a fifty-fifty split for proficient handicappers such that $5,000 will be allocated to straight betting, another $5,000 to combination betting. "Proficient" handicappers are defined as those who have demonstrated the ability to win 30 percent of their plays at average odds of 2.75 to 1. For those handicappers, as will be seen shortly, it's far better to proceed simultaneously on both fronts, as opposed to emphasizing win wagering or exotic wagering intermittently, depending on events or the size of the bankroll. Besides, it mimics reality; that's what most handicappers do.

So important is the $10,000 amount to begin that proficient handicappers who do not have those resources but expect to win substantial money for the season are urged to do what investors in other fields always do—borrow it! After all, racetrack betting for proficient handicappers is an investment program; speculative, high-risk, to be sure, but an investment program nonetheless. So those who qualify for the loan should borrow the difference to the extent that savings have been put aside for racetrack speculation. Borrow the capital investment. Pay the interest on a six-month loan. At 14 percent, or two above the present prime, the six-month interest on the full amount would be slightly more than $700, the cost of investing in one's handicapping knowledge and skill. Needless to say, incompetent handicappers need not apply. They will only lose the loan money.

Beyond financing, the borrowing serves several substantive purposes, the most important of which is that it lends to the proceeding, as squandering one's savings incrementally does not, a keener sense of the importance of money management. The awful money-management skills of many talented handicappers should improve immediately, a not unimportant matter.

Borrowing capital also has the desirable effect of forcing winning handicappers to display their true levels of proficiency. At season's end most will actually win money. How much depends on actual proficiency, which will have been determined exactly once and for all, and on the money-management methods employed. In the scheme of things the cost of the borrowed capital will be small potatoes.

What about the downside: losing the capital (loan money) at the track? Not to worry. Risk of actual loss will be minimal rather than high where handicapping proficiency can expect to select 30 percent winners at average odds of 2.75 to 1 from the two or three systematic method horses a day, which is no great shakes, and a Kelly optimal betting strategy is implemented. More on this momentarily.

If the capital investment cannot be borrowed, it must be saved, a painstaking task. Yet a clever alternative to borrowing or saving as much as $10,000 is plausible. Following investors in securities and commodities markets, handicappers can borrow or save half that amount, or $5,000, and invest it at a hypothetical 100 percent margin. That is, pretend the $5,000 is actually $10,000 and act accordingly. This is strictly psychological game-playing where the investment opportunity requires real cash, as at the races, but where handicapping proficiency has been highly reliable it offers psychological and real advantages as well.

With $10,000 of capital in hand, proficient handicappers can proceed to beat the races. To that great end, let's postpone discussing the advantages of exotic wagering briefly to take a hard look at straight wagering to win and the grind-it-out long-term road to success.

GRINDING IT OUT LONG TERM

This book by no means promotes the abolition of straight wagering to win with systematic money-management methods. In fact, it cries out for it. It pledges renewed allegiance to the grind-it-out approach in the win pool, but in the brand-new context, paradoxically, of exotic wagering. Both patterns of betting are appropriate to handicappers who prefer to combine recreational handicapping and wagering with profit motives, which includes practically all of us.

To explain.

Computer simulation studies have shown indisputably that the problems associated with systematic money management and win wagering are strictly pragmatic. Theoretically, proficient handicappers cannot lose when implementing a demonstrably effective systematic money-management method and should cling to it tenaciously. But the practical problems get in the way. Tedium and impatience are inherent in the systematic wagering pattern, a not unimportant impediment, precisely because the grind-it-out approach entails an extremely slow, lengthy process of profit-making. Computer studies reveal just how slow and gradual this process is. Early gains are so insignificant as to be almost inconspicuous. A full season's success is little more than a head start.

To illustrate the issues, consider the results of Bill Quirin's computer simulation studies of four systematic methods (Table 23) where the win percentage was a moderate 30.4 percent, the average odds on winners a realistic 2.75 to 1, the starting bankroll an agreeable $1,000, the number of bets 700, and bettors invested from 1 percent to 10 percent of the bankroll each time. Each series of 700 wagers was replicated 1000 times. Examine the tables.

The Kelly fixed percentage method invests *a percentage of bank-*

Table 23
Quirin's computer simulation studies of three systematic methods of money management

Bets range from 1 percent to 10 percent of a $1,000 starting bankroll for 1000 repetitions of a 700-race sample. The performance assumptions are 30 percent win proficiency at average win odds of 2.75 to 1.

FLAT BETS

Percent	*Average Bankroll*	*Largest Bankroll*	*# Tapouts*
1%	$1,984	$1,984	0
2%	$2,968	$2,968	10
3%	$3,735	$3,952	55
4%	$4,442	$4,936	100
5%	$4,748	$5,920	198
10%	$5,583	$10,840	485

"KELLY" FIXED PERCENTAGE

Percent	*Average Bankroll*	*Largest Bankroll*	*# Tapouts*
1%	$2,278	$2,278	0
2%	$3,855	$3,855	0
3%	$4,991	$4,991	0
4%	$5,054	$5,054	0
5%	$4,075	$4,075	0
10%	$70	$70	0

FIXED PERCENTAGE—MINIMUM

Percent	*Average Bankroll*	*Largest Bankroll*	*# Tapouts*
1%	$2,074	$2,799	0
2%	$3,297	$8,580	7
3%	$4,595	$17,366	49
4%	$5,544	$26,699	105
5%	$6,758	$122,190	211
10%	$4,954	$335,039	675

roll equal to the bettor's advantage over the game, which must be estimated rigorously and is calculated by the equation:

$$\% \text{ Advantage} = \text{Win } \% - \frac{(\text{Loss}\%)}{(\$ \text{ Odds})}$$

Handicappers unfamiliar with Kelly wagering propositions should acquaint themselves quickly. They will then understand that to bet aggressively, or more than your true advantage allows, *on a continual basis* is to guarantee losses, while to bet less than your true advantage allows means only that you will win less than you are capable of winning—a sad, but not desperate, state of affairs.

The method labeled "fixed percentage—minimum" is my own and is a combination of Kelly betting and flat betting. It invests a fixed percentage bet following win wagers, but only a minimum bet (the original fixed percentage bet) following a loss. The method is designed to minimize the erosion of accumulated profits during normal losing runs, which it does, but depends for its power on handicappers' tendencies to win and lose in clusters, not alternately. Computer simulations and real life, too, indicate that this clustering occurs reasonably well.

Examine the results of Quirin's simulation studies of the three methods again. What conclusions are supported here?

Handicappers should be greatly impressed that Kelly not only yields seasonal profits at every percentage level of investment but also that it *never* taps out. Without question, Kelly is the most reliable, most powerful of all systematic pari-mutuel wagering methods. Statisticians who conduct and interpret these studies are usually unbearable souls who will not tolerate tapouts. A single tapout incurs their wrath. Thus, statisticians and mathematicians fancy Kelly, because the method never taps out. Handicappers everywhere might concede the point, to be sure.

My own standards are less rigorous but not loose. I'll accept a five percent tapout rate, which equates to statistical significance in social science research. By this standard both flat bets and my fixed percentage—minimum method are acceptable up to 3 percent of the bankroll but not at the higher percents of investment. Two conclusions flow generously from these data.

First, flat betting, fixed percentage—minimum wagering, and *all other systematic money-management methods* that have been simulated by Quirin and others go bankrupt at unacceptable rates when 4 percent or more of the bankroll is bet continuously. At 5 percent of bankroll the methods tap out 20 percent or more of the time. Ten percent of bankroll taps out about one time in two.

Second, Kelly represents the most powerful long-term, grind-it-out method of systematic win betting *for all proficient handicap-*

pers. The more proficient the handicapper, the more powerful is the Kelly method. Thus, the method approaches the ideal.

In the simulations illustrated, the win percentage was 30 and the dollar odds 2.75 to 1, thus, the best estimate of true advantage is 4 percent (rounded down). Notice in the illustration that at 5 percent Kelly yields less profit than at 4 percent; Kelly is unforgiving to aggressive betting. Ten percent is a disaster; too aggressive a wagering pattern for a 30 percent handicapper. If the bettor's true advantage had been 8 percent or 10 percent, however, profits at those levels of investment would be much greater than at 4 percent.

Thus, if handicappers *know* their true advantage is greater than 4 percent, they can bet an equivalent percentage each time. Most handicappers do not know their real advantage over the game and will not do the spade work to find out, but many can perform at 30 percent winners at average odds at 2.75 to 1 on winners. This book, therefore, takes a firm position on systematic win wagering. Proficient handicappers should begin with the recommended bankroll of $5,000 and bet a 3 percent Kelly each time. The first bet will be $150. It would be more profitable to calculate a single race expectation each time and bet that amount, but handicappers simply will not bother to do that.

These bets are limited to supporting the prime selections of systematic handicapping methods; that is, the kinds of productive horses that repeat themselves all the time. If handicappers perform in accord with Quirin's performance assumptions, they will end with approximately $4,500 to $5,000 for a single long season of seven hundred bets, or a modest net of $3,500 to $4,500.

That reality explains why regular handicappers do not rely on systematic money-management methods. It takes too long to earn too little.

Yet what most handicappers do not appreciate sufficiently is that the long-term possibilities for skillful players are astounding. The most significant finding by far of computer simulation studies is the sensitivity of the bankroll to the win percentage. Handicappers who can win 34 percent of their wagers on prime selections, for example, at average odds on winners of just 2.6 to 1, can take home a small fortune in two short seasons if they simply persist in their systematic handicapping and betting patterns on those two or three key races a day that produce prime selections.

Consider the following two computer simulations of 3 percent

Kelly wagering for small and large numbers of bets at varying levels of win proficiency. The first is a blast of fresh air:

KELLY			
PW=.34	0=2.6	N=250	R=20
3% of capital			Capital $5,000
Average return			21,708.64
Win percentage			.34
Longest win streak			7.00
Longest lose streak			19.00
Average bet			375.16
Number of busts			0.00

By betting 3 percent of a $5,000 capital for 250 wagers, approximately one-third of the races carded during ninety days of racing, or a single season if you will, handicappers who win 34 percent of the bets, at average odds of 2.6 to 1, can fairly expect to earn $21,708.64, which is a nice season indeed. Notice that the number of busts following twenty replications (R=20) is zero. Moreover, the betting pattern tolerates a losing streak of nineteen, which sounds ghastly. The average bet is $375, which handicappers can tolerate, as can practically all major racetrack pari-mutuel pools. So this is the recommended approach to win wagering on prime selections, those two or three key horses a day that emerge from tried-and-true systematic handicapping methods and usually go postward at acceptable odds, or at what I like to call "low-priced overlays." Low-priced horses that are properly bet or are underlays, of course, are forgotten.

Significantly, the same Kelly pattern of investment at 30 percent win proficiency returned just $8,523 on average. That represents approximately $13,200 less in profits across a mere 250 wagers when the difference in win percentage is four points. At a difference of two points, or 32 percent winners, the average return was $11,695. Handicappers should appreciate that winning two additional races of each one hundred played means an additional profit of approximately $10,000.

To repeat, in systematic win wagering by Kelly the margin of profit for a season and any longer duration is most sensitive to the win percentage. Thus, the most important statistic under the handicapper's control is the win percentage. The average odds take care of themselves and are not very elastic.

To enlarge the possibilities fantastically, look at what happens to our 3 percent Kelly wagering method when the number of bets is increased to one thousand, which represents four short seasons or perhaps two desperately long seasons of continuous action.

KELLY

PW=.34	0=2.6	N=1000	R=20
3% of capital			
Average return			$277,890.57
Win percentage			.34
Longest win streak			7.00
Longest lose streak			22.00
Average bet			1,327.99
Number of busts			0.00

How about that! At .34 win proficiency, average win odds of 2.6 to 1, the return across a full two years would approximate a quarter million or more. Longest losing run was twenty-two straight. The constraint for handicappers outside of New York and southern California, of course, is the average bet size of $1,327, but these amounts would not distort the pools at the lead tracks on both coasts.

To dramatize further the case regarding win percentages, the same one thousand bets at .33 win proficiency nets $198,765 on average; at .32 the net drops to $161,000; at .31 it plummets to $76,300; and at Quirin's threshold of .30 proficiency it sinks to $26,100. The dollar difference in profits for four percentage points of win proficiency approximates a quarter million dollars!

Comparisons between Kelly and other systematic methods at .34 win proficiency across one thousand wagers reveals still further the power of Kelly, as no other method approaches the same return. My method, for instance, at .34 winners, even though never busting during the twenty replications, returns just $78,852 on average.

What prevents competent handicappers from implementing Kelly for the long haul are precisely the practical and psychological problems associated with systematic win wagering on two or three daily prime selections delivered up at acceptable odds by systematic handicapping methods. Boredom sets in post haste. The racing day is a long day. Yet the tedium can be controlled by allocating a second $5,000 bankroll to exotic wagering, conducted intuitively, as recommended here. Experienced, informed, skillful handicappers can now pursue exotic racetrack profits in the short run as actively and vigorously as win profits are pursued cautiously and deliberately in the long run, and at the same time they can profit on both tracks.

Other practical problems associated with systematic win wagering are less amenable to high-level action in the exotic pools. These problems get in the way relentlessly.

The handicapper's money must remain capital for a long time and not be translated into income. No profit-taking is allowed until the one thousand bets have been consummated. Other funds will be needed to support one's play as well as one's life. The races must be attended daily or at least regularly. That means full-dress handicapping becomes a persistent chore, a job. The skill must remain sharp, as we have noted the severe financial penalty attached to a single-point drop in win proficiency. The routine becomes one of handicapping, attending the races, and bookkeeping, leaving about four hours of the waking moments five days a week to the other activities of life, not to mention one's loved ones, who might become suspicious of the entire venture, notably when win proficiency slackens or has been gravely overestimated to begin with.

Grinding it out is just what the handle suggests, a long hard grind. But it is entirely feasible and can repay the capital investment loan before a season elapses, and nothing that follows is meant to intimate that solid professional handicappers cannot make important money by resorting to systematic money-management methods and the straight pools. They can.

INTUITIVE WAGERING

Approximately 70 percent of the races on the American calendar do not lend themselves to grind-it-out, long-term wagering with systematic money management. Either too competitive or too unreliable, the races do not yield a single systematic selection at acceptable odds. They deliver instead various outcome scenarios, alternative decisions, and final decision-making problems. This majority is the main concern of this book's money-management guidance. Where contenders have been replaced by scenarios and selections by alternative decisions, it follows that systematic money management should be replaced by intuitive wagering.

Not surprisingly, intuitive wagering decisions are characterized by the same *additive weighting* of relationships characteristic of intuitive handicapping. The two processes are hardly distinct; they overlap. Early in the handicapping, players develop a first sense of the relative odds that will be offered on horses that figure. This sensibility intensifies toward the end of the handicapping. Final decisions, in fact, result from a delicate balancing act in which estimates of winning probabilities are weighed against public odds.

The key consideration is value. Value refers to the relative worth of the bet. If three horses figure reasonably well and are judged comparable, but the first is 6 to 5, the second is 2 to 1, and third is 8 to 1, the third would be *worth* a bet, but the first and second would not. Significantly, if horses have a low probability of winning, high odds do not redeem them.

Whereas in systematic win wagering on prime selections the strict definition of value as a fair return on money invested is acceptable, because the competition has been judged clearly inferior, with exotic wagering it is not. As the risk of investing has increased, the notion of fair value is replaced by that of good value. What is good value? Whatever, the implication of that concept of wagering on low-priced horses in competitive races is plain: Short-priced horses must often be coupled in exotic wagers in ways that enhance their value dramatically.

A value orientation toward intuitive betting applies to win pools as well as to exotic pools. But because the win probability will be lower and the risk greater, the odds must be higher. Handicappers do not harvest good value from riskier win wagers when the odds hover below 7 to 2 or perhaps even below 5 to 1. As a consequence, these horses are best pursued in the exotics. Research supports the assertion, showing that handicappers are mainly spinning wheels with win bets on low-priced selections.

Alternatively, the concept of value does not mean that handicappers should automatically support the longest-priced horses. The trick is to assess the probable chances of winning and relate that to the size of the odds. This is the crux of intuitive wagering.

The broad guidelines apply when wagering to win intuitively with a value orientation. If a horse's chances look reasonably strong among two or three probable outcomes, odds should fall between 5 to 1 and 12 to 1. Most of these possibilities will actually be bet at between 4 to 1 and 8 to 1. If horses chances are best characterized as outside possibilities, the odds should range upward from 13 to 1.

In reality, most complex races will result in outcome scenarios and alternative decisions that entangle low-priced horses (odds below 7 to 2) with horses at moderate and higher odds. In these situations the value of the low-priced horses can be enhanced successfully only by exotic wagering. Certain tools are available to handicappers by which they can estimate whether exotic wagers offer good value or not, and all handicappers should certainly avail themselves of these. More on this briefly.

The rest of this chapter concerns itself with the exotics. Because among proficient handicappers one of the two central purposes of exotic wagering is to add value to horses overbet in the straight pools, I consider exotic wagering as a concept interchangeable with *value-added wagering*. The other purpose is not less important: to provide outlets for the sensible, advantageous use of all that handicapping information.

THE BY-LAWS OF INTUITIVE WAGERING

Laws are set in concrete and should not be violated. Besides, severe penalties are usually attached. The laws of pari-mutuel wagering, for instance, include the axiom, never bet underlays, as a series of wagers on true underlays guarantees losses. By-laws, in contrast, are merely regulatory. They are rules adopted by members of a special constituency, à la handicappers, to assure the orderly conduct of their affairs.

Here then I have drawn for handicappers eight by-laws of intuitive wagering, especially in the exotics. They should not be violated without cause.

1. *The capital investment for the season must be $5,000 or greater.*

   I shall not belabor this point anew but merely assert that the size of the original bankroll is a crucial determinant of success. The $5,000 figure has not been pulled from midair. It provides the financial sustenance needed at the track. A $2,000 bankroll is not enough. The money, to repeat, is neither income nor savings but capital, money to be invested for profit.

2. *The decisive concept of intuitive wagering is* value, *and not just fair value, as with systematic betting on prime selections, but good value.*

   Where value does not exist, no bets are placed. Where reasonably strong probabilities of horses' winning exist, good value falls between 5 to 1 and 12 to 1. Where only outside possibilities linger, good value means 13 to 1 and upward.

3. *Intuitive wagering emphasizes short-term gain, not long-term gain.*

   Finding value, intuitive handicappers intend to maximize profit for a race, a card, a week, a month, or a season. A season is a short term for investment purposes. Bets are scaled accordingly. On a given day, for example, a wager might be extraordinarily high and another the bare-bones minimum.

4. *The size of the wager should accord with the value of the betting opportunity.*

   In this regard, sound advice has been handed down to handicappers by the author with the most intuitive wagering experience, Andrew Beyer. Beyer recommends that handicappers establish a personal ceiling, or maximum, an amount they can comfortably risk when the prime selection and all the attending circumstances of the race appear to be ideal. All other bets are scaled downward from this threshold, as handicappers intuit horses' real chances and the relative values of the odds. Most bets, of course, will be far below the ceiling in size, perhaps 5 percent to 10 percent of the maximum.

   Beyer's personal ceiling is $3,000, an amount he says he invests a few times a season. He cautions that this figure is unrealistically high for most handicappers, which it decidedly is. Yet this allows Beyer to invest substantial amounts in exotics, an important consideration, as exotic wagering intends to cover *all* the real possibilities having real value, and there may be several.

   The point is that handicappers' betting ceilings cannot be artificially low or the game has been lost again.

   A perfectly legitimate way to establish a personal ceiling is to equate it with the largest bet invested in the past two seasons. By that standard my own ceiling would be $1,000, which was invested once during 1983, never during 1984, and only three times in my lifetime so far. A more pragmatic ceiling for me would be $500, an amount invested a few times a season.

Handicappers might likewise ask: What is the largest amount I invest a few times each season? Ordinary intuitive wagers in exotic pools will amount to between 5 percent and 15 percent of the ceiling. Other, more-favored exotic bets will range from 20 to 50 percent of the maximum; some bets will amount to 75 percent to 80 percent of the maximum. It depends on the strength of each decision, as experience—mainly successful experience—will show.

In regard to sizes of wagers, all texts on systematic money management urge handicappers to bet more when winning and less when losing, which is sound economics, of course, and psychologically comforting as well. The same precept holds for intuitive wagering, but here the sizing variable is not the incremental increases of bankroll resulting from a series of profitable wagers but rather the inherent value of the opportunity. That is, handicappers should bet more when value is greater, less when value is lower.

5. *A bare-bones wagering pattern is mandatory during losing runs.*

As experience with intuitive wagering grows, handicappers will become more acutely aware of winning and losing runs. When losing, intuitive wagering allows handicappers to cut losses to the bone. Cut back. All the way back. Skillful intuitive bettors do this instinctively and reasonably early in the loss pattern. It has taken me years to develop the instinct, but lately I have practiced the cutback skill successfully. This important intuition can save time, money, and tons of psychological energy.

In 1978, the only year my income depended exclusively on handicapping profits, during a three-week period at Hollywood Park, I won just one of twenty wagers. Actually, as absurd as this sounds, I handicapped quite well during the streak. Fifteen of the losers finished second. Another two finished third. Only two ran out. Nevertheless, there it is, a losing streak; 1 for 20. When this happened, I had been playing for five consecutive months and had amassed a huge profit. During the losing run, however, I stuck to my systematic money-management method, which invested five percent of an original $5,000 bankroll on each race. I therefore lost 80 percent of a five-month profit to the losing streak. If a similar losing run occurred today, I could recognize the telltale patterns, reduce the wagers to minimum size, and lose approximately 20 percent of any accumulated profits. Highly intuitive handicappers can sharpen this skill fairly well—and must.

A much more difficult intuitive wagering skill to develop and refine relates to the losing streak's opposite number, an extended stretch of winning. These happy times befall all proficient handicappers a few times each season. Most are unable to capitalize. This talent is arguably more important than all others. It represents our next by-law.

6. *Handicappers should be prepared to "crush" the races when winning.*

The "crushing" concept is borrowed, of course, from Andy Beyer, who alone among handicappers of my acquaintance has developed and refined this skill to its utmost. Its application in this book, however, is vastly different from what Beyer has proposed. This notice of crushing deserves to be treated separately, at some length, and will be in the next section.

7. *The intuitive wagering portfolio should remain relatively unstructured, to include Win Wagers, Doubles, Exactas, Quinellas, Triples, and even Pick-Six cards, but with the capital amounts proportionate to each unspecified.*

Handicapping and wagering values exist where intuitive handicappers find them. Unstructured exotic portfolios permit handicappers to relate the bets to values found. If structure must be placed on capital amounts, it is best accomplished by assigning separate pools of capital to the various types of exotic wagers. I have already demanded that handicappers create two separate wagering pools, one for systematic straight wagers, another for exotics and intuitive win bets, and each containing $5,000 or more to begin. As Triples and Pick-Six bets represent relatively costly investments involving the highest speculation and risk, separate betting pools for these games are also advisable. The capital investment to begin goes up, *dramatically*. Most of us will be left out. We don't have the necessary capital. Regardless, records of all exotic wagers should be kept on computerized spreadsheet applications, and the unsuccessful types of wagers eliminated from the portfolio.

8. *Rational scrambling techniques in exotic wagering must cover* all *the real possibilities offering good value.*

   Handicappers who emphasize intuitive wagering in exotics but try to finesse the games by covering some possibilities, and not others, are doomed to misery. Toward the end of the intuitive handicapping process handicappers will regularly confront various outcome scenarios and alternative decisions offering varying values. Poor value choices can be forgotten, but all probable combinations returning good value must be included in the investment. If that is not financially feasible, the races should be passed. Moreover, if adequate coverage of all the real possibilities would reduce the value of several unacceptably, it's best to pass altogether. The tendency otherwise is to omit some combinations having the greatest potential return, which is a mistake. Full coverage of the probables at good value is the operative guideline to successful exotic wagering.

THE "CRUSHING" CONCEPT

The "crushing" concept of intuitive wagering refers to the penultimate maximization of profit during the short run. In the normal case, skillful handicapping, good fortune, and inordinate pari-mutuel value coincide for a time. The short run can mean a race, a card, a week, a couple of weeks, or maybe, in the best of worlds, a month.

As promoted by Beyer, the crushing concept is reduced, artificially I believe, to "killing" a single race. Or more precisely, to killing a few races a season. Beyer has found that winning players get that way from occasional windfalls resulting from sizeable wagers on a few select, well-understood races. The races offer tremendous value and are perfectly understood that way by handicappers.

This book takes a different stance. Crushing the races now means escalating the bets substantially during those brief periods when sharp handicapping, good fortune, and good values coincide. Handicappers have already begun to win, and they suddenly sense a situation ripe for a killing. These short periods last from two to three weeks, not a single race. The recommended escalation is from the usual 5 to 15 percent of the ceiling bet to 25 to 50 percent of the ceiling. During the run a maximum wager might be made. Some wagers will be at 75 to 80 percent of maximum. It's winning time, as Magic Johnson likes to say.

Handicappers should understand that the crushing strategy is the ultimate key to big-money success in intuitive wagering. It is the absolutely most powerful concept of the intuitive approach. The problems many handicappers—myself included—will experience with Beyer's advice to attack a specific race, however, are triple-edged: rational, practical, and psychological. No matter what the apparent advantage offered by the handicapping probabilities and betting values, anything can happen in a single race. Too often it does. The error factor inherent in racing is high. Horses and jockeys make mistakes all the time; trouble occurs; accidents happen. Handicappers can lose any race. I happen to know—because he told me so—that the single largest bet Beyer has ever risked, he lost. During the past Santa Anita season I risked a maximum amount four times, and lost three times. The error factor is a strong deterrent to the single-race crushing instinct.

The practical constraint regards the size of the wager. Beyer can tolerate huge investments financially and does. His limit, to recall, is $3,000. Few of us can imitate him on the point. To "crush" a given race requires an enormous one-shot outlay. Few handicappers can afford it. None who cannot should try, as Beyer himself warns.

Moreover, if Beyer loses, he comes back whole. The huge loss does not affect his judgment, does not alter his behavior. The talent qualifies as the ultimate credential of the expert intuitive handicapper and bettor, but it is not well distributed. If extraordinary amounts are at stake and handicappers lose, the backlash can be more than demoralizing; it can be destructive.

Developing the capacity to escalate bets when handicapping skill, good fortune, and wagering values combine to advantage for the short term is the most dexterous, most advanced art of intuitive handicapping and wagering. It derives from extensive, thoughtful experience, a certain degree of success, and the capacity to take well-calculated, yet speculative, risks. Yet it is imperative to the kind of success that value-added, intuitive wagering is intended to bring about. Intuitive handicappers who want to maximize profits during the short run can take it for granted they must learn to escalate the bets decisively when winning.

I confess to a personal shortfall on the matter. Having ascribed to systematic money management since my initiation into handicapping, I have found the melodramatic escalation of wagers when winning a hard psychological barrier to penetrate. It is a skill to master for me, yet it is among the most decisive.

EXOTIC WAGERING TECHNIQUES

A glaring omission in the literature of handicapping is the systematic investigation of exotic wagering. Few meritorious studies are available to report. Perhaps the omission is more obvious than glaring. Until Exacta wagering appeared at Hollywood Park in 1969, exotic wagering had little visibility and no legitimacy in this sport.

Since then the Exacta has been widely instituted and has been followed in turn by the Quinella, Twin Q, the Triple, the Late Double, and the Pick-Six, the carryover aspect of the last invention offering handicappers their first million-dollar pools in history. Moreover, until recently most serious literature has characterized exotic wagering as gambling, or guessing, a teaching soon to collapse, even as allegiance to systematic handicapping methodology is slowly yielding ground to information management. Where key selections of systematic methods are emphasized, exotic wagering makes little sense. Combinations betting in fact has been skewered exactly for its capacity for turning winning key selections into losing propositions.

In consequence, not many serious attempts have been launched to disentangle the intricacies of exotic wagering. This soon will change. In the meantime, handicappers have been flying by the seat of their pants in the exotic pools, and what they understand about combination betting has accrued largely through the lore of experience. Perhaps the most common practice has been a reliance on the *key* horse. This horse figures first or second by systematic handicapping, and thereby represents the *key* to hitting the exotic combinations. In the usual case the key horse is coupled with other contenders. If the key horse performs as expected and the race's contention has been doped out properly, handicappers have an excellent chance of cashing combination tickets.

This book seeks to change that practice. It asks handicappers to replace the concept of the key horse with that of multiple keys. Where various outcome scenarios and alternative decisions predominate, as they do, multiple keys make sense, as a single key does not. In effect, the probable winner of each outcome scenario represents a potential key horse. Multiple keys are not necessarily linked to one another but rather to other horses that loom largely in each scenario.

By this approach the scrambling of horses in exotic combinations can become exaggerated. Thus, the concept of value takes precedence. Where key horse combinations offer value, they are included. Where they do not, they are dropped, or as Beyer proposes, used more lightly, as "savers" of capital. Means are available to handicappers who want to understand exotic values more exactly, and these should certainly be utilized. Most important are the projections by racetracks themselves to the probable payoffs of combination bets, as shown on television monitors. Without these predictions, to be sure, exotic betting is fool's play. When projected payoffs are compared to expected payoffs, based on odds probabilities, handicappers can decide where the true values reside or where they do not. These judgments are particularly vital in Exacta wagering. As exotic wagering has increased, so have the underlay combinations in those pools. On the other hand, the Exactas offer fantastic overlay combinations regularly, and thus have become the most generous source of profits for modern handicappers who are information managers. When Royal Heroine stole the show during the inaugural Breeder's Cup Event Day, November 10, 1984, at Hollywood Park, by winning the Breeder's Cup Mile on turf in a world-record 1:32$\frac{3}{5}$ seconds, the filly was favored to win and paid $5.40; the Exacta paid $654.40.

Needless to say, whenever handicappers achieve crystal-clear understanding of a race offering Exacta wagering, they are invited to entertain Beyer's rejoinder about "killing" a race. This is the time to plunge. After all, the single Exacta race obliges the fundamental money-management principle of maximizing gain at minimal risk. Two conditions are necessary: (1) the race must be clearly understood and (2) Exacta values must be extraordinary. On the latter point, handicappers are now advantaged by betting tools that crystalize Exacta overlays. Among the best is the Exacta-Perfecta Gauge.

In northern California resides an accountant named Andy Anderson who beats the races there regularly, unless he resorts to handicapping, in which case he loses. He wins by betting Exacta combinations mathematically. To this end Anderson employs his own invention, which he calls the Exacta-Perfecta Gauge. The gauge is an arithmetical slide rule that reveals the break-even points to one dollar for all the odds combinations in the Exacta pools. Anderson bets on combinations that are overlaid by 10 percent or more.

For instance, the gauge shows handicappers that the following common odds combinations should return the amounts indicated for each dollar invested:

Combination (top half at left)	Break-even payoffs to $1.00	$2.00 Exacta	$5.00 Exacta
8–5 and 5–2	8.70	17.40	43.50
2–1 and 5–2	11.00	22.00	55.00
5–2 and 3–1	15.00	30.00	75.00
3–1 and 4–1	22.00	44.00	110.00
7–2 and 5–1	31.00	62.00	155.00
8–5 and 6–1	17.00	34.00	85.00
9–5 and 10–1	31.00	62.00	155.00
5–2 and 7–1	30.00	60.00	150.00
5–1 and 10–1	80.00	160.00	400.00
8–1 and 12–1	150.00	300.00	750.00

I list the odds combination of 8–1 and 12–1 at the bottom because Anderson's gauge keyed the only racetrack wager for which I collected more than $5,000 in a single swoop.

It happened at Santa Anita in 1983. The handicapping depended on information management, not systematic method analysis, and the wager depended on Anderson's betting tool. The horse I liked was returning to action in a middle-distance route in two weeks following an awful but well-disguised trip, and was dropping from a $20,000 claiming race to $16,000 race. It was sent out besides by leading claimer trainer Mike Mitchell and was ridden by champion jockey Chris McCarron, who had experienced the troubled trip and stayed on. Minutes before post the odds to win were 8 to 1.

At the price I had decided to bet to win only, but en route to the window I stopped to examine the Exacta projections. A horse up in claiming price today from a recent "big win" at $12,500 would be exiting from the twelve-post in a full field. The odds to win were 12 to 1.

On the television monitor the 8–1 to 12–1 combination was projected for a $5 Exacta to pay $1,155, a large overlay. Instead of betting only to win as planned—I was not using separate pools of capital at the time—I bet $60 to win and two $20 Exactas from the trip horse at 8 to 1 to a pair of Exacta overlays, including the recent $12,500 winner at 12 to 1.

The 8 to 1 and 12 to 1 horses finished one-two. The $5 Exacta paid $1,160. The net on the $100 investment totaled $5,104, a value of 50 to 1. The Exacta Gauge keyed the killing. It informed me that an Exacta combination of interest to me was being offered as a 53 percent overlay.

A perfectly practical application of Anderson's gauge is to spot Exacta underlays. The application supports a principal purpose of Exacta wagering, which, to recall, is to convert straight pool underlays to exotic overlays. This purpose fastens on numerous low-priced horses and favorites. Handicappers are well-advised not to cover short-priced horses in Exactas without first comparing projected and expected payoffs. Many projected payoffs will identify conspicuous underlays. Avoid these.

On a recent social Saturday at Pomona Fair, a bullring near Los Angeles, the most likely winner on the card was a high-figure sprinter named Dancing Ribot. Its adjusted times were far and away superior to anything in the field, thus the speed boys figured to render the horse an underlay in the win pool. The price near post time was 3 to 2. The Exacta was available, but Pomona's information system had been spurting off and on all day, and for this race no Exacta probables were projected.

Tossing out an overbet second choice at 2 to 1, as urged cleverly by author Steve Davidowitz years back, I bought $10 Exactas from Dancing Ribot to four outsiders. Happily, one finished second to the solid favorite at 9 to 1. The 8 to 5 and 9 to 1 combination was expected by Anderson's gauge to return $25 to $1, or $125 for each five dollars, the cost of Exactas at Pomona. The actual payoff was $62.50, exactly a 50 percent injustice. My $40 investment netted approximately two dollars for each invested; a 2-to-1 shot. The dollar net was $85. Such a triumph!

As a rule, hardened handicappers have long since learned to avoid the short price in the straight pools, but the same hardened veterans make an analogous mistake every day in exotic pools. They back far too many exotic underlays. That guarantees losses as well, as all can appreciate.

As for rational scrambling techniques in exotics, in 1975 turf writer and handicapper Gordon Jones, of southern California, completed one of the few systematic longitudinal studies of exotic wagering on record. For five major race meets Jones studied his own exotic wagers to determine how well they performed in relation to straight bets. Jones' data are therefore subjectively biased, and they do not represent the national racing scene or calendar well. Yet the data can be considered hypothetical. They reveal interesting trends, to be confirmed or disconfirmed in the future by other, more rigorous, more representative samples. Similar studies can be conducted individually, of course, in the manner Jones proceeded, and handicappers are fairly urged to replicate the following in their personal situations.

In *Smart Money,* Jones described how key horses between odds of 5 to 1 and 20 to 1 that returned profits when bet to win should be combined in the Exacta and Daily Double to maximize the rate of return on the invested dollar. Below are the scrambling techniques and dollar profits for each combination. All key horse combinations were boxed, or reversed, with the other horses.

The Exacta	Rate of profit
Key horse to top contender	106%
Key horse to top two contenders	84%
Key horse to top three contenders	49%
Box top three horses	47%
Key horse top wheel	41%
Key horse bottom wheel	40%
Key horse to place	16%

The Daily Double:	Rate of profit
Key horse to top three in other half	69%
Key horse to top two in other half	46%
Key horse wheel	29%
Key horse to top one in other half	19%

In Exacta wagering, interestingly, key horses at 9-to-2 odds or less showed meaningful profits when combined top and bottom with the second choice only.

INFORMATION AND EXOTIC WAGERING

An assertion is in order. The richest overlays and greatest success in exotic wagering will belong to the most modern of handicappers who have access to the most information. On any given program, handicappers who are information managers will be found coupling speed and class horses that are visible to many with the trip horses, trainer pattern horses, track bias horses, body language horses, and other specialty types that are visible to only a few. These combinations will present informed handicappers with the overlays that represent repeated opportunities for optimal gains from small risks.

By this reasoning the handicappers at decisive advantage in the exotic games will be owners of microcomputers. Only high-tech handicappers can collect the bulk of information, manage it efficiently, and use it wisely. Successful exotic wagering and high-tech handicapping go hand in hand in the information age.

THE ROLE OF RACETRACKS IN THE AGE OF EXOTIC WAGERING

Since the Exacta appeared in 1969, racetracks by and large have ushered in the age of exotic wagering cautiously and unenthusiastically. Fair enough. The lack of enthusiasm traces to a deep-rooted tradition-bound fear that the public will perceive the tracks as encouraging gambling. That attitude qualifies today as an anachronism, but it persists. The citizens of California in 1984 passed a statewide lottery proposition by a fantastic majority. Other than to resist them, racetracks do little to compete effectively with these developments.

The resistance traces as well to a disturbing history of ignorance among racetrack executives as to the fundamental principles and procedures of the pari-mutuel wagering games they offer their customers. It is no less than rare to find a racing executive who really understands and appreciates the relations between pari-mutuel wagering and their customers' information needs. Many just don't want to know, as long as the bettors continue to pay their salaries, fund purse schedules, and subsidize track operations. The ignorance and primitive attitudes among track officials about pari-mutuel wagering and handicapping information is absolutely the bottom line as to why the tracks have never been able to attract the young-adult market it seeks—and needs. Whether younger racegoers will become loyal participants in the pari-mutuel games offered by racetracks depends ultimately on their ability to handicap and wager knowledgeably. Young adults tend instinctively to avoid participating in games in which they take a shellacking. So they stay away from racetracks, where they receive no tutoring in handicapping and betting.

To the good fortune of informed handicappers, exotic wagering has expanded and succeeded throughout the industry by a process of natural selection. In a vague, global, intuitive sense racing officials have learned that racetrack patrons enjoy exotic wagering, though not understanding why. As a clue, it is *not* because they are hell-bent to gamble. More to the point, as the general inflation roared along in the seventies, the value of straight win wagers to typical racetrack customers, especially small bettors, diminished to the breaking point.

Regular customers also need the kinds of betting opportunities by which they can use the multifaceted information they now possess. Tracks in competitive markets realized they could remain competitive for the gaming dollar only by expanding their customers' exotic wagering opportunities. For medium-sized and smaller tracks, promoting exotic opportunities became mandatory. So much so that today only isolated racing enclaves, such as Oaklawn Park in Hot Springs, Arkansas, or Longacres, near Seattle, can stay open without a considerable capitulation to exotic wagering.

In hotly competitive markets even the advent of on-track exotic wagering has not been enough and lately has been bolstered by the kinds of exotic, high-tech betting opportunities provided by simulcasting, inter-track wagering, and off-site betting. Without simulcasting and inter-track wagering, for one desperate example, Atlantic City would be closed down. So would other plants.

The contemporary, futuristic charge to racetracks everywhere is to recognize the priority of exotic wagering in the information age.

Loyal racetrack customers cannot use the multifarious handicapping information now available to them without exotic wagering opportunities. A crucial role of racetracks is to provide those opportunities. Genuinely progressive racetracks also will want to provide the wide array of handicapping information services that stimulate exotic wagering; successful exotic wagering.

My personal situation illustrates the modern handicapper's problem. I limit my play now to the winter season at Santa Anita, which abides by its reputation as *The Great Race Place* in most ways; excels as few tracks do in marketing, customer service, and overall operations; and offers its patrons an unmatched setting and unmatched atmospherics for playing the races. But Santa Anita has been comparatively dense when it comes to pari-mutuel wagering. Track management there has resisted the expansion of exotic wagering from the start but has been dragged along by the successful innovations of its neighbor and public enemy, Hollywood Park. As late as Oak Tree during the autumn of 1984, Santa Anita offered three Exactas as standard fare in the fifth, seventh, and ninth. When the feature features a short field of five or six, which happens far more often than it should, Santa Anita presents a fourth Exacta not in the interests of its customers but to boost its handle on an otherwise noncompetitive race. All other racetracks do the same.

Besides the poverty of intelligent thought that the practice of adding Exactas to short fields reflects, the gimmickry indicates how racetracks tend to attach policy to short-term self-interests and simultaneously disassociate it from the long-term financial health of their customers, particularly their best customers. Racetrack customers prefer exotic wagering opportunities in the competitive races where overlays abound, not in noncompetitive races, where Exacta underlays penalize the bettors as cruelly as do underlays in the straight pools. That argument does not go far with racetrack executives. For one thing, most of them do not understand it. Thus, they cannot appreciate its implications for the healthier long-term survival of numerous customers who are recreational bettors only, not to mention the interests of regular handicappers and customers. Presenting racetrack customers with pari-mutuel underlays as a regular diet only guarantees the customers' larger loss, a purpose more quickly achieved by tracks that promote Exacta wagering in short, noncompetitive fields.

For a second, modern marketing policy at racetracks is overly concentrated on the occasional uncommitted fan, and by methods that assure these people are likely to remain occasional uncommitted fans. Regular customers are supposed to need little or no attention, a dangerous perspective that is widely recognized among intelligent patrons of the sport for the contemptible attitude it reflects, and one that can eventually result in backlash.

One reason many visitors to racetracks who otherwise might not return frequently do is because the participative aspects of going to the races is neither marketed nor serviced. People go to racetracks to handicap and to bet on their selections. It's difficult to think of another industry that treats its best, most loyal customers as badly.

Following the 1984 Santa Anita winter season, to continue in this critical vein for a while longer, the publicity department released a memo purporting to show the relative advantage of small bettors vis-a-vis group syndications in Pick-Six wagering. A grandmother on her first excursion to the racetrack had submitted a two-dollar Pick-Six card and later walked out with $289,249.20, the sole winner that day, which involved an extra large pool and the heavy syndication betting that larger pools attract. Santa Anita's memo showed that across fifty-five days racing fans who had bet just $2 to $10 on the Pick-Six had won 31 percent of the pots. In contrast, bettors who had wagered $1,000 or more in the Pick-Six had won only 10 percent of the pots. From these data the publicity memo implied that $2 bettors had as distinct a chance of hitting the Pick-Six as do syndicates of big bettors.

To be frank, this is nonsense.

Had publicity reported the percentage of Pick-Six tickets that had been purchased for $2 to $10, and divided that percentage into the percentage of winners resulting from those amounts—the appropriate statistical procedure—Santa Anita would have learned that the probability of the $2 to $10 bettors' winning the Pick-Six is smaller than low. In fact, it is infinitesimal. On the other hand, the percentage of Pick-Six cards purchased for $1,000 or more is so minute, if those bettors win 10 percent of the pots, in relative terms, the probability of any big-betting syndication winning the Pick-Six is fantastically high. In fact, small syndications of bigger bettors who are handicappers, four to six perhaps, represent the only feasible, defensible approach to seeking riches by Pick-Six betting.

We can all agree that the peculiar, distinctive charm of the Pick-Six is that it makes possible a grandmother's betting $2 on a six-horse parlay on her first trip to the track and beating the syndications to the glory. No one regrets that. The news is a healthy hype for racetrack participation. But the Santa Anita memo reminds handicappers how naive racetracks can be about the pari-mutuel games they offer, and are supposed to manage.

Andrew Beyer recounted a story at the First National Conference on Handicapping in Los Angeles about betting the Pick-Six in Florida. He and three friends labored long and hard about the final horses to include on a $2,500 Pick-Six card. The final elimination was a nicely priced horse in the sixth, because it would be ridden by a green apprentice. The Pick-Six cards were deposited. Now comes an announcement just before the fifth race. Ladies and gentlemen, there will be a jockey change in the sixth race. Replacing the green apprentice on that interesting longshot will be Angel Cordero, Jr. Naturally, the horse won.

Beyer told how he went charging and screaming into the racing office, only to be met there by a nice fellow who didn't have the slightest concept of the significance of what had transpired.

All regular handicappers surely have their personal tales to tell, and I well remember mine.

As a relative newcomer at Hollywood Park in the early seventies, I decided to couple two favorites I liked in a $200 win parlay. The first horse won at 2 to 1. I put the $600 on the second half, an 8-to-5 shot, which figured to sit comfortably behind an early pace duel that should weaken the other contender before taking over in the stretch. Just after the announcer said the horses had reached the starting gate came a second dreadful announcement. The frontrunner that figured to burn up my main competition became a vet scratch at the gate.

I rushed to the window to sell my ticket. It was too late. They did not postpone the start. The lines were long. In textbook manner, my $600 parlay horse chased the lone remaining frontrunner from wire to wire and was beaten a length. I was furious. I demanded to see the mutuels supervisor and even went to operations; I actually wanted my money back. I got long looks, faint smiles, veiled insinuations that it was all my fault, and I should be manly enough to accept the defeat.

It was my first bad taste of racetrack insouciance toward the bettors.

In wild imaginings, I sometimes see a consumer movement that would force racetracks to come to grips intelligently and fairly toward the bettors on matters such as gate scratches, runoffs from the post parade, horses that break through the starting gate, unpardonable late jockey changes, and the like, all those unforeseen accidents for which the tracks have much responsibility, but do not accept it, preferring instead that their bettors suffer the consequences.

To be fair to my favorite racetrack, Santa Anita in 1985 began a concerted program of providing information services to handicappers, including nightly replays of all races from start to finish, trackside publication of horses receiving Lasix for the first time, head-on shots of race replays at the track, and even daily simulcasting to off-site betting parlors (Las Vegas).

The role of racetracks in the age of exotic wagering, however, is far less executive than the regulation of the sport in ways that protect the interest of bettors. It is merely administrative: to provide the legal wagering opportunities that permit customers to use the handicapping information that they bother to collect and to bet as they prefer. Ideally, customers should be entitled to play the races as they please, using whatever types of information and pari-mutuel wagers as have been sanctioned by the state racing boards.

Many tracks are moving aggressively toward that ideal; others are crawling along. A survey of major tracks in the United States indicates that the exotic wagering opportunities extended to customers on each race are generally plentiful. Most tracks already offer adequate opportunities in the exotics.

The survey reveals the progressive standing of my hometown track *Keystone* in the age of exotic wagering. *Keystone* offers the Daily Double and Exacta on each of the first two races; the Exacta on the third; the Exacta and Triple on the fourth; the Exacta on the fifth; the Late Double plus the Exacta on the sixth and seventh; the Exacta on the eighth; and only the Triple, alas, on the finale.

The tracks in arrears have been Churchill Downs, Longacres, Oaklawn Park, and Santa Anita Park.

Chapter 13

SCENES FROM THE RACETRACK

Indeed, if there is one pervasive influence on the handicapping experience, track bias comes close to filling the bill.

STEVE DAVIDOWITZ
Betting Thoroughbreds,
E. P. Dutton, 1979

THE THREE RACES reviewed here occurred in a week's span during the first six sessions of the 1984 Oak Tree at Santa Anita season, but they might just as readily have taken place during any week at any major track. In analyzing the three, the common denominator is the handicapper's need ultimately for information not contained in the past performance tables or results charts. The three also feature illustrations of this book's crucial concepts, including database information, intuitive thinking, making decisions (not selections), value-added wagering through the exotics, and multiple keys. It's the information management approach applied in three dramatic scenes.

SCENE I

We start with a middle-distance claiming affair that was manageable only by resorting to multiple keys and value-added wagering in the Exacta. Let's cycle through the past performances, looking for the data items and information to support each horse's case.

Up more than the customary notch off a claim, *Golden E.* may not have won in two seasons, but its form looked knife-sharp now, and young trainer Ron Ellis wins a high percentage. Its adjusted time September 15 at Pomona was among the quickest of the meet-

9th Santa Anita

1 1/16 MILES. (1.40⅘) CLAIMING. Purse $23,000. 3-year-olds and upward. Weights, 3-year-olds, 118 lbs.; older, 122 lbs. Non-winners of two races at one mile or over since July 24 allowed 3 lbs.; of such a race since September 1, 5 lbs.; since July 24, 7 lbs. Claiming price $32,000; for each $2,000 to $28,000 allowed 1 lb. (Claiming and starter races for $25,000 or less not considered.)

Golden E. — B. g. 4, by Golden Eagle II—Ginger Cove, by Kauai King
Br.—Smith P (Cal) — 1984 11 0 2 3 $17,625
Own.—Delaplane E E — 114 — Tr.—Ellis Ronald W — $30,000 — 1983 12 1 1 4 $26,700
Lifetime 23 1 3 7 $44,325 — Turf 9 0 1 3 $15,125

22Sep84-10Pom 1 1/16 :48⁴ 1:13⁴ 1:45¹ft *9-5 116 31½ 21½ 11 2½ Hansen R D7 c20000 85-17 GoldenFriend,GoldenE.,CrystlDrops 8
15Sep84-9Pom 1 1/16 :48¹ 1:13¹ 1:44⁴ft 2 118 44½ 34 21½ 23 Hansen R D4 Aw17000 85-15 Runway Ahead, Golden E.,OurLarry 6
26Aug84-3Dmr 1 1/16 Ⓣ:48⁴1:12³1:43²fm 2½ 117 615 511 57½ 57 ShoemkerW4 Aw17000 85-10 GeoStidham,QuotationMarks,LeCid 7
16Aug84-5Dmr 7½f Ⓣ:22³ :46¹1:29¹fm 5½ 117 67 63½ 53½ 31½ Pincay L Jr9 Aw18000 92-07 TooMchForT.V.,WshIhdMl'n,GldnE. 9
16Aug84—Lugged in stretch
3Aug84-7Dmr 1 1/16 Ⓣ:48²1:12⁴1:43²fm 14 117 810 85½ 78 65¾ Ortega L E8 Aw18000 86-10 HevenlyPlin,Mngione,EpsonDowns 8
21Jly84-5Hol 1 1/16 Ⓣ:48 1:12 1:41³fm 3 114 65 43½ 57 45 Hawley S2 Aw24000 86-10 Dancebel, Eterno, Mangiore 7
21Jly84—Hopped in air
8Jly84-5Hol 1⅛ Ⓣ:47³1:11²1:48²fm 16 114 54 64 42½ 3½ Hawley S7 Aw22000 87-09 Dalby, Socratic, Golden E. 7
24Jun84-5Hol 1 :46³ 1:11¹ 1:37¹ft 24 118 41 85¾ 913 915 Meza R Q4 Aw22000 65-20 EmprdorAlNort,Alpino,ExcssProfit 9
9Jun84-7Hol 1 :45³ 1:10³ 1:37²ft 16 112 64 63½ 54½ 32¾ Meza R Q6 35000 76-19 Just Arrived, Lost Creek, GoldenE. 8
31May84-8Hol 1 1/16 Ⓣ:46⁴1:10⁴1:42 fm 17 115 88¾ 88½ 814 811 CastanedaM5 Aw22000 78-11 Swift Message,BestLook,Dancebel 10
Oct 2 SA 4f ft :53 h — Sep 9 Dmr 5f ft :59 h — Sep 3 Dmr 3f ft :35⁴ h — Aug 11 Dmr 5f ft 1:00 h

One Eyed Romeo — Ch. g. 4, by Romeo—Tahitian Chant, by Distinctive
Br.—Manderly Farm (Ky) — 1984 5 0 2 1 $6,927
Own.—Benon-Berman-Soriano — 115 — Tr.—Soriano Morris — $32,000 — 1983 13 2 3 0 $25,550
Lifetime 21 2 7 1 $37,902

22Sep84-8Pom 6f :21³ :45¹ 1:11 ft 5½ 118 21 23½ 22 2½ Black K6 25000 88-17 Rstg,OnEydRomo,HomCourtRuling 8
6Sep84-9Dmr 1 1/16 :46³ 1:11² 1:43³ft 6½ 116 1½ 11½ 1hd 21 Sibille R1 20000 81-18 BOnTim,OnEydRomo,SqurYourHt 10
27Aug84-1Dmr 6f :22¹ :45¹ 1:10¹ft 6 116 97½ 78 56 32¾ Sibille R6 c12500 84-19 Trxton'sDbl,W.C.Shcky,OnEydRm 11
27Aug84—Broke in a tangle, bumped hard start; wide final 3/8
11Aug84-2Dmr 6½f :22¹ :45¹ 1:16¹ft 6½ 116 42½ 54 47 67½ DelahoussayeE1 16000 — — Restge,WickedHitter,LVrn'sBigMc 11
29Jly84-2Dmr 6f :22² :45⁴ 1:11²ft 6½ 116 1hd 22 23½ 54¾ McCarron C J7 16000 76-17 CndyInCort,WckdHttr,JnnyJmpUp 9
6Nov83-1SA 6f :21⁴ :44⁴ 1:10⁴ft 5 117 75¾ 710 79¾ 46 Pincay L Jr5 25000 78-18 LghtnngSwft,CndyStor,AustnsC.W. 7
6Nov83—Wide into stretch
21Oct83-9SA 1 1/16 :46¹ 1:11² 1:45²ft 3½ 117 76 97½ 914 1021¾ Pincay L Jr10 25000 52-20 BombyBrtndr,Egl'sBk,SidiBouSid 11
21Oct83—Veered in start
15Oct83-1SA 6f :22² :46 1:11³ft *2½ 118 51½ 31½ 21 2¾ Black K9 25000 79-17 Recind,OneEyedRomeo,JetMneuver 9
15Oct83—Wide 3/8 turn
18Sep83-2Dmr 6f :22¹ :45¹ 1:09³ft 3½ 121 74¾ 711 714 511½ McCarron C J3 32000 78-15 HndsmPckg,PlyngTWn,Swsh'sWnd 9
18Sep83—Wide into stretch
31Aug83-5Dmr 6f :22⁴ :46 1:10⁴ft *3 118 76 58 55½ 52¾ Black K5 40000 81-19 Rlr'sEnvoy,PlyngToWn,TMchFrT.V. 8
31Aug83—Bumped 5 1/2,1/16
Oct 3 SA 5f ft 1:00⁴ h — Sep 18 SA 5f ft 1:00¹ h — Sep 4 Dmr 5f ft :59 h — Aug 7 Dmr 5f ft :59³ h

ing, and the four-year-old had been parked outside on that bullring's turn last out. Ellis was sufficiently impressed to double-jump the gelding.

One Eyed Romeo looked similar in certain respects to *Golden E.*, with significantly better early lick. Very sharp form. Up in class—for the third consecutive race. Recent claim; improvement. It also had six speed points and a plus form factor last out. Like *Golden E.*, *One Eyed Romeo* could be counted upon mainly to finish second or third, instead of winning, especially against the tougher class.

Handicappers need to appreciate that *Sheriff Muir* continually works out well but races badly. The horse also resents dirt racing.

***Sheriff Muir** — B. g. 4, by Try My Best—First Round, by Primera
Own.—Longo I S — 108 5 — Br.—Stirling Col W (Ire) — Tr.—Velasquez Danny — $28,000
1984 8 0 0 0 $2,250
1983 4 1 0 0 $5,427
Lifetime 17 2 2 1 $15,615 — Turf 15 2 2 1 $15,615

10Sep84-5Dmr 6½f :221 :452 1:161ft 21 116 1½ 52¾ 719 726½ Fernandez A L5 32000 — — Jam Shot, L'Natty, Rye at Sea 7
10Sep84—Lugged out at 3/8
1Sep84-7Dmr 6½f :22 :443 1:162ft 55 116 41½1019112511127¼ Meza R Q1 50000 — — Embolden, RecordCatch,StandPat 11
29Apr84-9Hol 1 ①:4621:1011:342fm 22 114 32 31½ 78 711¼ Delgadillo C7 70000 85-06 Dare You II, Phosphurian, Mufti 7
23Apr84-5SA a6½f ①:221 :4441:151fm 12 116 1hd 1½ 54 67 Steiner J J7 62500 76-19 Midford, Prosperous, Azaam 8
23Apr84—Bumped start
8Apr84-7SA a6½f ①:211 :4341:153fm 10 112 12½ 12½ 44 79½ Garcia J A1 70000 72-18 Anstruther, Overoly, Golden Flak 7
30Mar84-5SA 1⅛ ①:4611:1111:491fm 20 116 35½ 23 21½ 42½ Garcia J A9 62500 79-19 Rusty Canyon, Lunar Ray,Broadly 10
4Mar84-7SA a6½f ①:211 :4341:152fm 47 116 2hd 64¼121312171¼ Steiner J J9 80000 64-19 Mr.PrimMinistr,GrtEstrn,Stlinctiv 12
27Jan84-8SA a6½f ①:213 :4411:154fm 41 114 12 11 65 814¾ Meza R Q2 Aw30000 65-20 EbonyBronze,FstPssge,Cloonwillin 8
30Dec83-8SA a6½f ①:211 :4421:153fm 39 113 87¾ 89¼ 99¼107¾ Guerra W A7 Aw30000 73-19 Shecky Blue, Azulino, Bruckner 12
30Dec83—Wide onto main track
23Jly83♦1Ascot(Eng) 1 1:432fm 10 129 ① 1¾ CrvlhM Tiffany Diamond ShrffMr,TndrSovrgn,LordProtctor 21
Oct 2 SA 5f ft 1:004 h Sep 27 SA 5f ft 1:00 h Sep 22 SA 5f ft 1:004 h ●Sep 17 SA 6f ft 1:123 h

Further, it spends itself on the front running wildly, then stops abruptly. It is one of three horses in this field of nine with no information support. Its brief rapid early speed might cause headaches for others that contest the pace, but that's all. Out.

Noble Air looks hopeless and deserves no further attention.

Following an eight-month layoff and two allowance tune-ups, the four-year-old *Bemidgi* is being dropped into the cheapest claim-

Bemidgi — Dk. b. or br. c. 4, by Honest Pleasure—Box Supper, by In Reality
Own.—Tartan Stable — 115 — Br.—Tartan Farms Corp (Fla) — Tr.—Lukas D Wayne — $32,000
1984 3 0 1 0 $4,800
1983 19 5 2 5 $71,725
Lifetime 26 6 4 6 $80,075 — Turf 6 0 0 2 $7,200

9Sep84-3Dmr 1 1/16 :481 1:12 1:413ft 11 117 56¼ 57½ 613 612½ ValenzuelPA4 Aw20000 79-16 Epson Downs, Decontrol, Eterno 7
9Sep84—Veered in, bumped start; wide 3/8 turn
26Aug84-3Dmr 1 1/16 ①:4841:1231:432fm 8½ 117 515 612 611 610 ValenzuelPA7 Aw17000 82-11 GeoStidham,QuotationMarks,LeCid 7
15Jan84-4SA 1⅛ :48 1:122 1:501ft 4½ 121 43 42 33 23 Valenzuela P A2 50000 75-16 So Goes, Bemidgi, Val De Roi 8
26Dec83-9SA 1 1/16 :464 1:11 1:422ft 4½ 115 914 811 712 79½ ValenzuelPA6 Aw21000 80-16 SwapTheGlass,DncingKin,Dr.Relity 9
7Dec83-9Hol 1 1/16 :461 1:102 1:424ft 2¾ 116 711 68½ 33 1¾ Valenzuela P A1 62500 81-18 Bmidgi,R.J.'sOrphn,Rufui'sNghtOut 7
7Dec83—Lost whip at 3/8
24Nov83-9Hol 1 1/16 ①:4731:1141:42 fm 10 116 912 96¾ 54½ 53 Valenzuela P A7 80000 86-11 GoodFinish,RondsDJmb,DrcLDuc 10
17Nov83-6Hol 1 1/16 :452 1:10 1:44 ft 2¾ 119 617 512 34 12½ Valenzuela P A7 50000 75-22 Bemidgi, Los Portales, Bold Treaty 7
5Nov83-9SA 1 1/16 :464 1:114 1:444ft *3-2 117 64¾ 64½ 63¾ 32 Pincay L Jr5 50000 75-20 Air Pocket, Bold Treaty, Bemidgi 7
5Nov83—Bumped start, wide into stretch
9Oct83-9SA 1 1/16 :461 1:111 1:432ft *2½ 118 812 68½ 56½ 47 Valenzuela P A3 50000 77-16 RondsDeJmbe,Rob'sMyChnce,ZcK. 9
9Oct83—Checked start
12Sep83-9Dmr 1 1/16 :471 1:111 1:424ft *6-5 116 45 46 33½ 1¾ Valenzuela P A2 50000 86-18 Bemidgi,Rob'sMyChance,ByTheSea 5
12Sep83—Lugged in stretch
Sep 26 SA 5f ft 1:002 h Sep 19 SA 5f ft 1:00 h ●Aug 10 SLR tr.t 6f ft 1:144 h

***Noble Air** — B. g. 7, by Royal Noble—Lynaire, by Inmone
Own.—Jackman J (Lessee) — 115 — Br.—Setchell P (NZ) — Tr.—Cox Robin — $32,000
1984 12 0 0 0 $1,075
1983 5 0 1 0 $714
Lifetime 37 4 2 1 $13,303 — Turf 28 4 2 1 $12,228

5Sep84-5Dmr 1 1/16 ①:47 1:1041:413fm 83 116 1222121812181214 McGurn C1 Aw22000 87 — Gordian, Dancebel, Matafao 12
18Aug84-9Dmr 1 1/16 :464 1:11 1:412ft 98 116 810 711 613 511½ McGurn C8 40000 81-13 Dcontrol,Lghthwyholm,Sl'sRylDrm 9
18Aug84—Broke slowly, lost whip at 1/16
11Aug84-9Dmr 1 1/16 :453 1:102 1:423ft 80 116 814 812 711 611½ Lipham T4 50000 75-12 Just Arrived, Norbet, Kilauea 8
2Jun84-7Hol 1 1/16 ①:4721:1121:414fm 75 116 98¾107¼ 89½ 98¾ Aragundi E7 50000 81-11 Kilauea, CheerOn,MyFriendMilton 10
20May84-6Hol 1 1/16 :462 1:103 1:431ft 78 116 76½ 78 87½ 77¾ Aragundi E5 50000 71-17 MusiclScor,Jimbo'sAc,GoodFinish 11
15Apr84-7SA 1 1/16 :463 1:11 1:423ft 59 114 59½ 812 814 815¾ Aragundi E7 Aw33000 72-15 DrkAccent,RiversideArtist,Conrton 8
15Mar84-5SA 1⅛ ①:4641:1131:491fm 59 116 69 63½ 68 68 Aragundi E8 80000 75-19 Phosphurian, Dare You II,ItemTwo 8
4Mar84-5SA 1⅜ ①:47 1:36 2:013fm 98 114 67 44½ 58 77½ Aragundi E6 Aw32000 71-19 Favoloso, TopCompetitor,Chivalry 11
26Feb84-9SA 1 1/16 :47 1:104 1:424ft 108 115 78½ 812 710 610½ Aragundi E5 Aw30000 76-15 No Hyst, Riverside Artist, Secret 9
17Feb84-8SA 1⅜ :474 1:364 2:013ft 82 115 54 68½ 613 617½ Aragundi E1 Aw30000 64-17 River Of Kings,Chivalry,PairOfAces 6
17Feb84—Bobbled, bumped after start, lugged out throughout
Oct 1 SA 6f ft 1:14 h Sep 24 SA 5f ft 1:002 h Aug 30 Dmr 5f ft 1:01 h

ing race of its career by trainer D. Wayne Lukas. It improved last out and should improve still more today. Will it beat the $32,000 claimers?

Bemidgi handled three-year-old claimers at $50,000 and better from far back as a three-year-old. The class drop is certainly appropriate following the long layoff and graduation to the older division. The workouts since the September 9 effort have been sharp, and handicappers know that the third race following a vacation is often the best. Those familiar with *Bemidgi*'s form cycle at three will recall a gradually improving horse, not an explosive one. Its wins were earned by long, steady late runs. *Bemidgi* should need more prepping before a best effort, in my opinion. When the colt was bet 3 to 1, I assigned it to the two-hole in any combinations bets I might try.

The two above represent two interesting outcome scenarios, although the first is well hidden.

Jimsel Ro. g. 7, by Lt Stevens—B Natalie, by Determine
Br.—Double HJ Stable (Ky) 1984 16 2 2 1 $23,723
Own.—Solar Stable 115 Tr.—Webb George H $32,000 1983 12 2 1 1 $37,985
Lifetime 53 9 3 6 $180,503 Turf 11 3 0 1 $47,125
12Sep84-2Dmr 1 1/16 :464 1:11 1:43 ft 12 116 55½ 12 13½ 15 Lamance C10 16000 85-17 Jimsl,StudntbodyLft,LVrn'sBigMc 11
5Sep84-3Dmr 1⅛ :464 1:114 1:504ft 13 115 109 108½ 85¾ 73½ Lamance C1 12500 72-21 Wenatche, Indian O.,ChargingStar 12
26Aug84-1Dmr 1 1/16 :46 1:112 1:434ft 6½ 116 45½ 22½ 1hd 1½ Lamance C11 10000 81-13 Jimsel, Cosumel, Denali Ridge 11
21Jly84-1Hol 1 1/16 :471 1:114 1:44 ft 11 115 64½ 32 31 2hd Sibille R3 10000 75-20 Preservative, Jimsel, North Line 7
12Jly84-9Hol 1⅛ :473 1:113 1:494ft 12 115 87¾ 75½ 710 68¾ Pedroza M A9 12500 70-20 GoldnFrind,MstrCrst,NturlSplndor 10
28Jun84-9Hol 1 :451 1:11 1:372ft 6 115 98¾ 76½ 97 96½ Lipham T9 c10000 73-19 Leaders Blade, First Blade,Cajolery 9
6Jun84-2Hol 7f :22 :452 1:232ft 8½ 115 910 911 65¾ 44 Lipham T9 10000 76-20 Incursion, Herbie Quayle, Dakar II 11
6Jun84—Wide
12May84-1Hol 1 1/16 :464 1:11 1:424ft 3½ 115 710 57 48½ 511¾ Meza R Q3 12500 69-15 NturlSplendor,GlxyPrince,Alcthous 8
25Apr84-1Hol 1 1/16 :462 1:104 1:433ft 16 115 710 78½ 65½ 41¾ Drexler H8 20000 75-17 Egle'sBek,IndinO.,TooMuchForT.V. 8
14Apr84-5SA 1 1/16 :451 1:11 1:444ft 9 116 1022 912 109 108 Lipham T6 20000 63-16 NturlSplendor,‡Cosumel,Egle'sBk 11
Oct 2 SA 5f ft 1:011 h Sep 24 SA 6f ft 1:124 h Sep 2 Dmr 4f ft :48 h Aug 19 Dmr 5f ft 1:002 h

***Groszewski** Gr. c. 4, by Godswalk—Claddie, by Karabas
Br.—McCalmont H (Ire) 1984 10 0 0 3 $6,805
Own.—Lebovitz H 115 Tr.—West Ted $32,000 1983 11 2 1 1 $11,627
Lifetime 30 4 3 4 $26,419 Turf 25 4 3 2 $21,814
26Aug84-9Dmr 1 1/16 :463 1:111 1:422ft 11 117 107¾ 96½ 68¾ 32¾ Lamance C10 32000 85-13 ItemTwo,GlintMinded,Groszewski 12
26Aug84—Extremely wide 3/8 turn
18Aug84-7Dmr 6½f :224 :451 1:161ft 12 115 76½ 612 610 55¾ Valenzuela P A6 40000 — — HandsomePckge,Wilder,CrimeFree 8
15Jun84-2Bel 1⅝ Ⓣ:50 1:4012:051fm 9 117 2hd 55½ 68¾ 56¾ Murphy D J8 50000 61-18 MonkeyBread,VIDeLMeuse,Brodly 10
9Jun84-3Bel 1 1/16 Ⓣ:4541:0941:402hd 43 119 87¾ 90 912 811 MrpDJ1 ⒷBlueLarkspr 83-07 NtiveRid,RemnddII,Who'sForDinnr 9
9Jun84—Run in divisions
23May84-2Bel 1 1/16 Ⓣ:4521:0941:414fm 8½ 117 1019 1017 919 1019 Murphy D J3 75000 68-12 QuickDip,ClrinetKing,RouletteWhl 11
23May84—Slow start
11May84-8Bel 7f Ⓣ:224 :46 1:231fm 4 121 2hd 11 52½ 69 Terry J2 Aw26000 79-17 Wilhelm, ChargingThrough,Eskimo 7
9Apr84-9GP a1 Ⓣ 1:391sf 7 115 68¾ 52½ 43 31¾ Terry J3 Aw20000 82-21 Tarheel Lad, DavisJoy,Groszewski 10
24Mar84-6GP 7f :224 :451 1:214ft 36 117 84¾ 64¾ 43½ 32 Terry J4 50000 93-18 NorthernOcn,BgO'Bucks,Groszwski 9
17Mar84-7GP 6f :223 :451 1:093ft 93 115 109 1016 918 915 Terry J5 Aw18000 76-17 Compo'sTmp,NmbrOnSpcl,ThsEys 10
6Mar84-4Hia 7f :224 :45 1:222ft 114 112 104½ 79½ 89¾ 68¾ Terry J3 70000 82-17 RightOnRed,FstReson,Aristocrticl 12
Sep 30 SA 7f ft 1:28 h Sep 24 SA 4f ft :482 h Sep 13 Dmr 3f ft :364 h Sep 7 Dmr 6f ft 1:133 h

Jimsel has back class. It has handled much better en route to its earnings of $106,503. The big win at Del Mar on September 12 smacked of a return to former heights. At least trainer Webb thought so, doubling the horse's claiming value today. *Jimsel* had looked dull throughout most of 1984, a jaded racer at age seven, but had won two of its last four, while beaten a head once and devastated the field last out while moving substantially ahead in class. *Jimsel* figured for me as a key horse. It might win big; it might fizzle.

The victim of a troubled trip from the ten-hole at peculiar Del Mar on August 26, *Groszewski* had nonetheless finished fast, earning an adjusted time and pace rating superior to the animals it would face today, notwithstanding the same claiming tag. Just as important, the Irish colt had flashed much improved form for trainer Ted West on its second start on the southern California circuit. It had previously been an in-and-out flop in Florida and New York. Groszewski was facing the cheapest field of its career today, always a plus for improving four-year-olds. It figured to be charging at a weakening pace in the stretch today and figured as a key horse prospect in a come-from-behind scenario.

***Jumbler**
Ch. h. 6, by Ex-Libris—En Fraude, by Mysolo
Br.—Haras de la Pomme (Arg) 1984 5 0 1 1 $3,332
Own.—Barrera L S **115** Tr.—Barrera Albert $32,000 1983 10 5 1 0 $10,501
Lifetime 29 8 3 5 $31,001 Turf 14 4 1 2 $17,911
9Sep84-9Dmr 1 ①:47¹1:10⁴1:35⁴fm 35 116 6⁵ 8¹⁵ 8²⁰ 8¹⁷¾ Castaneda M⁶ 62500 77-05 Valais,EmperdorAlNorte,Timberjck 8
5Aug84-5Sar 1⅛ⓣ:48⁴1:12⁴1:50 fm 12 1107 7⁹ 8¹⁰ 8¹³ 7¹⁴ Delgado G⁸ 75000 76-13 ElFntsm,RedBrigde,BlueEmmanuel 8
21Jly84-9Bel 1 1/16:46 1:11 1:43³sy 14 1107 3⁶ 3⁶ 3⁷ 3⁸ Delgado G¹ 75000 76-21 King's Swan, Beagle, Jumbler 8
7Jly84-4Bel 7f :22³ :45¹ 1:22⁴sy 28 122 7¹¹ 8¹⁶ 8¹⁴ 8¹⁵¾ Delgado G⁶ 100000 72-16 Sky Falcon, Irish Waters, Huckster 8
10Feb84♦10Hipodromo(Arg) a1 2:05²hy 4¼ 130 2⁶ CepedR Cl Rico Mnt H Brady, Jumbler, Scarface 9
22Dec83♦5Hipodromo(Arg) a1⅛ 2:04 ft 9-5 124 11½ Cepeda R Pr Mitla H Jumbler, El Fantasma, Wizzo 7
25Nov83♦6Hipodromo(Arg) a1⅛ 1:49¹ft 10 119 1² Cepeda R Pr A Lvllo H Jumbler, Wizzo, Prophet 8
28Sep83♦12SanIsidro(Arg) a7f 1:24⁴sf 3¼ 126 ① 8¹⁰ ZapatO Pr Atomvl Clb H Wizzo, Big Safety, Halley 10
11Sep83♦9SanIsidro(Arg) a7 1:23 fm 2¾ 133 ① 2ⁿᵏ Zapata O Cl Mlgrjo Wina Red, Jumbler, Hckautre 5
28Aug83♦10Hipodromo(Arg) a7f 1:25²hy 4¼ 126 4¾ ZapataO Pr Trumphl H Con Rulo, Austerlitz, Hckuatre 11
Oct 2 SA 5f ft 1:03 h Sep 22 SA 5f ft :50⁴ h Sep 5 Bmr 5f ft 1:01 hg Aug 31 Bmr 4f ft :50 h

Zac K.
B. g. 4, by Sturmvogel—Kriskinkoel, by Restless Wind
Br.—Kelley R W (Ariz) 1984 1 0 0 0
Own.—Nor F **115** Tr.—Nor Fabio $32,000 1983 9 2 3 1 $32,545
Lifetime 10 2 3 1 $32,545
12Sep84-7Dmr 7f :22⁴ :45 1:22⁴ft 13 117 4⁴ 5⁵½ 9¹⁵ 9¹²½ Olivares F³ Aw19000 — — Rosie's K. T., Nitro, MarkInTheSky 9
12Nov83-3SA 1 :46² 1:11³ 1:38 m *2 116 12½ 12½ 13½ 16½ Olivares F³ 40000 78-24 Zac K., Eagle's Beak,SingleSceptre 7
9Oct83-9SA 1 1/16:46¹ 1:11¹ 1:43²ft 5½ 116 2ʰᵈ 1ʰᵈ 2³ 3⁶ Olivares F¹ 50000 78-16 RondsDeJmbe,Rob'sMyChnce,ZcK. 9
14Sep83-7Dmr 6f :22¹ :44⁴ 1:09 ft 3½ 114 31½ 3³ 4⁵ 47¾ Meza R Q⁴ Aw17000 85-14 Expressmn,StffCommndr,SadDiggr 9
2Sep83-9Dmr 1 :45¹ 1:10 1:36²ft 14 1115 1⁵ 1⁵ 12½ 2¾ Estrada J C³ 62500 85-19 Staff Commander, Zac K., Bemidgi 8
10Jly83-3Hol 1 :45⁴ 1:10⁴ 1:37¹ft 6¾ 109 31½ 4⁴ 57¾ 49¾ Hawley S⁸ Aw22000 70-21 KhalDave,NatomsExchnge,Sirluovt 8
10Jly83—Lugged out 7/8 turn
4Jly83-7Hol 1 :45⁴ 1:10² 1:36⁴ft 33 1125 13½ 12½ 11½ 22½ CapitineNM² Aw22000 79-17 Stabilized, Zac K., Upper Court 7
8Jun83-7Hol 1 1/16:46³ 1:11² 1:45¹ft 9½ 115 2ʰᵈ 3½ 7⁷ 98¾ Pierce D² Aw22000 61-20 Negotiate, Gigawatt, Bemidgi 9
26May83-2Hol 6½f :22 :45¹ 1:17³ft 10 114 1½ 12½ 1⁵ 1⁶ Olivares F¹⁰ M35000 82-21 ZacK.,YaYaGooGoo,RoylSt.George 12
26May83—Bumped in rear quarters at start
17Jan83-5TuP 5½f :22¹ :45¹ 1:04³ft 16 119 6³ 43½ 3⁴ 2¾ Guerra V⁹ Mdn 84-14 Cuervo, Zac K., Dreamer's Alibi 11
Oct 1 SA 6f ft 1:13 h Sep 26 SA 1 ft 1:43³ h Sep 20 SA 5f ft 1:01⁴ h Sep 9 Bmr 4f ft :46⁴ h

Jumbler looked awful. The precipitous drop should not help this Argentinian six-year-old.

Zac K. had beaten maiden claiming horses and three-year-old winners in the mud for its pair of lifetime wins, both while unmolested on the front. It would not be lonesome up there today. More important, *Zac K.* had experienced one sprint in eleven months. Probability statistics indicate that the most ineffective stretch-out pattern following a long layoff is a single sprint. That's two strikes, and strike three is the low odds promised from several public selectors' suggesting *Zac K.* can take this field wire-to-wire.

The handicapping process resulted in two outcome scenarios, a class scenario featuring *Jimsel* and a figure-pace-form scenario with *Groszewski. Golden E., One Eyed Romeo,* and *Bemidgi* filled the runners-up slots in each scenario. The more I ruminated about the possibilities, I decided *Jimsel* would likely win or run out. I decided not to link the multiple keys together in Exacta combinations but couple them independently with the others. *Groszewski* proved to be a short-priced prospect and could not be supported otherwise. The odds near post looked like this:

Golden E.	7-1
One Eyed Romeo	7-1
Bemidgi	3-1
Jimsel	5-1
Groszewski	5-2

I took Jimsel to win ($80) and concentrated on the Exacta with *Groszewski.*

The combinations with *Bemidgi* were undervalued; pass.

So I bought $30 Exactas from *Groszewski* to *Golden E.* and *One Eyed Romeo*, which returned fair values on the monitor. The investment totaled $140.

In this way four horses of nine in the field were covered, with multiple keys bet both to win and in four Exacta combinations, each offering fair value to handicappers. *Jimsel* and *Groszewski* were *not* linked to each other. If *Jimsel* won big, the win net would be $400; $60 would be lost to the Exacta. If *Groszewski* won and *Jimsel* disappointed, a winning Exacta would net approximately $600; $80 to win would be sacrificed.

The result proved beneficial indeed, as it so often does by the intuitive approach. The winning $5 Exacta paid $127. The profit on the race was $622.

NINTH RACE
Santa Anita
OCTOBER 6, 1984

1 1/16 MILES. (1.40⅘) CLAIMING. Purse $23,000. 3-year-olds and upward. Weights, 3-year-olds, 118 lbs.; older, 122 lbs. Non-winners of two races at one mile or over since July 24 allowed 3 lbs.; of such a race since September 1, 5 lbs.; since July 24, 7 lbs. Claiming price $32,000; for each $2,000 to $28,000 allowed 1 lb. (Claiming and starter races for $25,000 or less not considered.)

Value of race $23,000; value to winner $12,650; second $4,600; third $3,450; fourth $1,725; fifth $575. Mutuel pool $266,126. Exacta Pool $411,106.

Last Raced	Horse	Eqt.	A.	Wt	PP	St	¼	½	¾	Str	Fin	Jockey	Cl'g Pr	Odds $1
26Aug84 ^{9}Dmr^{3}	Groszewski		4	115	7	3	$3^{1/2}$	4^{hd}	4^{1}	3^{3}	$1^{1\frac{1}{2}}$	Valenzuela P A	32000	2.50
22Sep84 ^{8}Pom^{2}	One Eyed Romeo		4	117	2	1	1^{1}	1^{1}	1^{hd}	1^{1}	2^{nk}	Black K	32000	7.60
9Sep84 ^{3}Dmr^{6}	Bemidgi		4	116	4	4	7^{3}	9	7^{1}	$5^{1/2}$	3^{2}	Delahoussaye E	32000	3.10
22Sep84 ^{10}Pom^{2}	Golden E.	b	4	114	1	2	6^{4}	6^{6}	$3^{1/2}$	4^{1}	4^{2}	Meza R Q	30000	7.10
12Sep84 ^{2}Dmr^{1}	Jimsel	b	7	115	6	9	9	$8^{1/2}$	8^{1}	6^{2}	5^{nk}	Sibille R	32000	5.10
10Sep84 ^{5}Dmr^{7}	Sheriff Muir	b	4	108	3	7	4^{1}	3^{hd}	$5^{1\frac{1}{2}}$	7^{7}	6no	Dominguez R L^{5}	28000	26.10
12Sep84 ^{7}Dmr^{9}	Zac K.		4	115	9	5	2^{1}	2^{1}	$2^{1\frac{1}{2}}$	$2^{1/2}$	$7^{6\frac{1}{2}}$	Olivares F	32000	4.70
5Sep84 ^{5}Dmr^{12}	Noble Air		7	115	5	8	8^{1}	7^{1}	9	8^{3}	8^{3}	McGurn C	32000	69.60
9Sep84 ^{9}Dmr^{8}	Jumbler	b	6	115	8	6	$5^{2\frac{1}{2}}$	5^{2}	$6^{1\frac{1}{2}}$	9	9	Castaneda M	32000	31.80

OFF AT 5:25 Start good. Won driving. Time, :23, :46⅘, 1:10⅘, 1:36⅘, 1:43 Track fast.

$2 Mutuel Prices:			
7-GROSZEWSKI	7.00	4.40	2.60
2-ONE EYED ROMEO		6.80	4.00
4-BEMIDGI			3.20

$5 EXACTA 7-2 PAID $127.00.

Gr. c, by Godswalk—Claddie, by Karabas. Trainer West Ted. Bred by McCalmont H (Ire).

GROSZEWSKI found room along the rail once straightened for the wire and scooted through to overtake ONE EYED ROMEO. The latter carved out all the pace and gave way grudingly. BEMIDGI fell far back midway, then finished strongly but was too late. GOLDEN E. loomed boldly from the outside at the quarter pole and hung. JIMSEL failed to reach serious contention. SHERIFF MUIR had no rally. ZAC K. wide on the clubhouse turn, forced the pace and stopped.

Owners— 1, Lebovitz H; 2, Benon-Berman-Soriano; 3, Hatley M E; 4, Delaplane E E; 5, Solar Stable; 6, Longo I S; 7, Nor F; 8, Jackman J (Lessee); 9, Barrera L S.

Trainers— 1, West Ted; 2, Soriano Morris; 3, Lukas D Wayne; 4, Ellis Ronald W; 5, Webb George H; 6, Velasquez Danny; 7, Nor Fabio; 8, Cox Robin; 9, Barrera Albert.

Overweight: One Eyed Romeo 2 pounds; Bemidgi 1.

Groszewski was claimed by Yasuda G; trainer, Palma Hector O; Zac K. was claimed by Siegel Jan & Samantha; trainer, West Ted.

The gross was $762, or a $622 net on a typically decipherable overnight race that was unbeatable by conventional win wagering alone.

SCENE 2

This one tasted personally sweet for three reasons: (1) the payoff was outstanding, (2) the decision was intuitive, not logical-deductive, and (3) the critical factor is track bias, until mid-1984 a weak underdeveloped aspect of my game that had been improving steadily.

The third on October 8 was a six-and-a-half furlong sprint for $20,000 three-year-old fillies. In a full field only two horses—Bulig Z. and Dynamometer—possessed a lick of early speed. Their records appear on page 361.

To set the stage, the Oak Tree main track had been favoring inside speed for three days, a bias that had picked up momentum the day before and repeated itself loudly in the first today. The first was a $10,000 claiming sprint for older Cal-breds. The winner sped wire-to-wire in 1:09 1/5 seconds, some twelve lengths faster than par for the class at Santa Anita, and looked like this in the past performances:

Lucky Buddy — Ch. g. 4, by Gaelic Dancer—Do Good Honey, by King's Balcony
Br.—Old English Rancho (Cal) — 1984 9 1 1 2 $11,7
Own.—Bernstein-Block-Forman et al 118 Tr.—Bernstein David $10,000 — 1983 4 1 0 0 $9,7
Lifetime 13 2 1 2 $21,520

14Sep84-10Pom	6f	$:21^2$	:45	$1:10^1$ft	*3-5	117	1½	1½	$6^{8½}$	7^{15}	Mena F[1]	10500	78-16	New Stradegy, Pre Book, Papa To
29Aug84-2Dmr	6f	$:21^4$	$:44^4$	$1:11^3$ft	2¾	116	$1^{3½}$	$1^{2½}$	1^1	3^1	Castaneda M[4]	12500	79-19	WthLbrty,Shrly'sStvR.,LuckyBddy
20Aug84-2Dmr	6f	$:22^1$	$:45^2$	$1:10^3$ft	3¾	116	$1^{1½}$	1^3	1^4	1^4	Pedroza M A[2]	10000	85-17	LckyBddy,WndyScott,RoylSt.Gorg
29Jan84-1Hol	7f	$:22^1$	$:44^4$	$1:22^3$ft	6	115	2^1	4^4	5^{11}	$6^{16½}$	McCrronCJ[4]	c16000	67-18	The Bagel Kid, Wantazee,MajorBi
2Jan84-3Hol	6f	:22	:45	$1:10^2$ft	4¾	115	3^{nk}	1^{hd}	2^{hd}	$6^{7¾}$	McCarron C J[1]	20000	77-16	Khalidon, Major Bill, Chris's Lad
12May84-6Hol	6f	:22	$:44^2$	$1:09^3$ft	2¾	117	$2^{1½}$	$3^{3½}$	3^4	$7^{5¾}$	McCarron C J[3]	25000	83-15	Jojohnick,Mr.Unbelievable,MjorBi
5May84-4Hol	6f	:22	$:44^4$	$1:10^3$ft	9½	116	3½	3^3	$2^{2½}$	2no	Black K[1]	20000	84-16	Hard Hit, Lucky Buddy,JurgleHaz
15Apr84-5SA	6½f	$:21^4$	$:44^3$	$1:16^2$ft	*2½	117	1^{hd}	3^4	$6^{9½}$	$7^{16½}$	Pincay L Jr[2]	32000	72-15	L'Natty, Major Bill,Mr.Ambassado
25Mar84-3SA	6f	$:21^4$	:45	$1:10^2$ft	5½	111^5	2^1	$2^{1½}$	$2^{2½}$	$3^{2½}$	Fuentes F P[4]	32000	84-17	Calabonga, Hard Hit, Lucky Budd
22Apr83-1SA	6f	:22	$:45^1$	$1:10^3$ft	*4-5	118	2^1	1^{hd}	1½	$1^{1½}$	Pincay L Jr[5]	M40000	85-20	LuckyBuddy,CallEddie,BostonMgi

Oct 4 SA 3f ft $:36^2$ h   Sep 29 SA 5f ft 1:02 h   ●Sep 23 SA 4f ft :47 h   Sep 8 Bmr 4f ft :47 h

Lucky Buddy could obviously outrun $10,000 Cal-breds when left alone up front, but this gelding had never approached 1:09 1/5 seconds in its life. The bias accounted for the final time.

Add to this the result of yesterdays's fourth, a six-furlong sprint for $32,000 horses, three-and-up. The field contained the classiest late-running sprinter in the barns at the $32,000 level. It also contained a speedball from the Longacres-Yakima circuit, the only front speed in the race. The two looked like this:

Both were low-priced propositions, *Che* at 5 to 2 and catch it. *Che* had backed off from rapid pace duels in its four Longacres races and had been pulverized twice at Del Mar, so badly last out it had to be eased in the final sixteenth. It was dropping 50 percent in claiming price today. The Longacres sprint on April 15 was particularly bothersome, as *Che* raced untouched through slow early fractions at the short five-and-a-half furlongs but managed nevertheless to tire badly enough to finish third. Its Yakima victories in two minor stakes were at five and five-and-a-half furlongs. Was this six-year-old simply a short-winded horse? Did it figure to last at Santa Anita, track bias notwithstanding? At 5 to 2, I passed the opportunity, but speed handicappers I respect did not. They backed *Che* confidently.

Che won wire-to-wire in a blazing 1:09 flat, all fractions sizzling. *Calabonga* raced powerfully this day too, but despite gaining fast in

Che — Dk. b. or br. g. 6, by Windy Tide—City Siren, by Prince Alert
Own.—Sherry J & Helen — 116
Br.—Sherry Helen & J (Wash) — 1984 9 2 1 1 $9,021
Tr.—Eidson Daniel J — $32,000 — 1983 13 4 3 0 $23,960
Lifetime 40 11 4 5 $65,681

12Sep84-3Dmr 6f :22 :45 1:10 ft 9 116 2hd 2½ 53½ 616¾ DelahoussayeE5 62500 71-17 LughingBoy,Polly'sRulr,RivlsFctor 6
12Sep84—Eased final 1/16
12Aug84-7Dmr 6½f :22¹ :45¹ 1:16¹ft 19 116 2hd 2hd 31½ 68¼ Hawley S2 Aw30000 — — Proof, Dedicata, Noble Fury 7
20May84-9Lga 6f :21³ :44¹ 1:09 ft 31 116 2hd 1hd 1hd 87 Baze D3 Speed H 84-21 ColemanCreek,Markell,Hgley'sNest 8
29Apr84-9Lga 5½f :22 :45³ 1:04 ft 20 118 32 61½ 3nk 52¾ Aragon J6 Renton H 89-19 NewBroom,FlyingBob,Sam'sTable 11
15Apr84-8Lga 5½f :22¹ :45³ 1:04 ft 3 119 11½ 12½ 12½ 3½ Aragon J6 Aw9500 91-19 Sailalong, Iron Billy, Che 6
8Apr84-9Lga 5½f :21³ :44⁴ 1:04 gd *3-5 120 31 44 56½ 68¾ FrzirB3 [S]Lewisclarkh 83-16 FlyingBob,RaceyChamp,Sam'sTble 8
17Mar84-9YM 5½f :22¹ :46 1:04 ft *1-2 121 1½ 11 11 12 Frazier B5 Baze H 95-14 Che, Iron Billy, Dilone 5
10Mar84-9YM 5f :22¹ :46 :58 ft 1 118 2hd 11½ 11½ 11 Frazier B4 E Rodeo H 90-17 Che, Hotcha Brown, Doc Barrie 5
26Feb84-8YM 4½f :22³ :46¹ :52 ft *1-3 119 3 2hd 2½ 24½ Frazier B5 Aw1400 90-09 Muckledeedun, Che, SplendidSpirit 5
16Oct83-9Lga 1⅛ :46² 1:11 1:43¹ft 30 116 1hd 1hd 49 822 AragonJ9 [S]Wash Chp 61-23 Moonltly,FlyingLightly,Jimbo'sAc 11
●Sep 29 SA 4f ft :48³ h — Sep 23 SA 4f ft :48¹ h — Sep 3 Dmr 6f ft 1:12⁴ h — Aug 27 Dmr 5f ft 1:00 h

Calabonga — B. g. 6, by Forceten—T V Quiz, by Victoria Park
Own.—Alesia F & Sharon — 116
Br.—Asbury C A & T H (Ky) — 1984 14 5 1 4 $61,650
Tr.—Ippolito Steve — $32,000 — 1983 12 1 1 6 $24,380
Lifetime 48 8 5 17 $136,130 — Turf 2 0 0 0 $550

3Sep84-2Dmr 6½f :22² :45² 1:16¹ft 3½ 116 88¾ 52 2hd 1hd Ortega L E3 25000 — — Calabonga,RomanRelm,WorldRuler 9
25Aug84-2Dmr 6½f :22² :45² 1:16¹ft 6 114 1211 1111 85½ 55½ Ortega L E4 30000 — — Dansacha, Meteorite, Jam Shot 12
25Aug84—Bumped start
9Aug84-5Dmr 6f :22² :45² 1:09³ft 3½ 116 55 53½ 34½ 33 Ortega L E7 32000 87-18 Cryptarch, Meteorite, Calabonga 10
22Jly84-4Hol 6½f :21⁴ :44³ 1:16⁴ft *3-2 116 610 58½ 54½ 3½ Ortega L E5 32000 85-15 BlueSeas,SeniorSenator,Calabonga 7
4Jly84-7Hol 7f :22 :44⁴ 1:22¹ft *2½ 116 1114 88½ 66½ 34 Ortega L E4 32000 82-18 ToughEnvoy,AFastPeace,Calbong 12
4Jly84—Wide 3/8 turn
10Jun84-6Hol 7f :21⁴ :44¹ 1:21³ft 7½ 118 109¾ 1110 68½ 46¾ Ortega L E10 Aw20000 82-18 ExplosivePssr,ExcssProfit,U.S.Stl 12
10Jun84—Lugged out at 3/8
28May84-3Hol 6f :21² :44 1:09¹ft 7 116 711 713 45 2nk Black K1 32000 91-19 Mr.Unbelievbl,Clbong,HndsomPckg 9
3May84-5Hol 6½f :22² :45 1:16¹ft 2½ 112 56 54¾ 33½ 32¾ Ortega L E2 35000 86-19 UnbeknownsttoM,Cryptrch,Clbong 5
8Apr84-2SA 6f :22 :44³ 1:09³ft 4 116 816 611 46 42 Ortega L E5 40000 88-16 HndsomePckge,LughingBoy,SkiRcr 8
8Apr84—Wide into stretch
25Mar84-3SA 6f :21⁴ :45 1:10²ft 5 116 711 55½ 44½ 12 Ortega L E7 32000 86-17 Calabonga, Hard Hit, Lucky Buddy 8
25Mar84—Broke slowly, wide into stretch
Oct 1 SA 6f ft 1:13⁴ h — Sep 25 SA 5f ft 1:00² h — Sep 18 SA 5f ft 1:01³ h — Sep 12 Dmr 3f ft :36¹ h

the last strides proved no real threat. The positive speed bias had taken over in Oak Tree's sprints.

In the third on October 8 only *Dynamometer* and *Bulig Z.* showed good early speed. The two shared 13 of the 25 speed points allotted to the entire twelve-horse field; *Dynamometer* had 28 % of the speed points, and *Bulig Z.* had 24 %.

The odds near post were as follows:

Bulig Z.	7-5
Dynamometer	22-1

The race shaped up as a classic example of underlay-overlay propositions. Moreover, *Bulig Z.* could be tossed out quickly on more than just its ridiculous price, which could be attributed in large part to the combination of its own early speed and the presence of jocky leader Chris McCarron in the saddle. *Bulig Z.* last won for $12,500, in ordinary time, and had not exercised since September 9, which was twenty-nine days past. The filly rarely worked out, true, but did for September 5 for new trainer Eddie Truman shortly after the August 20 claim. But not again, or since its first start for Truman on September 9.

Now examine *Dynamometer*. Start by throwing out its first, where it obviously was beaten at the gate.

Its three-year-old Longacres-Yakima record is consistently impressive. Never clear while fighting the early pace, the nicely bred Forceten filly had stopped once, its first try versus winners. Its latest two races on the lead are the best. On August 26 at Yakima, in the Miss Soph Handicap, while losing by a neck at 11 to 1 *Dynamometer* earned an extraordinarily high adjusted speed rating of of 117, when

3rd Santa Anita

6½ FURLONGS. (1.14) CLAIMING. Purse $13,000. Fillies. 3-year-olds. Weight, 121 lbs. Non-winners of two races since July 24 allowed 3 lbs.; of a race since then, 5 lbs. Claiming price $20,000; if for $18,000 allowed 2 lbs. (Races when entered for $16,000 or less not considered.)

Dynamometer
Own.—Kessler & Tannenbaum **116**
Dk. b. or br. f. 3, by Forceten—Bold Dynastic, by Dynastic
Br.—Kessler & Tannenbaum (Cal) 1984 6 1 3 0 $6,310
Tr.—Vienna Darrell $20,000 1983 2 M 0 1 $323
Lifetime 8 1 3 1 $6,633

21Sep84-8Lga 6½f :22 :45 1:17 ft 7¾ 116 10[18] 10[22] 8[15] 8[9¾] Aragon J4 Ⓕ 25000 75-21 Rocco'sRockt,BllyJn,Mlcforthght 10
9Sep84-7Lga 6f :213 :443 1:10 ft 11 115 2hd 2hd 2hd 2no James M5 Ⓕ 25000 86-12 Merinated,Dynamometer,SilkenMrk 8
26Aug84-9YM 5½f :22 :453 1:043ft 3 119 2½ 1hd 1hd 2nk ArgnJ8 ⒻMiss Soph H 92-25 Merinated,Dynamometer,Shhgrnde 9
19Aug84-4Lga 6½f :231 :463 1:172ft 5½ 118 1½ 1hd 2½ 2½ James M1 Ⓕ 28000 81-15 Mry'sMove,Dynmometer,MillStLss 6
29Jly84-6Lga 6f :22 :452 1:101ft 7¾ 112 21 1hd 3nk 76¼ James M2 Ⓕ 25000 79-17 Splndcous,SwftMrcl,Mlcforthought 7
13Jly84-4Lga 6f :221 :453 1:11 ft 4½ 118 1hd 2hd 2hd 1¾ James M2 ⒻM20000 81-18 Dynmometr,Riny'sDstiny,GmofRoni 7
4Dec83-3YM 5½f :22 :47 1:05 ft *9-5 120 64 59 44½ 411¾ Aragon J3 ⒻMdn 78-12 MissBarbie,‡RainSpirit,BelleBtteur 8
30Oct83-3YM 5½f :214 :462 1:054ft 10 120 47 42½ 31 31 Aragon J3 Mdn 85-14 JustAirTime,TbleHeirss,Dynmomtr 10
Oct 4 SA 4f ft :482 h Aug 16 Lga 4f ft :49 h Aug 9 Lga 4f ft :51 b

Bulig Z.
Own.—Leisuretime Stable **116**
Dk. b. or br. f. 3, by Delaware Chief—Mandalay, by Bounding Main
Br.—Bisbines R L & Patricia (Cal) 1984 9 2 3 1 $22,550
Tr.—Truman Eddie $20,000 1983 0 M 0 0
Lifetime 9 2 3 1 $22,550

9Sep84-1Dmr 6f :22 :442 1:10 ft *9-5 116 22 23½ 23½ 23½ McCarron CJ3 Ⓕ 20000 84-16 BobHutsonEsquir,BuligZ.,DstntJul 11
20Aug84-1Dmr 6f :214 :444 1:094ft *3-2 116 33 25 25 25 Garcia J A6 Ⓕ c16000 84-17 Fan Club, Bulig Z., Vedalia 12
20Aug84—Broke against bit, off slowly
10Aug84-1Dmr 6f :221 :453 1:113ft *3½ 111 1hd 1hd 13 14½ Garcia J A10 Ⓕ 12500 80-20 Bulig Z., Pochola, Tear Pad 12
21Jly84-2Hol 6f :22 :45 1:111ft 9e 116 3½ 64½ 1011 1213½ Castaneda M8 Ⓕ 40000 67-20 HttieWing,SpringBid,ShdyHostess 12
12Jly84-5Hol 6f :22 :443 1:10 ft 20 112 31 47½ 815 817½ Meza RQ1 ⒻⓈAw20000 69-20 Fleet N' Irish, La Fonteyn,FairAma 8
29Feb84-6SA 6f :22 :453 1:104ft *9-5 117 22½ 22½ 11½ 15½ McCrronCJ7 ⒻM50000 84-15 Bulig Z., Crownlet, Society Riva 8
17Feb84-4SA 6½f :213 :452 1:182ft *9-5 117 1hd 11 1½ 32½ McCrronCJ3 ⒻM40000 75-17 What Magic, Wine Kiss, Bulig Z. 9
17Feb84—Veered in, bumped after start
3Feb84-2SA 6f :214 :451 1:11 ft 6¾ 117 1hd 1hd 1hd 21½ DelgadilloC3 ⒻM40000 81-15 PssiontPldg,BulgZ.,Gmblr'sDughtr 12
20Jan84-4SA 6f :214 :45 1:112ft 17 114 2hd 2hd 22½ 46¾ DelgadilloC5 ⒻM40000 74-19 MusicalBall,AlotaGlitter,PollyHigh 11
Sep 5 Dmr 4f ft :473 h

the form speed rating and variant are added. Back at Longacres on September 9, *Dynamometer* lost to the same winner of the Miss Soph following a bitter speed duel all the way, this time by a nose. Notably, the fractions on September 9 are considerably faster than the filly had seen before. It stayed in there, battling. This claiming filly is fast and competitive, and now has moved to trainer Darrell Vienna's barn at Santa Anita.

It looked to me that *Dynamometer* would be ahead of *Bulig Z.* at the first call by two lengths perhaps, but that scenario proved wrong. Under McCarron, *Bulig Z.* broke the gate sharply, while *Dynamometer* came out flat-footed. Jockey Kenny Black urged *Dynamometer* immediately and put a head in front a quarter mile along. The two raced neck and neck for another quarter mile and into the upper stretch. The race set up now for a latecomer, but nothing really closed. In slow final furlong time *Bulig Z.* hung, and *Dynamometer* lasted.

THIRD RACE
Santa Anita
OCTOBER 8, 1984

6 ½ FURLONGS. (1.14) CLAIMING. Purse $13,000. Fillies. 3-year-olds. Weight, 121 lbs Non-winners of two races since July 24 allowed 3 lbs.; of a race since then, 5 lbs. Claiming price $20,000; if for $18,000 allowed 2 lbs. (Races when entered for $16,000 or less not considered.)

Value of race $13,000; value to winner $7,150; second $2,600; third $1,950; fourth $975; fifth $325. Mutuel pool $367,394.

Last Raced	Horse	Eqt	A.Wt	PP	St	¼	½	Str	Fin	Jockey	Cl'g Pr	Odds $1
21Sep84 ^{8}Lga^8	Dynamometer	b	3 117	8	3	1^{hd}	1^{hd}	$1\frac{1}{2}$	1^{nk}	Black K	20000	18.60
10Sep84 ^{1}Dmr^5	Cream Pocket		3 118	10	4	7^6	4^1	3^3	$2^{2\frac{1}{4}}$	Delahoussaye E	20000	4.50
9Sep84 ^{1}Dmr^2	Bulig Z.	b	3 116	11	1	2^4	2^4	2^3	3^{hd}	McCarron C J	20000	1.20
10Sep84 ^{1}Dmr^7	Miss Via Magnum		3 116	3	7	$4\frac{1}{2}$	$5^{1\frac{1}{2}}$	5^1	4^{nk}	Meza R Q	20000	36.10
29Aug84 ^{1}Dmr^6	Merry Headliner		3 109	1	11	9^1	9^2	$6^{1\frac{1}{2}}$	$5^{1\frac{3}{4}}$	Dominguez R L^5	18000	98.60
25Sep84 ^{10}Pom^4	Taidy	b	3 111	7	6	$5\frac{1}{2}$	6^2	7^2	$6^{1\frac{1}{2}}$	Lozoya D A^5	20000	88.30
9Sep84 ^{1}Dmr^6	Winning Gold		3 118	9	2	6^{hd}	$7^{2\frac{1}{2}}$	9^1	$7\frac{1}{2}$	Sibille R	20000	57.20
25Aug84 ^{1}Dmr^6	Awkward Age		3 117	5	9	$10\frac{1}{2}$	10^4	10^5	$8\frac{1}{2}$	Pincay L Jr	20000	5.00
25Sep84 ^{10}Pom^2	Time To Rule		3 114	2	8	8^1	8^1	$8\frac{1}{2}$	9no	Garcia J A	18000	23.00
21Jun84 ^{1}Hol^4	Passionate Pledge		3 116	6	5	3^1	$3\frac{1}{2}$	4^1	$10^{3\frac{1}{2}}$	Valenzuela P A	20000	8.20
29Aug84 ^{1}Dmr^1	La Femme Natural		3 116	4	10	11	11	11	11	Hawley S	20000	7.60

OFF AT 2:08 Start good. Won driving. Time, :21⅘, :45⅕, 1:10⅘, 1:17⅘ Track fast.

$2 Mutuel Prices:

8–DYNAMOMETER	39.20	14.00	5.60
11–CREAM POCKET		5.60	3.20
12–BULIG Z.			2.80

Dk. b. or br. f, by Forceten—Bold Dynastic, by Dynastic. Trainer Vienna Darrell. Bred by Kessler & Tannenbaum (Cal).

DYNAMOMETER made every pole a winning one, discouraging a constant challenge from BULIG Z. until the furlong marker and then held over CREAM POCKET in game fashion. The latter began to move to the leaders from the outside turning for home, was cutting into the winner's margin rapidly at the end but ran out of real estate. BULIG Z. forced the issue outside the winner the entire trip and lacked renewed energy in the final furlong. MISS VIA MAGNUM made some progress from the rail. AWKWARD AGE was never dangerous. PASSIONATE PLEDGE was an early factor and stopped. LA FEMME NATURAL lost contact midway. LOVE'S VENTURE (9) WAS SCRATCHED AT THE GATE ON ADVICE OF THE TRACK VETERINARIAN. ALL REGULAR WAGERS ON HER WERE ORDERED REFUNDED WHILE ALL OF HER PICK SIX SELECTIONS WERE SWITCHED TO THE FAVORITE, BULIG Z. (1).

Owners— 1, Kessler & Tannenbaum; 2, Bernstein-Guiliano-Guiliano; 3, Leisuretime Stable; 4, Universal Stable; 5, Knight G L & R G; 6, Deerwood Stock Farm; 7, Corwin-Fuller Stable-Stoke; 8, Rowan & Whitney; 9, Worley Barbara B & W; 10, The Ladies' Confederacy; 11, Housego J W.

Trainers— 1, Vienna Darrell; 2, Bernstein David; 3, Truman Eddie; 4, Mulhall Richard W; 5, Harrell Cecil; 6, Mazzone Paul A; 7, Robbins Jay M; 8, Canney William T; 9, Tinsley J E Jr; 10, Mitchell Mike; 11, Harte Michael G.

Overweight: Dynamometer 1 pound; Awkward Age 1.

Dynamometer was claimed by Sulli J; trainer, Luby Donn; Cream Pocket was claimed by Miller F G; trainer, Brooks L J.

Scratched—Love's Venture (9Sep84 ^{2}Dmr^6); Tear Pad (28Sep84 ^{8}Pom^2); Martha Elizabeth (18Sep84 ^{8}Pom^1).

The win mutuel paid $39.20. Aware of the existing speed bias in sprints, I had approached the October 8 card on the lookout for a bias opportunity, but I did not expect the overlay I found here. The morning line rated *Dynamometer* 12 to 1 and I expected that to be sliced in half by the speed handicappers. But too many fastened onto McCarron and *Bulig Z.*, despite the low price. Influenced strongly by the evidence on track bias, I intuitively took the other track and bet $100 to win on *Dynamometer.* I realized the track bias would move the filly along. *Bulig Z.* did not seem much to beat. The scenario proved correct, resulting in the most generous profit I've yet to receive from a strong track bias: $1,860.

As a postscript, when *Dynamometer* and horses like it return to action they no doubt become instant underlays and will very probably lose. The horses will be raised slightly in class, and the speed bias that moved them forward will likely have disappeared. Track bias involves a multitude of handicapping ramifications. At Oak Tree in 1984, for instance, frontrunning horses that had stopped against negative speed biases at both Del Mar and Pomona, the prior stops on the circuit, moved up dramatically. Late-runners that had shined brightly at Del Mar and Pomona had scarcely a hope of repeating at Oak Tree, and did not. Handicappers who record daily track biases in the database will be able to query the database on the matter weeks or months later. Not infrequently, bias information retrieved in that way will explain good and bad performances that will be vastly dissimilar today.

SCENE 3

The Autumn Days Handicap opens the Oak Tree fall season at Santa Anita. It's an open stakes, ungraded, unlisted, and contested down Santa Anita's unique European-style hillside turf course. In any open stakes, class handicappers scout for horses that have performed well against stronger stakes competition, preferably in graded or listed features. If these stars look in form and suited to the distance, footing, and probable pace, the horses shape up as automatic best bets, provided the odds are fair to favorable. Oddly, they often are. Class appraisal apparently remains one of the most difficult skills for regular handicappers to develop.

Moreover, a new type of class standout has been running away with numerous stakes and allowance races on the United States calendar. The horses represent one of the best, most generous sources of overlays in contemporary handicapping for players who underscore the significance of class advantages in the stakes divisions and who know what to look for. Precisely that kind of classy article found its way into the Autumn Days at Oak Tree in 1984. The horse was a standout. The mutuel was a standout too. The past performance tables of the entrants appear below. Try to find the predictable winner.

To facilitate the information management approach, below is an itemized list of stakes information not trapped in the past performances. The information items have a chronology similar to the order of appearance of the horses in the past performances:

- The Vacaville Stakes at Solano is a minor forgettable stakes.
- The Ramona handicap is a $125,000-added Gr. 2 stakes for three-year-olds and up.
- The Rancho Bernardo Handicap is a listed stakes, purse of $50,000-added, for three-year-olds and up.
- *Bara Lass* won multiple sprint stakes of high caliber and a bankroll of half a million but has also looked unwilling in the late stages of its last several sprints, and it has never triumphed on turf, winning only $9,000 of its earning on the grass.
- The Anoakia is a $60,000 Gr. 3 sprint for two-year-olds. The Railbird is a $50,000 Gr. 3 stakes for three-year-olds. *Percipient* has been far more effective as a sprinter, but it has not yet raced on turf and is not bred for that surface.
- The Gr. 3 pair of stakes won by *Lina Cavalieri* in Italy have purses of $30,000 and $35,000, respectively, with the Royal Mares open to older horses. The Gr. 1 Tesio Stakes has a $70,000 purse and is open to four-year-olds. The Verziere is a listed stakes for three-year-olds (purse not available).
- Two additional data items are pertinent: *Our Native*, the sire of *Betty Money*, is one of the most promising young turf sires in the nation. Its latest impact value is 2.25, and its winners have returned healthy profits on a series of $2 bets.
- *Lyphard's Princess* is the heavy prerace favorite at 7 to 5 ; *Percipient* is 3 to 1.

8th Santa Anita

ABOUT 6 ½ FURLONGS. (Turf). (1.11⅘) 16th Running of THE AUTUMN DAYS HANDICAP(2nd Division). Purse $45,000 added. Fillies and mares of all ages. By subscription of $50 each, $600 additional to start, with $45,000 added, of which $9,000 to second, $6,750 to third, $3,375 to fourth and $1,125 to fifth. Weights Friday, September 28. Starters to be named through the entry box by the closing time of entries. A trophy will be presented to the owner of the winner. Closed Wednesday, September 26, with 34 nominations.

Awesome Promise *

Dk. b. or br. m. 5, by What Luck—Sharp Pencil, by Olden Times
Br.—Forrester Mrs Geri (Ky) 1984 3 1 1 0 $21,500
Own.—Hughes B W 113 Tr.—Vogel George 1983 7 3 1 0 $34,050
Lifetime 12 5 3 0 $63,585

7Sep84-3Dmr 6½f :22 :444 1:17 ft 39 1105 1½ 1² 1³ 1nk Fox W I Jr5 ⓕAw32000 — — AwesomePromise,BttyMony,Circulr 7
7Sep84—Bumped start
9Feb84-8SA 1 :46 1:103 1:353ft 23 1105 1½ 2½ 58½ 616 FuentesFP4 ⓕAw40000 74-17 Adored, Elusive, Past Pleasures 7
13Jan84-8BM 6f :223 :452 1:11 ft 2½ 114 3½ 3½ 3² 2½ LamanceC1 ⓕAw20000 84-26 NoMorBlus,AwsomPromis,AGftAgn 7
14Dec83-8BM 6f :224 :47 1:133m *3 115 55½ 56½ 78½ 811 Munoz E4 ⓕAw20000 61-39 Joi'skiToo,BigDrems,CheryleKing 11
14Dec83—Broke slowly, steadied after break
8Aug83-8Dmr 1 :462 1:111 1:362ft 6½ 115 1½ 67½ — — Sibille R1 ⓕAw30000 — — Skillful Joy, Winsome Eris, Kippy 6
8Aug83—Eased; Broke in a tangle
16Jly83-11Sol 6f :222 :444 1:092ft *1 118 1hd 1hd 11½ 1¹ Munoz E5 ⓕVacaville 97-07 AwsomProms,Lt'sImg,Shoto'Luck 10
16Jly83—Bumped break
27May83-6GG 6f :214 :444 1:101ft *4-5 114 1½ 1hd 1³ 1⁵ Munoz E4 ⓕAw16000 88-22 AwsomPrms,QckStdy,UnchnMyHrt 8
13Apr83-3SA 6f :221 :453 1:102ft 3½ 120 11½ 1¹ 2½ 2no DlhoussyE3 ⓕAw24000 86-23 GloccMorr,AwsomPrms,FmsPrfrmr 6
16Mar83-5SA 6½f :212 :441 1:162ft 5½ 120 11½ 1hd 46½ 714½ RomeroRP2 ⓕAw23000 73-17 RiseDPi,SunnyRidg,FmousPrformr 8
27Feb83-5SA 6f :213 :452 1:104gd 5 113 1⁴ 11½ 1² 12½ RomeroRP8 ⓕAw19000 84-15 AwsomPromis,DvinLook,Comprhnd 8
Sep 28 SA 5f ft 1:012 h Sep 23 SA 4f ft :483 h Sep 16 Dmr 3f ft :381 h Sep 4 Dmr 3f ft :374 h

Circular

Ch. f. 3, by What a Pleasure—Convenience, by Fleet Nasrullah
Br.—Glen Hill Farm (Fla) 1984 8 3 3 1 $52,800
Own.—Glen Hill Farm 112 Tr.—Proctor Willard L 1983 0 M 0 0
Lifetime 8 3 3 1 $52,800

7Sep84-3Dmr 6½f :22 :444 1:17 ft 7-5 113 42½ 32½ 3³ 3nk McCrrnCJ3 ⓕAw32000 — — AwesomePromise,BttyMony,Circulr 7
24Aug84-8Dmr 6f :221 :451 1:092ft 9-5 117 1¹ 1hd 2½ 22½ Pincay LJr1 ⓕAw21000 88-18 Percipient, Circular, Alabama Nana 6
10Aug84-7Dmr 7f :221 :444 1:224ft 3-2 117 11½ 1hd 2hd 2½ VlenzulPA1 ⓕAw22000 — — Inquisition, Circular, Allusion 6
6Jly84-5Hol 6f :213 :442 1:093ft *4-5 115 21½ 2½ 12½ 11½ VlenzulPA6 ⓕAw22000 89-15 Circular,BrightOrphan,PatriciaJmes 7
6Jun84-8Hol 7f :212 :44 1:221ft 4½ 114 2hd 2² 42½ 57½ ShoemkrW3 ⓕRailbird 78-20 Mttrnd,Gn'sLdy,LuckyLuckyLcky 10
16May84-7Hol 6f :22 :441 1:09 ft *4-5 120 1½ 1½ 2hd 1hd ShoemkrW2 ⓕAw20000 92-18 Circular, Sligo Town,DreamFeather 6
31Mar84-6SA 6½f :212 :443 1:171ft *8-5 117 1³ 1⁵ 1⁶ 14½ Valenzuela PA1 ⓕMdn 84-18 Circular, Valegal, Bobbinette 12
11Mar84-3SA 6f :214 :452 1:102ft *4-5 117 1¹ 1² 1¹ 21½ Valenzuela PA6 ⓕMdn 84-18 LoveSmitten,Circulr,QuenOfMirdor 9
11Mar84—Broke slowly
Sep 29 SA 5f ft 1:003 h Sep 23 SA 4f ft :501 h Sep 5 Dmr 4f ft :501 h Aug 22 Dmr 4f ft :49 h

Lyphard's Princess

B. f. 4, by Lyphard—Avum, by Umbrella Fella
Br.—PreskuerMrsG&VkngFmsLtd (Ky) 1984 11 3 2 1 $77,500
Own.—Northwest Farms 117 Tr.—Fanning Jerry 1983 9 0 3 1 $23,691
Lifetime 24 4 7 2 $119,923 Turf 21 4 7 2 $118,548

9Sep84-8Dmr 1⅛Ⓣ:4841:1241:482fm 10 115 73½ 41½ 32½ 43½ Sibille R8 ⓕRamona H 93-05 FlagDeLune,RoylHeroine,SltSpring 9
31Aug84-7Dmr 1 Ⓣ:4631:1041:352fm*2-3 115 22½ 1½ 13½ 12½ Toro F1 ⓕAw35000 96-04 Lyphrd'sPrincss,PolitRbuff,Milgros 6
17Aug84-8Dmr 1⅛Ⓣ:4811:1141:421fm 2½ 115 65½ 4³ 2² 1² Toro F6 ⓕAw24000 99-06 Lyphrd'sPrincss,Agigl,RflctToGlory 8
4Aug84-7Dmr 1 Ⓣ:48 1:1221:363fm *2½ 116 54½ 53½ 33½ 2¹ Toro F7 ⓕAw24000 89-08 AuntieBetty,Lyphrd'sPrincss,ClrTlk 9
4Aug84—Extremely wide into stretch
28Jly84-5Dmr 1⅛Ⓣ:4821:1231:45 fm*8-5 117 2hd 1¹ 1¹ 52½ Toro F1 ⓕⓈOsunitas 81-16 RdyForLuck,ContinntlGrl,AgttdMss 8
28Jly84—Run in divisions
19Jly84-8Hol 1 Ⓣ:47 1:1041:353fm 2½ 116 43½ 5⁵ 4³ 2no Toro F7 ⓕAw32000 90-10 ContinntlGirl,Lyphrd'sPrincss,Agigl 7
2May84-7Hol 1⅛Ⓣ:47 1:11 1:412fm 4½ 114 1⁶ 13½ 1hd 34½ Fell J1 ⓕAw30000 87-08 FnnyRogh,RdEmbr,Lyphrd'sPrncss 8
23Apr84-3SA a6½fⓉ:221 :4441:151fm *2½e114 63½ 4⁴ 2hd 11½ Fell J5 ⓕAw32000 83-19 Lyphrd'sPrncss,Aggl,PtchPtchPtch 7
30Mar84-7SA 6f :22 :452 1:104ft 9½ 114 73½ 62½ 4³ 52½ Fell J1 ⓕAw30000 81-18 Gatita, Princess Lurullah,RedRaisin 8
1Mar84-5SA 6½f :214 :442 1:164ft 4 114 3¹ 42½ 87½ 8⁹ Olivares F7 ⓕAw28000 78-21 Centavos,RedRisin,PrincessLurullh 8
1Mar84—Checked at 3/4
Oct 1 SA 4f ft :464 h Sep 25 SA 5f ft 1:022 h Aug 25 Dmr 4f ft :471 h Aug 15 Dmr 4f ft :471 h

Bara Lass

B. m. 5, by Barachois—Rama Lass, by Jaipur
Br.—Stevens S E (Tex) 1984 15 3 4 2 $291,013
Own.—Stevens S E 119 Tr.—Lukas D Wayne 1983 20 7 1 3 $226,200
Lifetime 56 17 9 10 $532,362 Turf 5 0 0 1 $9,000

31Aug84-8Dmr 6f :221 :444 1:083ft 7½ 121 1½ 3½ 1hd 4⁴ BlckK1 ⓕRch Brnrd H 91-16 PlsurCy,LovlirLind,PridOfRoswood 8
13Aug84-8Dmr 6½f :221 :452 1:16 ft 2½ 122 2½ 21½ 1½ 21½ McCrrnCJ4 ⓕAw35000 — — Pleasure Cay, Bara Lass, Centavos 7
13Aug84—Checked stretch
3Aug84-8Dmr 6½f :22 :444 1:16 ft *1 122 2hd 1hd 1hd 2nk VlenzulPA5 ⓕAw30000 — — Filomena Galea, BaraLass,Centavos 8
18Jly84- Hol 5f :221 :452 :572ft — 115 2hd 2hd 3½ 32½ VlenzulPA5 Hol Exp H 88-19 Commemorate,Premiership,BarLss 5
18Jly84—Lugged in, steadied late; No wagering
7Jly84-7Hol 5f Ⓣ:22 :441 :562fm 9½ 118 41½ 52½ 71½ 6⁵ McCarronCJ1 Aw40000 92-06 NightMovr,‡Commmort,GminiDrmr 8

30May84-8Hol 6f :214 :44 1:091ft 2 124 1½ 21 3½ 5¾ Black K5 ⒻSlvr Spn H 88-22 HolidyDncr,NughtyMdm,MdmForbs 6
2May84-8CD 7f :233 :461 1:233ft *4-5 123 1hd 2hd 21 44 BrfldD4 ⒻBlue Delight 85-13 RegalValley,FirstFlurry,SpaceAngel 7
2May84—Run in divisions; Hesitated start
18Apr84-7Kee 6f :22 :454 1:113m *3-2 126 4nk 77 79 614¾ MpleE5 ⒻThrobrd Clb 69-20 BidsAndBlds,LssTrump,GrcnComdy 8
30Mar84-8SA 1 :452 1:101 1:354ft 3½ 117 11 1½ 1hd 2no McCrrCJ4 ⒻSusns Grl 89-18 Princess Rooney, BaraLass,Capichi 4
21Mar84-8SA 5½f :214 :444 1:031ft 4½ 117 42½ 33 34 32¾ HawleyS6 El Conejo H 92-15 NightMover,Dave'sFriend,BaraLass 7
21Mar84—Run in divisions
Sep 26 SA 4f ft :483 h Sep 17 Dmr 4f sy :504 h Aug 22 Dmr 4f ft :484 h

Percipient
B. f. 3, by Topsider—Naughty Intentions, by Candy Spots
Br.—Hibbert R E (Ky) 1984 5 2 1 0 $30,900
Own.—Hibbert R E 116 Tr.—Manzi Joseph 1983 4 2 0 1 $79,537
Lifetime 9 4 1 1 $110,437
14Aug84-8Dmr 6f :221 :451 1:092ft *4-5 115 21 2hd 1½ 1¾ VlenzulPA3 ⒻAw21000 91-18 Percipient, Circular, Alabama Nana 6
5Aug84-7Dmr 6½f :222 :452 1:16 ft *6-5 115 31½ 31 1¾ 16 VlenzulPA3 ⒻAw19000 — — Percipient,SpringLoos,KlondikKuti 7
8Jly84-8Hol 1⅛ :451 1:094 1:491ft 21 121 21 44½ 916 921¾ VlnzulPA6 ⒻHol Oaks 61-16 MomntToBy,Mttrnd,LckyLckyLcky 9
6Jun84-4Hol 7f :212 :441 1:232ft *4-5 114 53 44¼ 3½ 2hd McCrrnCJ3 ⒻAw22000 88-16 AgittedMiss,Percipint,BrightOrphn 6
6Jun84-8Hol 7f :212 :44 1:221ft 10 119 62¾ 42½ 32½ 46¾ PincayLJr10 ⒻRailbird 80-20 Mttrnd,Gn'sLdy,LuckyLuckyLcky 10
6Jun84—Wide 3/8 turn
12Nov83-8SA 1 1/16 :463 1:113 1:442sy 6½ 115 32½ 42½ 36½ 317½ GuerrWA8 ⒻOak Leaf 61-24 Life's Magic, Althea, Percipient 9
26Oct83-8SA 7f :222 :451 1:231ft 11 117 1hd 1hd 1½ 11½ GuerraWA1 ⒻAnoakia 84-20 Percipient, Althea, PersonableLady 4
15Oct83-6SA 6f :22 :452 1:104ft 3½ 117 11½ 11½ 12 12 Valenzuela PA5 ⒻMdn 84-17 Percipient, Mallorca, Stake Lady 5
5Sep83-6Dmr 6f :224 :462 1:113ft 17 116 86½ 85½ 56½ 46½ Sibille R10 ⒻMdn 74-17 LovebleMiss,PersonbleLdy,StkLdy 12
5Sep83—Broke slowly, bumped after start, wide into stretch
Oct 2 SA 3f ft :351 h Sep 28 SA 6f ft 1:15 h Sep 23 SA 4f ft :504 h Sep 9 Dmr 4f ft :484 h

***Lina Cavalieri**
B. f. 4, by Star Appeal—Grey Invader, by Dorring-Do
Br.—Razza Ascagnano (Eng) 1984 2 0 1 0 $3,247
Own.—Moss Mr-Mrs J S 116 Tr.—Frankel Robert 1983 9 6 1 0 $90,660
Lifetime 17 7 3 1 $98,407 Turf 17 7 3 1 $98,407
21Apr84♦5StCloud(Fra) a1 5/16 2:24 yl 26 130 Ⓣ 915 DtteriS ⒻPxCorrida(Gr3) Fly Me, Marie de Litz, Estrapade 11
25Mar84♦1Rome(Italy) a1⅛ : gd — 116 Ⓣ 21 Dettori S ⒻPrBerlino Retrousse, Lina Cavalieri, LaVerna 4
25Mar84—No time taken
30Oct83♦3Milan(Italy) a7f 1:25 sf — 129 Ⓣ 9 DttoriS PrChiusura(Gr2) Nandino, Sinio, Baran 9
16Oct83♦6Milan(Italy) a1 1:40 sf 4½ 121 Ⓣ 1½ DettoriS ⒻPr Bgtta(Gr3) Lina Cavalieri, Sedra, Soigneuse 12
2Oct83♦5Rome(Italy) a1⅛ 2:06 sf — 122 Ⓣ 26 DettoriS ⒻPr Tesio(Gr1) RightBank,LinaCavlieri,Retrousse 11
26Jun83♦4Turin(Italy) a1 1:362gd 5 121 Ⓣ 1hd DttorS ⒻPrRylMares(Gr3) LinaCvlieri,SummerReview,Delices 11
29May83♦1Milan(Italy) a1 1:454sf 9-5 117 Ⓣ 1¾ Dettori S ⒻPrVerziere Lina Cavalieri,LidaPerelli,Benguela 5
●Sep 30 SA 5f ft :59 h Sep 25 SA 6f ft 1:13 h Sep 19 Dmr 5f ft 1:01 h Sep 11 Dmr 5f ft 1:001 h

***Love Me True**
B. m. 5, by He Loves Me—Lady Lambourn, by Habitat
Br.—Kanasashi T (Eng) 1984 10 1 1 3 $30,900
Own.—Kanasashi T 113 Tr.—Doyle A T 1983 11 2 3 3 $47,825
Lifetime 23 3 4 6 $98,375 Turf 17 2 2 5 $76,850
2Sep84-7Dmr 1 1/16 Ⓣ:47 1:11 1:422fm 12 116 712 87¼ 86 86¼ DlhoussyE4 ⒻAw24000 90-04 FncyWings,ClerTht,ReflectToGlory 9
19May84-5Hol 1⅛ Ⓣ:463 1:10 1:472fm 14 115 58 58 57¾ 510½ Meza R Q4 ⒻⓈCnvnce 82-08 FnnyRogh,PrdOfRoswod,Nc'NPrpr 6
11May84-8Hol 1 1/16 Ⓣ:491 1:132 1:424fm *2¼ 117 32½ 32 2½ 62¾ Pincay LJr2 ⒻAw40000 82-15 Sedra,FactFinder,WeekendSurprise 6
14Apr84-4SA 1⅛ Ⓣ:464 1:104 1:482fm 3 115 54 64½ 53½ 21 DlhoussyE3 ⒻAw36000 84-16 SilkPyjms,LoveMeTrue,Nic'NPropr 6
31Mar84-8SA 1⅛ Ⓣ:46 1:10 1:481fm 29 114 59½ 58½ 46 34 GurrWA4 ⒻⓈSt Lca H 82-15 Nn'sDncer,AngelClre,LoveMeTrue 11
31Mar84—Bumped late
17Mar84-8SA a6½f Ⓣ:211 :432 1:143fm 8½ 115 911 810 68½ 44 McCrrCJ2 ⒻMonroviH 82-17 Tangent,IrishO'Brien,FriedaFrame 10
8Mar84-8SA 1⅛ Ⓣ:47 1:121 1:502fm 4 114 711 65½ 34 34 ShoemkrW3 ⒻAw35000 71-25 Angel Clare, Caferana,LoveMeTrue 7
8Mar84—Bumped, jostled at 1/8, steadied at 7/8
24Feb84-8SA 1⅛ Ⓣ:464 1:12 1:494fm 8½ 115 75 74 54 34½ DlhoussyE2 ⒻAw35000 74-22 Brorita, Caferana, Love Me True 9
3Feb84-8SA 1⅛ Ⓣ:473 1:121 1:501fm 2¾ 114 21 31 27 413¾ McCrrnCJ5 ⒻAw35000 62-24 BidForBucks,Cafern,RinOnMyPrde 7
5Jan84-5SA a6½f Ⓣ:211 :433 1:16 fm 15 115 1112 915 69 11½ DlhoussyE4 ⒻAw27000 79-21 Love Me True, Brorita, Upatree 11
Sep 28 SA 6f ft 1:141 h Sep 22 SA 6f ft 1:142 h Sep 16 Dmr 6f ft 1:131 h Sep 9 Dmr 6f ft 1:123 h

Centavos
Ch. f. 4, by Scout Leader—Special Key, by Key to the Mint
Br.—Beddo & Miller (NM) 1984 9 2 0 4 $54,300
Own.—Beddo & Miller 112 Tr.—Moreno Henry 1983 2 1 0 1 $2,316
Lifetime 16 6 2 5 $89,933 Turf 1 0 0 0 $2,625
31Aug84-8Dmr 6f :221 :444 1:083ft 15 113 79¾ 66 64½ 76¼ ShrW6 ⒻRch Brnrd H 88-16 PlsurCy,LovlirLind,PridOfRoswood 8
13Aug84-8Dmr 6½f :221 :452 1:16 ft 6 115 42½ 43 31 35¼ ShoemkrW5 ⒻAw35000 — — Pleasure Cay, Bara Lass, Centavos 7
3Aug84-8Dmr 6½f :22 :444 1:16 ft 12 116 73½ 44 32½ 31¾ McGurn C7 ⒻAw30000 — — Filomena Galea, BaraLass,Centavos 8
7Jly84-3Hol 6½f :221 :441 1:16 ft 12 115 1hd 1½ 1½ 31¾ McGurn C1 ⒻAw45000 88-18 MdmForbs,DontstopThmsc,Cntvos 7
3Jun84-7Hol 6f :214 :442 1:10 ft 14 114 1hd 13 13½ 1nk McGurn C2 ⒻAw26000 87-21 Centavos,PrincessLurullh,HiYuLulu 7
5Apr84-8SA a6½f Ⓣ:213 :441 1:144fm 38 119 32½ 45½ 45½ 42¾ McGurn C2 ⒻAw35000 82-16 DncngDsplIII,RnOnMPrd,WddngFlr 6
5Apr84—Lugged out, steadied at 1/4
1Mar84-5SA 6½f :214 :442 1:164ft 24 114 43½ 32½ 1½ 1no McGurn C5 ⒻAw28000 86-21 Centavos,RedRisin,PrincessLurullh 8
17Feb84-5SA 6f :22 :443 1:094ft 6 114 21 32½ 42 43½ McGurn C6 ⒻAw25000 85-17 Bonbonaire,PrincessLurullh,Bullion 9
17Feb84—Lacked room 1/8
4Feb84-3SA 6f :22 :444 1:093ft 45 114 1hd 1½ 21 33 McGurn C6 ⒻAw24000 87-14 MadamForbes,Bonbonire,Centvos 10
31Dec83-9Sun 6f :222 :452 1:104ft 8-5 107 31 2hd 2hd 2½ MartinezOAJr5 Aw3200 87-21 ‡BourbonAndBls,Cntvos,MssJblton 9
31Dec83—Placed first through disqualification
Sep 30 SA 4f ft :484 h Sep 24 SA 6f ft 1:142 h ●Sep 18 SA 4f ft :471 h Sep 11 Dmr 4f ft :48 h

```
Betty Money                      Dk. b. or br. m. 5, by Our Native—Need No Proof, by Prove It
                                 Br.—Michael Mr-Mrs R Jr (Ky)          1984  7  1  2  2    $24,450
Own.—Hunt N B              115   Tr.—Whittingham Charles               1983 12  3  1  0    $58,281
                              Lifetime  43 11  7  5  $351,585          Turf  6  1  1  0    $22,826
7Sep84-3Dmr   6½f :22  :444 1:17 ft      9¾  1105 55  57  66¾ 2nk  Lozoya DA7 ⒻAw32000 — — AwesomePromise,BttyMony,Circulr 7
    7Sep84—Wide into stretch
9Jun84-9LaD   170 :454 1:113 1:431ft     3¾  116  511 55  31  2no  Kaenel J L7 ⒻAw16500 81-17 Lady D. II, Betty Money,Juliet'sPet 8
5May84-10LaD  7f  :222 :444 1:233ft     *2¾  116  3½  3½  43  32¼  Kaenel J L5 ⒻAw14500 104-13 CoastPtrol,Juliet'sPet,BettyMoney 7
11Apr84-8OP   6f  :22  :452 1:104ft      5¾  121  43  43½ 47  33¾  Kaenel J L6 ⒻAw20000 87-16 Lila Jean, Workin Girl, BettyMoney 7
24Mar84-9OP   1   :461 1:103 1:354ft     12  114  55½ 54½ 66½ 611½ Kaenel J L3 ⒻPippin — — Heatherten, Nizhanee, Quill Castle 7
14Mar84-9OP   1   :471 1:112 1:371ft     2¼  118  33½ 33½ 35  53¼  Kaenel J L4 ⒻAw20000 — — Picture Point, Frosty Tail, Svarga 9
17Feb84-8OP   6f  :214 :45  1:10 ft     *8-5 118  64¾ 53½ 32½ 1½   Day P6      ⒻAw17500 95-09 BttyMony,PicturPoint,PurPltinum 12
5Oct83-8Bel   7f  :232 :464 1:24 ft       9  1105 2hd 3nk 56  41½  Messina R2  ⒻAw37000 80-21 I'mInTime,Chieftn'sCommnd,VivSc 6
10Sep83-6Med  1⅟₁₆Ⓣ:461 1:093 1:411fm    34  113  64½ 56½ 611 713½ HernndzR5   ⒻViolet H 83-07 Twosome, Princess Roberta,Svarga 8
    10Sep83—Run in divisions
27Aug83-5Bel  1¼Ⓣ:473 1:364 2:013fm      12  113  1hd 23  78½ 813¾ Cordero A Jr1 ⒻHcp0 72-19 Geraldine'sStore,Twosome,Mistrett 8
  Sep 28 SA 6f ft 1:132 h      Sep 14 Dmr 3f ft :38 h      Aug 29 Dmr 5f ft 1:003 h      Aug 23 Dmr 5f ft 1:01 h
```

The information about the stakes races comes from the handicapper's database. *Daily Racing Form* does not provide it. Without a stakes division database, therefore, handicappers have no way of comparing the added money activities of the horses in featured competition. They thus have no reliable means of separating stakes competitors, which are best contrasted on their stakes records, plus the fundamentals of form, early speed distance, and pace, and not on final speed, trainer, trips, or any other factor. The stakes division database must provide handicappers with the grade designations, purse values, and eligible ages. Grade designations must include "L" for listed, the newest classification in the stakes hierarchy. Listed stakes supersede open stakes in importance, but are less important than Gr. 3 stakes. The database should contain those three data items for all United States stakes and stakes of Canada, England, France, Germany, Ireland, Italy, Argentina, and Chile.

Once the database displays the stakes information, handicappers need to interpret it smartly. Here are five basic guidelines that point the way:

1. Gr. 1 races are decidedly superior to Gr. 2 races.
2. Gr. 2 races are decidedly superior to Gr. 3 races.
3. Gr. 3 and listed stakes are usually superior to open stakes that are ungraded, unlisted, but are rather indistinct among themselves.
4. Open stakes that are ungraded and unlisted are best distinguished as to relative class by purse values, but the horses should also measure up on matters of pace, distance, footing, and current form.
5. Restricted stakes are inferior to open stakes and are often won either by runners-up in the open contests or by developing three-year-olds that are moving ahead.

These elementary guidelines clear the handicapper's path to profits in featured races such as the Autumn Days Handicap, assuming they have access to the telltale information. Dismissing *Bara Lass* on current form, by these standards the probables for the Autumn Days of 1984 were two: *Lyphard's Princess*, and *Lina Cavalieri*. The others lack the proper stakes credentials. *Percipient* was a Gr. 3 winner at two, but in many cases the two-year-old record does not translate well to age three. What has *Percipient* accomplished at three? Not enough in the Gr. 3 contest (Railbird) it tackled. She was slaughtered in the Gr. 1 Hollywood Oaks.

At 7 to 5, *Lyphard's Princess* could be discounted in this non-Exacta affair.

Now look closely again at *Lina Cavalieri*.

Excepting the Berlino Stakes, its first start at four, this filly has raced exclusively in graded and listed stakes, winning the listed Verziere and two Gr. 3 events. More impressive, at three the horse placed in the Gr. 1 Tesio Stakes, finishing ahead of older stakes horses. It lost badly once at three, to males in the Gr. 2 Chiusura Stakes. In sum, *Lina Cavalieri* is a multiple graded stakes winner that has placed in Gr. 1 competition, and no filly in this field can lay comparable claims against it.

Is *Lina Cavalieri* in form?

Examine the morning workouts at Santa Anita for September 30 below. *Lina Cavalieri*'s best-of-morning workout was among only five swifter than a minute on a day of relatively slow workouts.

What about the trainer?

Robert Frankel is one of the most talented horsemen in the country, a threat with every horse he sends to the post, including any stakes imports.

What about the distance and probable pace?

Of distance, handicappers must appreciate that excessive class advantages often overcompensate for short form and unfamiliar distances. At the same time, they often do not. With imports, these considerations can become puzzling. The foreign shippers often have not raced in months and just as often are entered at distances not among their most comfortable. The issue reduces itself to relative class. Where class advantages are pronounced, they can be counted upon to nullify disadvantages on form and distance.

Fortunately, the odds are regularly delightful on foreign horses, precisely because handicappers do not trust their current form, do not accept unusual distances, and do not really understand the implications of the class factor. *Lina Cavalieri* went postward at 8 to 1 in this race, when it should have been a low-priced favorite or second choice. The result chart testifies to the power of its victory.

Foreign horses stabled in southern California that have exited graded or listed stakes overseas and now challenge open, ungraded, unlisted stakes or allowance races here, have been winning approximately twice the rightful share of their starts. They rank among the best sources of overlays for the American horseplayer. In fact, the prices are consistently outstanding, and will be for a few seasons more.

Handicappers who construct personal databases are well advised to compile the recommended data on foreign stakes races.

Sunday, September 30, 1984

SANTA ANITA PARK – Track Fast

Three Furlongs—	**:32**	
AMomentinTime	:38⅘	h
Adam Jo	:36⅘	h
Beach's Rock	:36⅘	h
Eager	:35⅘	h
Eradicate	:36⅗	h
Explosive Passer	:38⅗	h
Finnway Lady	:36⅖	h
Glaze	:38⅘	h
Jimmy Zeke	:36⅘	h
Kapi Kat	:38	h
La Jolla Pearl	:39⅘	h
Marydke Place	:37⅘	h
Mon's Souvenir	:37⅖	h
Orchestra	:36⅖	h
Pirate's Frolic	:36⅖	h
Saintly Saint	:35⅘	h
Stanley Station	:35⅘	h
Total Woman	:36⅖	h
Why Zanthe	**:35⅖**	**h**
Four Furlongs—	**:46⅘**	
Alyanna	:48⅗	h
Amapola	:48⅘	h
Angelo G.	:53	h
Ascension	:48⅕	h
Barland	:49	h
Because It'sTrue	:50	h
Bold Polly	:49⅘	h
Centavos	:48⅘	h
Circle of Steel	:48⅗	h
Clear as Crystal	:48	h
Crimson Cameo	:48	h
Debonaire Junior	:49⅘	h
Del Indio	:50	h
Double Deficit	:47⅗	hg
Ed's Bold Lady	:49	h
Fastic	:49⅘	h
Fluke	:48⅗	h
Frivolous Nature	:50⅕	h
Hard Number	:50⅖	h
High Natural	:48	hg
In Natural Form	:48⅕	h
Johnotable	:51	b
Jojohnick	:49	h
Message toGarcia	**:47**	**h**
Mind Storm	:50⅕	h
Mr. Sensational	:47⅗	hg
Night Mover	:48⅖	h
Noble Dex	:52⅖	hg
Not the Regular	:50	h
OneO'ClockJump	:48⅘	h
Palestine Sun	**:47**	**h**
Pleasure Dome	:48	h
Post Flag	:48⅗	h
Pro Bowler	:48⅗	h
Ready for Luck	:47⅘	h
Romantic Roman	:49⅖	h
Secret Eagle	:50	h
Secret Fling	:50⅖	h
Spring Bid	:48⅗	h
Stand Pat	:48⅕	h
Total Departure	:48⅖	h
Woodland Way	:49⅘	h
Five Furlongs—	**:58⅗**	
Aglint	1:03⅖	h
Alola*	1:00	h
Ancient Rites	1:01⅖	h
Baby Grace	1:00	h
Blazing Irish	1:05	h
BothEndsBurnng	1:03	h
Brookings	1:01⅗	h
Bunch A Bunk	1:01	h
Captain A. R.	1:01⅘	h
Cold	:59⅗	h
Danzadar	1:00⅖	hg
Dare You II	1:00⅕	h
Duchess Petrone	1:02	h
Dynamo Doc	1:01⅘	h
Encargada	1:01⅘	h
Eterno	1:00⅕	h
Fighting Fit	1:00	b
FoolishIntntions	1:00⅕	h
Free Majesty	1:02⅘	h
Hot Princess	1:00⅘	h
I'm Sizzling	1:00⅘	h
Irish O'Brien	1:01⅗	h
Iron Leader	1:01⅘	h
Key to the Arc	1:01	h
Kilima Point	1:02⅗	h
Killora	1:01⅕	h
LaPaginaSportiv	1:02⅘	h
Lady Universal	1:01	h
Let's Get Raced	1:01	h
Lina Cavalieri	**:59**	**h**
Lucid Moments	:59⅘	h
Mark's Cove	1:03⅘	h
Morgan D.	1:00⅘	h
Naskarion	1:00	h
Order Control	1:00⅗	hg
Pax in Bello	1:01⅖	h
Phi Beta Kappa	1:00⅘	h
Pinball	1:04⅖	h
Pirate'sSerenade	1:03⅖	h
Pocasset	1:01⅗	h
Pretense d'Or	1:03⅘	h
Preysayetera	1:01⅘	h
Propertius	1:02⅗	h
Re Ack	1:01	h
Saint Cadvan	1:00⅗	hg
Saratoga Six	1:00⅕	h
Sari's Delight	:59⅖	h
Scalper's Price	1:02⅘	h
She's A Honey	1:02⅘	h
Simi Dancer	1:02	h
Small Habit	1:01⅗	h
SpecialKindaGuy	1:01⅖	h
Speed Spy	1:01⅘	h
Stickette	1:02⅗	h
Svetlana Epris	1:03⅗	h
Swivel	1:01	h
Sword Prince	1:01⅗	h
Take Petra	1:03⅕	h
Tank's Prospect	1:01⅖	h
The Babe	1:01⅖	h
TheRunningGrek	1:01⅖	h
Vital Force	1:01⅘	h
WtchYourMoney	1:03⅗	h
Wildglen Driver	1:02⅗	h
Woodcote	1:01⅕	h
World Ruler	1:01⅖	h
Zaheender	1:03	h
Six Furlongs—	**1:07⅗**	
Ackermann	1:14	h
Andrew 'n Me	1:13⅕	h
Bean Bag	1:12⅕	h
Byron	1:11⅘	h
Cachuma	1:17⅗	hg
Cupid Dancer	1:11⅘	h
Devastating Miss	1:16⅕	h
Durable	1:13⅕	hg
Dynamite	1:13	h
End Display	1:14⅕	h
Feasibility Study	1:12⅗	h
Flack	1:15⅗	h
Isan Emperor	1:17⅗	hg
Jet Royal	1:14⅗	h
Kimstep	1:17⅗	h
King's Waltz	1:14⅗	hg
Kingsider	**1:11⅗**	**h**
Kutati King	1:15⅖	hg
Lord Ben	1:16⅕	hg
Lord Protector	1:14⅗	h
Mountain Bear	1:14⅘	h
Neal's Style	1:14	h
Pol and Dic	1:17	h
Premier Coup	1:13⅘	h
Robersky	1:13⅕	h
Rock Softly	1:15	h
Rough Flight	1:14	h
Savannah Dancer	1:16	h
Shy Bride	1:13⅘	h
Snow Motion	1:13⅗	h
SpectacularLady	1:16	h
Star Pirate	1:16	h
Steelinctive	1:14	h
Summer Creek	1:14	h
Super Starlz	1:14	h
T. H. Lark	1:13⅖	h
Tammy Lu	1:17⅗	hg
Trakady	1:14⅗	h
Valais	1:12⅖	h
Veiled Sands	1:14⅗	hg
Vivian's Jade B.	1:13⅖	h
Seven Furlongs—	**1:20**	
Dandy Dispute	1:31⅗	h
Easy Easy	1:26⅗	h
Evening M'lord	1:27⅘	h
Full O Wisdom	1:28⅖	h
Groszewski	1:28	h
Klystron	1:28⅖	hg
Leanita	1:27	h
LoadtheCannons	1:26⅗	h
Michadilla	1:27⅗	h
Nuclear	1:26⅗	h
Prando	1:27	h
Serbian Princess	1:30⅗	h
She's Got Style	1:31	h
The Sali of Swat	1:28⅗	h
Valiant George	1:27⅗	hg
1 Mile—	**1:33⅗**	
FbulousSelection	1:42⅗	h
Item Two	1:44	h
Polly La Femme	1:42⅖	h
Ridgeline	1:44⅕	h
Watch Word	**1:39⅘**	**h**

When horses coming out of listed stakes in Europe, for example, land in classified and nonwinners allowance races here, alerted handicappers find themselves looking these proverbial gift horses in the mouth, and should not be caught napping.

Getting back to the Autumn Days, a $60 win wager on the Italian-based *Lina Cavalieri* yielded an even $480. An Exacta combining that graded stakes horse with the horse bred so nicely for turf, *Betty Money*, trained by the great Charlie Whittingham, no less, would have been sweeter still. Oak Tree mysteriously offered Exacta wagering on its Saturday feature opening week (full field), but on no other feature. No chance for an exotic score.

As a postscript, the two other sources of information needed to unravel this featured race knowingly, breeding I.V.s for grass and workouts, could also be trapped conveniently in the handicappers' database. In the age of information, to repeat once again for the emphasis it so deserves, no considerations are more central to successful handicapping than effective information management and the computerized relational databases that make that feasible.

EIGHTH RACE
Santa Anita
OCTOBER 3, 1984

ABOUT 6 ½ FURLONGS.(Turf). (1.11⅘) 16th Running of THE AUTUMN DAYS HANDICAP (2nd Division). $45,000 added. Fillies and mares of all ages. By subscription of $50 each, $600 additional to start, with $45,000 added, of which $9,000 to second, $6,750 to third, $3,375 to fourth and $1,125 to fifth. Weights Friday, September 28. Starters to be named through the entry box by the closing time of entries. A trophy will be presented to the owner of the winner.

Closed Wednesday, September 26, with 34 nominations.
Value of race $50,650; value to winner $30,400; second $9,000; third $6,750; fourth $3,375; fifth $1,125. Mutuel pool $510,783.

Last Raced	Horse	Eqt.A.Wt	PP	St	¼	½	Str	Fin	Jockey	Odds $1
21Apr84 ^{5}Fra^{9}	Lina Cavalieri	4 116	5	8	7^{3}	$7^{2½}$	5^{1}	1^{3}	Delahoussaye E	8.00
7Sep84 ^{3}Dmr^{2}	Betty Money	5 115	8	1	6^{2}	6^{2}	6^{3}	2^{hd}	Shoemaker W	8.40
24Aug84 ^{8}Dmr^{1}	Percipient	3 116	4	4	3^{1}	$2^{1½}$	$1^{½}$	$3^{1½}$	Valenzuela P A	3.20
7Sep84 ^{3}Dmr^{1}	Awesome Promise	5 113	1	2	2^{hd}	5^{3}	$3^{½}$	$4^{½}$	Lipham T	40.70
2Sep84 ^{7}Dmr^{8}	Love Me True	5 117	6	5	8	8	$7^{½}$	$5^{½}$	Pincay L Jr	8.20
9Sep84 ^{8}Dmr^{4}	Lyphard's Princess	4 117	2	6	$5^{2½}$	4^{hd}	4^{1}	$6^{1½}$	Toro F	1.40
31Aug84 ^{8}Dmr^{4}	Bara Lass	5 119	3	3	$1^{½}$	1^{hd}	2^{1}	$7^{5½}$	McCarron C J	5.40
31Aug84 ^{8}Dmr^{7}	Centavos	4 112	7	7	4^{hd}	$3^{½}$	8	8	Ortega L E	24.40

OFF AT 4:58. Start good. Won driving. Time, :22⅘, :45⅘, 1:08⅕, 1:14⅘ Course firm.

$2 Mutuel Prices:			
5–LINA CAVALIERI	18.00	9.00	5.80
9–BETTY MONEY		8.60	4.60
4–PERCIPIENT			4.60

B. f, by Star Appeal—Grey Invader, by Derring–Do. Trainer Frankel Robert. Bred by Razza Ascagnano (Eng).

LINA CAVALIERI, not away alertly, found her best stride crossing the dirt path, found room up the rail to make her bid and blew past the leaders to win going away. BETTY MONEY split horses and closed willingly. PERCIPIENT set or forced the issue from the outset and hung late. AWESOME PROMISE raced in tight quarters much of the trip along the rail, came out for the drive and lacked the needed rally while lugging out. LOVE ME TRUE was no threat. LYPHARD'S PRINCESS was well placed throughout and hung while being floated out near the end. BARA LASS had nothing left for the final furlong. CENTAVOS was not away alertly, rushed up and tired. LAOIS PRINCESS WAS A STAKES SCRATCH PRIOR TO THE START OF THE DAY'S WAGERING.

Owners— 1, Moss Mr-Mrs J S; 2, Hunt N B; 3, Hibbert R E; 4, Hughes B W; 5, Kanasashi T; 6, Northwest Farms; 7, Stevens S E; 8, Beddo & Miller.

Trainers— 1, Frankel Robert; 2, Whittingham Charles; 3, Manzi Joseph; 4, Vogel George; 5, Doyle A T; 6, Fanning Jerry; 7, Lukas D Wayne; 8, Moreno Henry.

Overweight: Love Me True 4 pounds.

Scratched—Circular (7Sep84 ^{3}Dmr^{3}); Laois Princess (17Aug84 ^{5}Dmr^{4}).

$2 Pick Six (6–7–3–11–2–5) Paid $62,632.00 for 6 Wins, 2 Tickets. 5 Wins Paid $813.40; 154 Tickets. Pool $313,474.

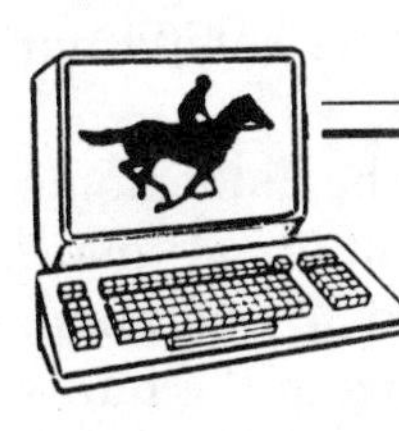

Chapter 14

THE MIS LIFE CYCLE

Which comes first—the computer or the computer program? Neither!

JAMES MARTIN

HANDICAPPERS WHO BECOME information managers do not accomplish the transformation overnight. The change occurs gradually; it takes two years at least. But it lasts a generation. This chapter describes the changeover as a series of four steps. It rests on the proposition that computerized information systems have a life cycle consisting of four major phases. Unlike other life cycles, however, the information system is neverending.

Personal computers and databases supply the management information that is needed to solve handicapping problems and make final decisions. Handicappers are viewed as managers of the information supply and the system that encodes, stores, processes, retrieves, and displays it. As managers of information, handicappers identify information needs, convert needs to objectives, organize resources, supervise implementation of the information flow, solve problems, and make decisions. Managers exercise control. The control concept is central. It prevents handicappers from becoming the handmaidens of computers, or worse, of computer salespersons.

As we have seen, the system that handicappers control is referred to as a management information system, a managerial strategy now used—and misused—extensively at the corporate level, but no less meaningful conceptually at the individual level. The popular acronym is MIS.

Handicappers should keep in mind two key points: (1) MIS development occurs in steps, and (2) managers and users of the information must work collaboratively with information specialists and computer technicians. In practice, the collaboration usually involves computer programmers and database design technicians. This is a cooperative adventure, integrating the logical representation of data and the physical representation of data. Handicappers should not intend to proceed alone.

To rely on academic terminology that should not be bothersome to handicappers, the four phases of MIS development are (1) planning, (2) analysis and design, (3) implementation, and (4) operation. We conclude our treatment of high-tech handicapping in the information age with pointed, practical guidelines for completing each phase. To stress a practical matter that dramatizes the importance of the conceptual and logical aspects of MIS development, the hardware and software *should not be purchased* until phase three, of implementation, after the brain work has been done. This consideration is so important that I wish to take it a step forward by suggesting that handicappers who already own personal computers suspend belief in those machines while engaging the chapter, and decide later whether the computer system they now own actually satisfies their information needs. It's a deadly mistake to sacrifice real information needs to the constraints of existing hardware. New computers can be bought to replace obsolete systems, and should be.

Before proceeding to the planning phase, let's examine the components of the physical system that should emerge. These include hardware, software, and noncomputerized information resources, even people.

THE PHYSICAL SYSTEM

The handicapper's MIS has six major components. All are integral.

1. *Hardware*, or a computer configuration, consisting in all likelihood of a microcomputer; secondary storage media, especially disks; and input-output devices, notably a keyboard, terminal, and printer. The computer network should also include a datacom (data communications) modem for dialing remote databases and communicating with other computers.
2. *Software*, or the programmed instructions that direct the computer's internal operations. High-tech handicappers will

collect two types of computer programs: applications software and systems software.

Applications software refers to the conventional programs that direct computers toward the stepwise completion of predefined processing tasks, perhaps a speed handicapping application that produces speed figures or a class handicapping application that produces class ratings. Most commercial programs on handicapping now produce selections or rankings, and can be distinguished from applications that provide information. This book favors applications programs that provide information and discounts applications that produce selections or rankings. High-tech handicappers should do their own handicapping. Computers are handicapping tools, not handicappers.

Word processing and spread sheets are highly visible commercial applications for the general market, and handicappers have been advised to obtain a spreadsheet application.

Systems software programs usually are embedded in microcomputers by manufacturers and control the central workings of the computer. Until now, an exception has been database management systems (DBMSs), software packages that have come from vendors and act as interfaces between operation systems, programmers, applications programs, query languages, users, and the actual database in physical storage. This book insists on relational DBMSs; no substitutes. The database management system will be the most important and most expensive package of software high-tech handicappers will buy. They should rely on professional expertise and guidance on the matter, accepting either the recommendations in the book or friendly professional consultation about new and better products that will absolutely flood the market soon enough. So indispensable are relational database management systems to achieving the characteristics of data independence and flexibility that manufacturers of personal computers can be expected to make them standard equipment on microcomputers soon, even as operating systems are standard now.

3. *Databases*, or the data resource. Computerized databases are nothing more than electronic file-keeping systems, but their

data management and processing power can be fantastic. In the ideal database world, where all handicapping data have been stored in relational databases and the key data relationships identified, no question would ever go unanswered. High-tech handicappers will be striving toward that ideal from now on.

As described in this book, computerized databases consist of tables, tables, tables, and tables: columns and rows of data items and data relationships. Database technology allows handicappers to manipulate the columns and rows in innumerable ways, either to process the instructions of applications programs or to respond to the spontaneous queries of users.

Databases can be noncomputerized, of course, or computerized but nonrelational, but neither situation really advances the cause. Manual or file structured databases are ultimately too inefficient and inflexible to facilitate information management. The consumption of time is inexcusable. The processing tasks become unbearable. Handicappers become frustrated. Many quit using the databases altogether. The more things change, the more they remain the same.

The handicapper's database ultimately includes all the data items and relationships that describe the handicapping process. But it begins in small manageable ways. High-tech handicappers are advised to set up a few tables of high interest and begin.

4. *Human resources,* or a network of handicapping associates and acquaintances that contributes various types of information. Some handicappers will own microcomputers; others will not. Human resources might supply the information otherwise provided by applications programs that handicappers do not yet have. Friends and acquaintances also might have access to other databases. Several handicappers in a network might develop computerized databases that vary in subjects and data item relations, and share the information resources among themselves.

   Human resources that are not high-tech types might be devoted trip handicappers with daily notes or body language experts who can distinguish sweating horses that are frightened from sweating horses that are sharp. Others might be expert speed handicappers who can identify the top figure horses on a card. Another might have compiled in-depth trainer statistics and act as the local authority on trainer patterns.

The point is that a management information system involves several handicapping specialists who share their expertise with the group. Informal links to racetrack and handicapping information sources become more formalized. A subsystem of people that facilitates a *quid pro quo* handicapping environment develops nicely, and it works.

Other essential human resources are computer programmers and systems design specialists, preferably people who are also handicappers or racegoers. The technicians are vital to MIS development. They convert identified information needs into alternative system designs, transform logical handicapping processes into applications programs, and translate a logical representation of the database into the physical representation of the data. They also instruct on hardware and software acquisition, and take the lead in the implementation, operation, and revision of the system. They are indispensable information management resources and should be courted carefully from the outset. Fortunately, many computer technicians who play the races are out there and just as eager to become part and parcel of a high-tech handicapping information system.

By means of collaboration and participation in the handicapping MIS, as suppliers of specialized information, handicappers who do not own microcomputers and do not have technical skills with computers can participate fully in the information system. In that way they earn access to the benefits of the microcomputer and of electronic databases.

5. *Off-line sources of information*, or the unscheduled observations of races and horses, as well as the everyday experiences trackside, including informal spontaneous contacts with insiders and other racegoers. High-tech handicappers are usually on their own at the racetrack, outside of the confines of the computer world and apart from other members of the information team. They are off-line. Yet information continually pours in. The major sources are trip observations, bias notes, body language inspections, odds board calculations,

and *ad hoc* conversations with friends and associates. Of the latter, conversations, handicappers should prefer the same output they expect from their personal computers, not selections or opinions, but information. All of this vital off-line information must be integrated with information obtained from the on-line system.

Much of the empirical information collected on the spot at racetracks will eventually be transmitted to the computerized database and the MIS, certainly trip and bias notes, probably body language, and possibly tote trends. At the track, however, sources of information will be empirical and soft, their utility dependent greatly on the savvy, alertness, and carefulness of the individual handicapper. High-tech handicappers who become information managers will be tuned in constantly to the incessant dribble of information at the racetrack.

By the way, on this point systematic method handicappers who arrive at the races with firmly held selections will not be as mentally prepared to take advantage of the information sources they find at racetracks as will handicappers who arrive with informational analyses, various outcome scenarios, and alternative decisions, but no final decisions.

6. *Information-oriented handicappers,* or individuals who prefer the information management approach to the exclusive reliance on systematic methods. In the motif of the times, such handicappers will proceed most effectively by applying everything they know to every race they analyze.

If *all* the components of the MIS are not present and in fine working order, the effectiveness of the information management approach will be compromised. That means the approach and handicappers who adopt it will be best serviced at the racetrack and not as well at simulcasting sites, inter-track wagering sites, off-site betting parlors, race books, or at home, betting by telephone. High-tech handicapping information systems must not be confused with data communications systems or simulcasting.

High-tech handicappers belong at the racetrack. The information displayed by microcomputers and television is not enough.

The work toward the development and implementation of the handicapper's MIS begins shortly after handicappers have made a conscious decision to follow the high-tech, information management directions. The first phase is the most crucial: the planning phase.

THE PLANNING PHASE

The planning phase is concerned with identifying the handicapper's information needs. The phase lasts approximately sixty days. Planning takes time. No computers or software should be purchased, or even considered as commercial products, at this point. Phase 1 is strictly a conceptual, logical planning process.

Assessing information needs is not overly complicated, but it does require a reasonable degree of mental clarity. A potential stumbling block is ignorance regarding the range of handicapping information that now exists. On that point handicappers might revisit Chapter 2 of this book and supplement that with other readings that have so far lagged. The annotated bibliography following this chapter refers handicappers to the books or articles that elaborate on the information summarized in the second chapter. Such review will assist handicappers in completing the first steps of the needs assessment model presented in Figure 19. I call it the Is-Should model, and handicappers are recommended to complete its assessment procedures carefully, step by step.

The Is-Should label denotes a discrepancy model whereby handicappers identify the gaps between the handicapping information they currently use (Is) and the information they would prefer to use (Should). Where no gaps exist, handicappers suffer from no discrepancies between actual information use and preferred information use. Such handicappers have no identified information needs. For those identifying gaps that are stifling their play, the planning purpose is to convert identified information needs into information objectives. The objectives specify new types of information to be collected and used.

The information objectives are next prioritized, according to personal values and tastes, and in recognition of the practical constraints hounding the development of personal information systems. Few of us can manage all of the handicapping information at once. As a practical matter and helpful hint, personal computerized databases should be relatively small and manageable at the beginning. They should grow larger and more complex over time.

Figure 19
Information Needs Assessment: An Is-Should Model

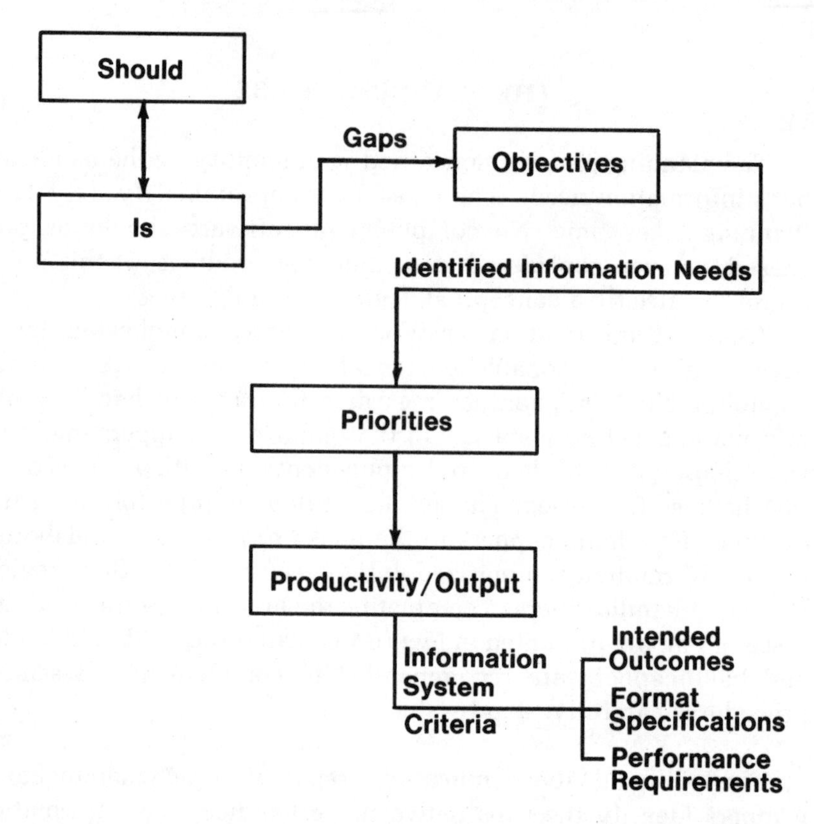

The principle of the planning process is to move from relatively broad definitions of information needs and objectives to relatively specific statements of the intended outcomes to be achieved by the information system. The final planning step converts information priorities into system output statements. These should be stated in written form. What are the intended outcomes of developing an information system? Speed figures? Class ratings? Trip notes? Bias information? Pace ratings? Stakes information? Trainer data?

List all the outputs. Be specific and descriptive about the desired characteristics of the information you need. What should the format of each output be? What should the information look like? How should it be displayed? Used?

Given the desired information outputs and formats, what should be the performance criteria of the information system that produces the information? What kinds of output devices must the system contain? What must be its processing capabilities? What are the input requirements? What are the storage requirements?

Another prime principle of effective planning is to involve in the planning process all the people who will be seriously impacted by the resulting plans. To handicappers who will develop individually tailored information systems and computer work stations, the principle does not apply. If human communications networks will be part of the system, however, all members of the network should participate in the planning. The point is so basic that no meaningful planning should proceed until the network membership has been identified and planning conferences conducted. To skip this participatory step or to postpone it until first planning decisions have already been made is to open doors to the interpersonal and human relations problems that typically plague information systems in the organization. By and large information managers do not like to implement plans that do not reflect their personal points of view, let alone engage in operations imposed upon them. Handicappers can be expected to feel the same way on the matter of participating in computerized information systems and handicapping communications networks.

The planning process is also cyclical. When personal values shift or the information base of handicapping gets broadened anew, planners can return to the Is-Should comparisons and renew the planning process in its entirety.

Planning is the most important phase of the MIS life cycle. It's a precondition for effective system design and implementation. The total system swings on the completeness and clarity of the plans. Where plans have been incomplete or unclear, all else will be problematic and ineffectual in various ways from the outset. That point holds for individuals as well as teams.

ANALYSIS AND DESIGN

The analysis phase of MIS development is intended to identify the sytems design best suited to satisfying the information needs and productivity outputs identified during the planning phase. As with planning, analysis is strictly a logical exercise. Though handicappers should be expected to study the dizzying array of hardware

Figure 20
Analysis: Identifying alternative system designs and choosing the most cost effective ones.

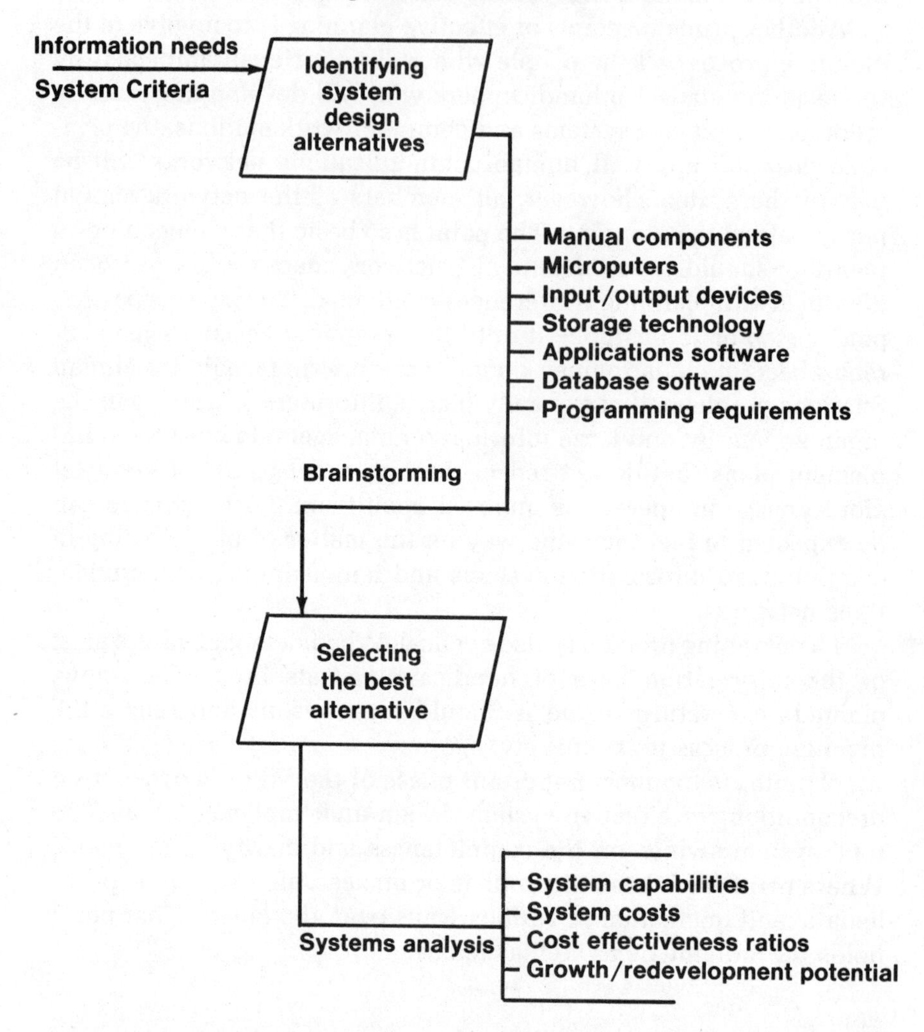

flooding the market, no computers are purchased. Though in-depth reviews of the available applications programs occur during this phase, software is neither purchased nor developed. Analysis lasts from ninety days to six months, perhaps longer. Shortcuts are not advised.

The analysis phase moves from identifying several alternative system designs to selecting the most cost-effective design for the stated purposes. The two analysis steps are outlined in Figure 20.

* * *

Analysis begins with a clear understanding of the identified information needs and proceeds with a series of freewheeling brainstorming sessions. What design components—manual, computerized, human—in the marketplace can satisfy the system performance criteria already identified (planning)? During brainstorming, cost constraints are not emphasized. Performance standards are taken into consideration, of course, but standards of excellence, so-called state-of-the-art system components, are discounted as well. The idea is to identify a number of system design configurations that simply satisfy the performance criteria. The brainstorming is greatly open-ended; it is nonjudgmental.

A first question regards which system components will be manual and which will be computerized. This discussion assumes a limited use of microcomputers at least. As Figure 20 indicates, a computerized system embraces not only computers but also input/output devices, storage technology, applications software, database software, and any new programming requirements. New programming, of course, involves collaboration with computer programmers who might perform the tasks. To repeat, at this point costs remain secondary to performance capability.

To conduct the analysis stage, handicappers, notably those with scarce computer background or experience, proceed best by consulting specialty magazines, vendors, and technical resource people who are friends and associates. Discussions can be wide-ranging, but should be focused at the start on the handicapper's identified information needs and the corresponding performance criteria of the information system. A number of alternative system designs, at least three, should be developed before proceeding to the selection step of the analysis phase. Each alternative system should be described in written form and portrayed in graphic form.

To select a best system design, high-tech handicappers relate system capabilities to costs and choose the design having the most impressive cost-effectiveness ratio. Collaboration with design and programming specialists continues during the selection process. It is important to keep in mind that the system design selected should be capable of accommodating the future growth and development of the handicapping information system. The original database will expand. Database management systems for microcomputers will proliferate. Numerous applications programs will be developed through the seasons. Data communications networks, from computer to computer, and from the home computer to regional and national databases, represent a futuristic certainty.

In conducting a system design analysis, a helpful hint is to ask vendors or computer scientists to identify the state-of-the-art components of computerized information systems, and next ask them to identify other design components that can duplicate or approximate the state-of-the-art. In turn, the recommended components can be evaluated by handicappers in terms of their personal identified information needs.

IMPLEMENTATION

During implementation the information system is purchased, installed, and tested. Bugs are detected and corrected, especially with the software. The physical database is constructed and data are input.

Implementation lasts another three to six months, mainly so that adequate testing can be conducted and the database constructed. As both will be needed to implement the system, hardware and software can be purchased concurrently. Purchasing contracts should specify provisions for services needed from vendors due to problems identified during installation and testing.

All components of the information system should be thoroughly tested during implementation. Testing should never, never be overlooked. The testing also tests the users (handicappers), who may require extensive hands-on training and practice with the system, especially with the system software. Database management systems will be especially tricky for users to implement.

Implementation also embraces the coordination of the human resources who will participate in the information system. Who will do what, how, and when?

A final implementation period of thirty days simulates the actual operation of the information system during live racing days. Procedural flaws will be identified and corrected at this time. In the end, the information system runs smoothly; the information flows as expected.

OPERATION

Technically, the operation phase is merely a continuation of the implementation phase. Presumably the information system continues to run smoothly; the information continues to flow as desired. Beyond normal operating procedure, however, operation implies evaluation. The operative question really asks: Does the system work as intended? That is, do handicappers obtain the information they need? Has problem-solving in handicapping been enhanced? Has decision-making been improved? Do the system managers (handicappers) identify more winners? Do they earn greater profits? If the answers prove negative, the system has proved inadequate and begs to be improved. The MIS life cycle is therefore repeated, perhaps indefinitely, until matters improve, the significant management information is produced, and the expected results have been achieved.

A final consideration. Operations must also accommodate the growth and redevelopment of the system. To be sure, more information about handicapping will be produced. Handicappers will subsequently desire to add applications programs and to enlarge the databases. New technology will also emerge. Flexible, expandable physical systems must be accompanied by adaptable, growth-oriented users. High-tech handicapping in the information age should not be viewed just as a specialized practice but also as a developmental process. The MIS life cycle of planning, analysis, implementation, and operation will be repeated several times as the information requirements of modern handicapping change, and high-tech handicappers discover they have vitally new information needs.

THE EDUCATIONAL REQUIREMENTS

Beyond the considerable changes and commitments this book already has requested of handicappers, its author also wants to send the interested readers back to school. A formal education consisting of three courses will graduate systematic method handicappers into high-tech handicappers who will become information managers. The education will confer the added benefit of preparing its consumers to enter the age of information in the home and broaden society with a long head start on the crowd.

The three courses will be found at most university extension schools, community colleges, and continuing-education programs. They should be taken in this sequence: (1) introduction to management information systems, (2) principles of database management, and (3) hands-on experience with microcomoputers. A fourth course, an elective, if you will, would offer hands-on experience with database management systems (software) for microcomputers. If the course work conincided with the planning, analysis, and implementation phases of the MIS life cycle, how convenient!

The complete education takes approximately a year and costs a few hundred dollars. But it lasts longer than a generation. It lasts throughout the age of information.

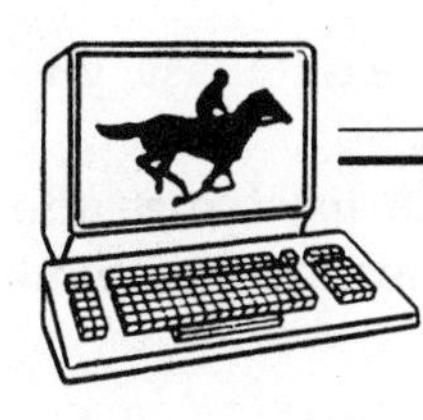

GLOSSARY OF HIGH-TECH TERMS

Applications Program A list of instructions that guides the computer to the solution of a problem or the completion of a task.

Basic A programming language for beginners; an acronym meaning *Beginner's All-purpose Symbolic Instruction Code.*

Binary Digit An electronic element of data having two states—on and off.

Bit An electronic element; it can be either *off* or *on,* its binary states; a contraction for binary digit.

Byte A group of bits representing a character.

Central Processing Unit (CPU) The transformation and control part of a computer; main memory.

Character A combination of on and off bits that are coded for computer storage.

Compatibility Correct communication between an operating system and applications software.

Computer Configuration A physical arrangement of computers, input/output devices, and secondary storage media.

Computer Schematic The basic architecture of a computer.

Concatented Key A key composed of two data items or fields.

Database An electronic record-keeping file.

Database Management System A software system to manage a database.

Database Software Computer program that interfaces with a database; DBMSs, query languages, communications.

Datacom A system of data communications that extends beyond the computer environment.

Datacom Networks A system by which a CPU and remote terminals communicate by means of modems and channels.

Data Description Language (DDL) A language for describing data items.

Data Dictionary A catalogue of all data items in a database, giving names, description, and location.

Data Independence The property of being able to change the overall logical or physical structure of the database without changing the programmer's view of the data.

Data Item The smallest unit of data that has meaning in describing information; synonymous with field.

Data Manipulation Language (DML) A language used by programmers to cause data to be transferred between applications programs and the database.

Decision Support System (DSS) An information-producing system aimed at a particular decision a manager must make.

Direct Access Storage An access mechanism that can move directly to a record stored on a disk.

Direct Access Storage Device A disk.

Disk Data recording tracks arranged in a vertical cylinder.

Distributed Data Processing The condition of dispersing computers, terminals, and data management responsibility throughout the organization.

Domain The columns in a database; a collection of data items of the same type.

Field The location of a data item in physical storage.

File A group of records on a subject.

Flat File A two-dimensional array of data items in rows and columns.

Floppy Disk A soft data recording cylinder of relatively low storage capacity, similar to 45-rpm records; diskette.

Hard Disk A hard data recording cylinder of relatively large storage capacity, similar to a stack of records.

Heuristic A principle of problem-solving that implies trial-and-error methods of finding solutions.

Index A table used to determine the location of a record in physical storage.

Input Devices Units employed to enter data into a computer; keyboards and voice-recognition units.

Key A data item used to identify or locate a record.

Kilobyte One thousand bytes.

Magnetic Tape The medium of sequential secondary storage; excellent medium of historical storage.

Management Information System (MIS) An organized method of providing information to decision-makers for purposes of planning, problem-solving, and decision-making.

Microcomputer An integrated electronic information processing system, consisting of a microprocessor, storage, and input/output units.

Modem A communications device that converts a digital signal to an analog telephone signal and vice versa; it means modulator-demodulator.

MS/DOS A high-level operating system developed by IBM.

Off-Line Not connected to the CPU of a computer.

On-Line Connected to the CPU of a computer.

Operating Systems Software that controls and directs the internal workings of a computer.

Output Devices Units employed to display processed data; terminal screens and printers.

Peripherals Input/output devices and secondary storage media.

Primary Key A data item that uniquely identifies a record.

Primary Storage The storage part of the CPU.

Query Languages Languages that communicate the *ad hoc* requests of computer users to a database.

Record A group of data items on a subject.

Relation A flat file; a two-dimensional array of data items.

Relational Database A database consisting of tables.

Schema The logical structure of a database.

Secondary Key A data item having a value that is not a unique identifier of a record.

Secondary Storage Storage units outside of the CPU; the main media are disks and tapes.

Semiconductor Chips The logic and arithmetic circuitry of a microcomputer.

Sequential Secondary Storage Data records arranged in sequence on a magnetic tape.

Software The array of computer instructions.

Subschema A programmer's view (application program) of a database.

UNIX A high-level operating system.

Word A number of bytes processed as a unit; several bytes.

BIBLIOGRAPHY

Ainslie, Tom. *Ainslie's Complete Guide to Thoroughbred Racing.* New York: Simon and Schuster, 1968, 1979. The classical concepts, principles, and practices of handicapping; comprehensive handicapping; the economics of racing.

———. *Ainslie's Encyclopedia of Thoroughbred Handicapping.* New York: William Morrow and Company, Inc., 1980. An alphabetical discussion of all the major concepts and topics of handicapping; descriptions of various methods.

Allswang, John M. "Expanding the capabilities of dBase II." *Interface Age,* April 1984. A basic discussion of the first database management system for microcomputers.

Beyer, Andrew. *Picking Winners.* Boston, Massachusetts: Houghton Mifflin Company, 1975. The definitive treatment of speed handicapping, projected times, and speed charts based on the concept of proportional time.

———. *The Winning Horseplayer.* Boston, Massachusetts: Houghton Mifflin Company, 1983. A discussion of trip handicapping; speed and trip handicapping in combination; modern money-management guidelines; trip notation.

Bjork, Robert. "The Role of Human Memory in the Age of Information." A technical paper delivered at the New England Telephone Seminars, October 5–6, 1983. Mnemonics; short-term and long-term memory.

Davidowitz, Steven. *Betting Thoroughbreds.* New York: E. P. Dutton, 1977. Trainer patterns; track biases; key races; speed and pace discussion; basic handicapping.

Davis, Fred. *Thoroughbred Racing: Percentages and Probabilities.* New York: Millwood Publications, 1974. The winning percentages and probabilities of numerous handicapping characteristics; the concept of impact values (I.V.s); computational procedures for establishing the handicapper's morning line.

Gabel, David. "Starting from scratch with a data base." *Personal Computing,* March 1984. The basic purposes, design features, and developmental steps of building computerized databases.

Heck, Mike. "A data base for experts." *Interface Age,* April 1984. A discussion of Knowledge Man, a database management system.

Jones, Gordon. *Smart Money.* Huntington Beach, California: Karman Communications, 1977. Combination betting; how to play the Daily Double and the Exacta; rationale for exotic wagering.

Lawlor, Greg. "Santa Anita trainer computer study, 1978–83." *JGL Enterprises,* San Diego, California, 1984. Trainer performance data and patterns of strengths and weaknesses for forty leading trainers at Santa Anita Park.

Langone, John. "They save horses, don't they?" *Discover,* May 1984. Advances in equine veterinary medicine; surgical procedures.

Ledbetter, Bonnie, and Ainslie, Tom. *The Body Language of Horses.* New York: William Morrow and Company, Inc., 1980. Equine body language; six profiles of body language and behavior at the racetrack; handicapping guidelines.

Mahl, Huey. "Money Management." A technical paper presented at Sports Tyme Handicapping Seminar, Dunes Hotel and Country Cub, Las Vegas, Nevada, December 12–13, 1979. Explanation of the Kelly criterion; application of the Kelly method to pari-mutuel wagering.

Martin, James. *Principles of Data-Base Management.* Englewood Cliffs, New Jersey: Prentice-Hall, 1976. An in-depth discussion of the concepts, principles, design features, applications, and functions of the database in the large organization by the leading expert in the field; relational databases; databases and management.

McLeod, Raymond. *Management Information Systems.* Science Research Associates, a subsidiary of IBM, Chicago, 1983. Information management; management information systems; the computer; the MIS life cycle.

Meyer, John. "The T.I.S. Pace Report." *The National Railbird Review,* Volume II, Numbers 8 and 9, San Clemente, California, 1981. A comparative study of four approaches to pace analysis; pace ratings.

Mitchell, Dick. *A Winning Thoroughbred Strategy.* Los Angeles, California: Cynthia Publishing Company, 1985. The concept of mathematical expectation; how to construct a handicapper's morning line, based on probabilities; money management methods; computer applications programs in coded language.

Mott, Roger E. "Horse Race Handicapping Made Easy . . . Maybe." *Software Supermarket,* May 1984. Evaluation of computer programs for making handicapping selections; comparison of five applications programs.

——— "Project Database." *PC Magazine,* August 21, 1984. Review of high-level database management systems; comparison of leading DBMSs on forty-eight dimensions.

Quinn, James. *The Handicapper's Condition Book.* Las Vegas, Nevada: GBC Press, 1981. Analyses of all major racing conditions; the class factor; elimination and selection profiles.

———. *The Best of Thoroughbred Handicapping.* New York: Casino Press, 1984. Essays on major ideas and methods in handicapping from 1965 through 1984.

Quirin, William L. *Thoroughbred Racing: State-of-the-Art.* New York: William Morrow and Company, 1984. Contemporary interpretive guidelines for reading the past performances; speed and pace figures.

———. *Winning at the Races: Computer Discoveries in Thoroughbred Handicapping.* New York: William Morrow and Company, Inc., 1979. Percentages and probabilities of all important handicapping characteristics; importance of early speed; leading grass sires; computer-generated multiple regression models of handicapping.

Roman, Steven A. "An Analysis of Dosage." *The Thoroughbred Record,* April 1984. Evolution of pedigree evaluation; dosage; the dosage index; research findings; applications.

———. "Dosage: a practical approach." *Daily Racing Form,* "Bloodlines" column, May 1981. Explanation of dosage; the dosage index; how to calculate the dosage index.

Roussopoulos, Nicholas, and Yen, Raymond T. "An adaptable methodology for database design." *Computer,* May 1984. Analyzing the database environment; system analysis; conceptual modeling; logical schema design.

Scott, William L. *How Will Your Horse Run Today?* Baltimore, Maryland: Amicus Press, 1984. Investigations of the form factor; form defects; form advantages; applications.

———. *Investing at the Racetrack.* New York: Simon and Schuster, 1982. Ability times; fully systematized handicapping.

Selesner, Gary. "Laying the foundation for futuretrack." *Gaming Business Magazine,* May 1983. Third-generation racetrack design principles; the Meadowlands, Garden State.

Shaughnessey, Joseph J. "Spreadsheets." *Run,* March 1984. Various spreadsheet applications for microcomputers; features.

Welles, Chris. "Teaching the brain new tricks," *Esquire,* March 1983. Cognitive psychology research on intuitive reasoning; mnemonics; how experts think; decision-making.

Ziemba, William T., and Hausch, Donald B. *Beat the Racetrack.* New York: Harcourt Brace Jovanovich, 1984. The Z-System of place and show betting; inefficiency-of-market principles; formulae; illustrations; wagering charts and graphs.

INDEX